P9-DZA-001

DISCARDED
FROM
UNH LIBRARY

LIQUID SCINTILLATION COUNTING

RECENT DEVELOPMENTS

ACADEMIC PRESS RAPID MANUSCRIPT REPRODUCTION

Proceedings of the
International Symposium on Liquid Scintillation Counting
Held 20-22 August 1973
Sydney, Australia

LIQUID SCINTILLATION COUNTING

RECENT DEVELOPMENTS

EDITED BY

Philip E. Stanley
Waite Agricultural Research Institute
Department of Agricultural Biochemistry
University of Adelaide
Glen Osmond, South Australia

Bruce A. Scoggins
Howard Florey Institute
of Experimental Physiology and Medicine
University of Melbourne
Parkville, Victoria, Australia

Academic Press, Inc. New York and London 1974
A Subsidiary of Harcourt Brace Jovanovich, Publishers

ACADEMIC PRESS, INC.
111 Fifth Avenue, New York, New York 10003

United Kingdom Edition published by
ACADEMIC PRESS, INC. (LONDON) LTD.
24/28 Oval Road, London NW1

LIBRARY OF CONGRESS CATALOG CARD NUMBER: 73-7444

PRINTED IN THE UNITED STATES OF AMERICA

Contents

CONTENTS

CONTENTS

Preface

The International Symposium on Liquid Scintillation Counting was held from 20th - 22nd August, 1973 at the Wentworth Hotel, Sydney, Australia. One hundred delegates attended the Symposium and discussed the current status of many recent developments in liquid scintillation spectrometry. The decision to hold a Symposium of this type in Australia was based on what was considered a need for a forum to discuss both the theoretical and applied aspects of liquid scintillation counting. In Australia, the technique of liquid scintillation spectrometry has had an important role in the majority of universities, laboratories and institutes involved in nuclear physics, biomedical science, geophysics, and agriculture. However, geographical isolation and the fact that scientists utilizing the same basic principles and equipment work in many different disciplines means that there is seldom an opportunity to exchange ideas. It was hoped that by inviting scientists who are research leaders in the varied and important aspects of liquid scintillation spectrometry overseas and within Australia, a suitable interchange of information could be achieved. In this regard our hopes were truly fulfilled.

We were pleased that Dr. J. L. Symonds, Acting Director of the Australian Atomic Energy Commission Research Establishment, could be present to open the Symposium. It was a special pleasure and privilege to have Dr. John Birks present at the Symposium on the occasion of his 25th Jubilee in the Realm of the Scintillation Process, and to present a plenary lecture covering his endeavors in the field. A different aspect of John Birk's ingenuity is seen in his closing remarks which so skillfully analyzed the "state of the art." The other plenary lecturers, Dr. Yutaka Kobayashi, Dr. Harley Ross, and Professor Eric Schram, were to provide expertise and a solid basis for discussion in three important areas of scintillation counting. Our colleague, Dr. John Coghlan, took the unenviable responsibility of organizing and chairing the Workshop Session. A panel consisting of Drs. Bransome, Kobayashi, Laney, Ross, and Schram ably discussed a number of contentious issues which arose during the Symposium. An edited synopsis of the Workshop Session is included in this volume. Also included in this book is the paper of Dr. Laumas (India) who regrettably was unable to attend the meeting because of illness. To our plenary lecturers, workshop participants, session chairman and to those contributing to the published proceedings, we extend our grateful thanks for making the Symposium a success.

We gratefully acknowledge the active participation of Mr. Peter Annand, Marketing Director of Searle Nucleonics, a Division of Searle Australia Pty. Ltd., who with us conceived the idea of holding the Symposium in Australia. As a member of the organizing committee his encouragement and moral support during the Symposium were of great value. Also, we wish to thank Searle Nucleonics for their generous financial support which enabled us to carry through a long gestation period and then to give birth to the actual meeting. Convention Organizers Pty. Ltd. handled most professionally the logistical aspects of the Symposium for us. Of their staff, we are greatly indebted to Mrs. Karen Forbes who carried out seemingly impossible tasks with efficiency and willingness. We wish to thank Miss Jan Peavy of the Waite Agricultural Research Institute for valuable secretarial assistance and especially for her preparation of the abstracts. The assistance of the secretarial staff at the Howard Florey Institute is also acknowledged and also that of the many other people who in various ways contributed to the Symposium.

Finally, we wish to record our sincere appreciation to Academic Press for their assistance and advice in preparing this volume for publication.

Philip E. Stanley
Bruce A. Scoggins

List of Contributors

Adu, J. K., Department of Agricultural Biochemistry and Soil Science, Waite Agricultural Research Institute, The University of Adelaide, Glen Osmond, South Australia 5064

Apelgot, S., Fondation Curie-Institut du Radium 11, Rue P. et M. Curie et 26, Rue d'Ulm. 75231, Paris

Baucom, Terry L., Department of Biochemistry, University of Georgia, Athens, Georgia 30602

Birks, J. B., The Schuster Laboratory, University of Manchester, U.K.

Bransome, E. D. Jr., Division of Metabolic and Endocrine Disease, Medical College of Georgia, Augusta, Georgia

Butkus, Aldona, Howard Florey Institute of Experimental Physiology and Medicine, University of Melbourne, Parkville, 3052, Australia

Calf, G. E., Australian Atomic Energy Commission, Lucas Heights, N.S.W. 2232

Carter, T. P., Department of Experimental Pathology, University of Birmingham Medical School, Birmingham B15 2TJ. England

Chemama, R., Fondation Curie-Institut du Radium 11, Rue P. et M. Curie et 26, Rue d'Ulm. 75231, Paris

Church, V. E., Australian Atomic Energy Commission Research Establishment, Lucas Heights, N.S.W.

Coghlan, John P., Howard Florey Institute of Experimental Physiology and Medicine, University of Melbourne, Parkville, 3052, Australia

Dodeja, Nancy, Endocrine-Metabolic Unit, Department of Medicine, Peter Bent Brigham Hospital and Harvard Medical School, Boston, Massachusetts

Downes, A. M., C.S.I.R.O., Division of Animal Physiology, Ian Clunies Ross Research Laboratory, Prospect, N.S.W., Australia

Ediss, C., Division of Bionucleonics and Radiopharmacy, University of Alberta, Edmonton, Alberta, Canada

Everett, L. J., Research Manager, Packard Instrument Company, Inc., 2200 Warrenville Road, Downers Grove, Illinois 60515

Faini, George, J., Department of Biochemistry, University of Georgia, Athens, Georgia 30602

Fardy, John J., Chemical Technological Division, Australian Atomic Energy Commission, Private Mail Bag, Sutherland, N.S.W. 2232, Australia

Fraser, I., The University of Sydney, Department of Physical Chemistry, Sydney, Australia

Frilley, M., Fondation Curie-Institut du Radium 11, Rue P. et M. Curie et 26, Rue d'Ulm. 75231, Paris

Gezing, Michael, Amersham/Searle Corp., Arlington Heights, Illinois

Gillespie, R., The University of Sydney, Department of Physical Chemistry, Sydney, Australia

Gresham, P. A., Department of Experimental Pathology, University of Birmingham Medical School, Birmingham B15 2TJ, England

Hartley, P. E., Australian Atomic Energy Commission, Research Establishment, Lucas Heights, N.S.W. 2232

Jackson, N., C.S.I.R.O. Division of Animal Genetics, Epping, N.S.W. Australia

Kaartinen, N., Consultant, Packard Instrument Pty, Ltd., 2-4 Vale Street, St. Kilda, Victoria, 3182 Australia

Kisieleski, Walter, Argonne National Laboratory, Argonne, Illinois 60439

Kobayashi, Yutaka, The Worcester Foundation for Experimental Biology, Shrewsbury, Massachusetts 01545

Kreveld, P., District Manager, Packard Instrument Pty, Ltd., 2-4 Vale Street, St. Kilda, Victoria, 3182 Australia

Laney, Barton H., Searle Analytic Inc., Des Plaines, Illinois

Laumas, K. R., Department of Reproductive Biology, All-India Institute of Medical Sciences, New Delhi 110016, India

Lee, John, Department of Biochemistry, University of Georgia, Athens, Georgia 30602

Lowenthal, G. C., Australian Atomic Energy Commission Research Establishment, Lucas Heights, N.S.W. 2232

Malcolm, Philip J., Biometry Section, Waite Agricultural Research Institute, The University of Adelaide, Glen Osmond, South Australia

Maudsley, David V., The Worcester Foundation for Experimental Biology, Shrewsbury, Massachusetts 01545

Maxwell, C. A., C.S.I.R.O., Division of Animal Physiology, Ian Clunies Ross Research Laboratory, Prospect, N.S.W., Australia

Murphy, Charles L., Department of Biochemistry, University of Georgia, Athens, Georgia 30602

Neary, Michael P., Geochronology Laboratory, The University of Georgia, Athens, Georgia 30602

Noakes, John E., Geochronology Laboratory, The University of Georgia, Athens, Georgia 30602

Noujaim, A. A., Division of Bionucleonics and Radiopharmacy, University of Alberta, Edmonton, Alberta, Canada

Oades, J. M., Department of Agricultural Biochemistry and Soil Science, Waite Agricultural Research Institute, The University of Adelaide, Glen Osmond, South Australia 5064

Painter, Kent, Amersham/Searle Corporation, Arlington Heights, Illinois*

Polach, Harry A., The Australian National University, Radiocarbon Dating Laboratory, Canberra, Australia

Rahman, S. A., Department of Reproductive Biology, All-India Institute of Medical Sciences, New Delhi, 110016, India

**Present address:* Micromedic Diagnostics, Inc.
P.O. Box 464
Fort Collins, Colorado

Rasmussen, G. T., Analytic Chemistry Division, Oak Ridge National Laboratory Oak Ridge, Tennessee 37830

Ross, H. H., Analytic Chemistry Division, Oak Ridge National Laboratory, Oak Ridge, Tennessee

Schram, Eric, University of Brussels, Belgium

Scoggins, Bruce A., Howard Florey Institute of Experimental Physiology and Medicine, University of Melbourne, Parkville, 3052, Australia

Sharpe, S. E. III, Division of Metabolic and Endocrine Disease, Medical College of Georgia, Augusta, Georgia

Sharry, L. F., C.S.I.R.O., Division of Animal Physiology, Ian Clunies Ross Research Laboratory, Prospect, N.S.W., Australia

Spaulding, James D., Geochronology Laboratory, The University of Georgia, Athens, Georgia 30602

Stanley, Philip E., Department of Agricultural Biochemistry, Waite Agricultural Research Institute, The University of Adelaide, Glen Osmond, South Australia*

Temple, R. B., The University of Sydney, Department of Physical Chemistry, Sydney, Australia

ten Haaf, F. E. L., Nuclear Applications Laboratory, N. V. Philips' Gloeilampenfabrieken Eindhoven, The Netherlands

Underwood, Richard H., Endocrine-Metabolic Unit, Department of Medicine, Peter Bent Brigham Hospital and Harvard Medical School, Boston, Massachusetts

Wiebe, L. I., Division of Bionucleonics and Radiopharmacy, University of Alberta, Edmonton, Alberta, Canada

**Present address:* Department of Clinical Pharmacology
The Queen Elizabeth Hospital
Woodville, South Australia 5011

TOWARDS AN UNDERSTANDING OF THE SCINTILLATION PROCESS IN ORGANIC MOLECULAR SYSTEMS

J. B. Birks
The Schuster Laboratory, University of Manchester, U.K.

Introduction. My interest in organic scintillators began just 25 years ago, and I have been interested in them ever since. Silver jubilees are rare occasions. I propose to celebrate this particular anniversary by departing from the usual pattern of review papers and restricting this talk to an account of my own personal voyage of discovery. There have been many other contributors to the field, and accounts of their work are to be found in my books [1-5] and other review articles [6-23] .

The story opens in 1948 in the Department of Natural Philosophy at the University of Glasgow where I was completing my Ph.D. studies on the microwave properties of ferromagnetic materials. All my colleagues were engaged in research related to nuclear physics, and Professor P. I. Dee suggested that I should seek a suitable project in this area and join the "team". In one of the laboratories I found an unusual instrument consisting of a large naphthalene crystal attached to a photomultiplier. Its designer, G. M. Lewis, explained that this was an organic scintillation counter of the type discovered the previous year by H. Kallmann. "Ionizing radiations impinge on the scintillator and produce light flashes or scintillations", he explained. "Why?", I asked. "We don't know", he replied. The next day I informed Professor Dee of my new project, the study of the scintillation process in organic molecular systems. 25 years later I am still working on it. It was a much tougher assignment than I realised, but en route there has been a lot of interesting physics. We've found answers to many of the original questions, and in turn we've uncovered many new questions that we didn't know existed.

Organic crystals. Kallmann's first organic scintillator (1947) was a naphthalene crystal, grown from moth-balls bought in a chemist's shop in war-ravaged Berlin. In 1948 P. R. Bell reported that crystalline anthracene had a scintillation yield about 5 times that of crystalline naphthalene. 25 years later crystalline anthracene remains the most efficient organic scintillator in general use. Unfortunately it cannot be grown in such large crystals as naphthalene.

It was to overcome this limitation that I decided to study the scintillations from naphthalene crystals containing different small concentrations of anthracene [24]. There is an efficient solvent-solute energy transfer process in organic crystal solutions due to exciton migration and transfer, and above a certain anthracene concentration the scintillation yield equals and even exceeds that of a pure anthracene crystal. Because of the elimination of the self-absorption which occurs in anthracene crystals [65] , mixed organic crystals have the highest scintillation yields of any known organic molecular systems [10] . Moreover they can readily be grown in volumes as large as pure naphthalene crystals.

Despite these virtues mixed organic crystals have found little practical application, because of the parallel discovery of two more convenient types of organic solution scintillator. M. Ageno, M. Chiozzotto and R. Querzoli discovered organic liquid solution scintillators in 1949, and in 1950 M. G. Schorr and F. L. Torney found the first organic plastic solution scintillators. Although organic liquid and plastic solutions have lower scintillation yields than organic crystal solutions, they suffer from no limitations of size and shape. Huge sheets and massive blocks of plastic scintillator are used in high energy physics and whole-body counting, and kilolitre tanks of liquid scintillator are used in cosmic ray and neutrino research. There is no need to remind this Symposium of the advantages of liquid scintillators for the direct incorporation and assay of radio-active specimens using the internal counting technique.

<u>Scintillation response</u>. The next topic which I decided to study was the relative scintillation response to different types of ionizing radiation. The scintillation response of an organic scintillator to 5 MeV α-particles is only about 10 per cent of that to β-particles of the same

energy. This effect, which is common to all organic scintillators, is an example of what is known as the L.E.T. (linear energy transfer) effect by the radiation chemists and biologists [31-2].

It was discovered [25] that for β-particles the scintillation response L is proportional to the particle energy E, but that for α-particles L is proportional to the particle range r. Expressed alternatively, the specific scintillation response dL/dr is proportional to the specific energy loss dE/dr when the latter is low as for β-particles, but dL/dr is constant and independent of dE/dr when the latter is high as for α-particles. A theory was proposed [27,28] to account for the effect, leading to a relation of the form

$$\frac{dL}{dr} = \frac{S(dE/dr)}{1 + kB(dE/dr)} \tag{1}$$

At low dE/dr, dL/dr = S (dE/dr), where S = dL/dE is the scintillation yield for β-particles. At high dE/dr, dL/dr = S/kB = constant, as observed. The excited molecules of specific concentration S(dE/dr), responsible for the scintillation, are considered to be quenched with a probability k by ionized or damaged molecules of specific concentration B (dE/dr) produced by the ionizing particle. The parameter kB was determined for an anthracene crystal from observations of the scintillation response to β-particles and α-particles, and relation (1) was used to predict the response to protons and other ionizing particles (Figure 1). Relation (1) applies to all types of organic scintillator, although the values of kB depend somewhat on the nature of the crystal or solvent.

It was thought that molecules permanently damaged by the ionizing radiation might be responsible for the ionization quenching, and a study was therefore made of the deterioration of the scintillation yield of anthracene crystals under prolonged α-particle irradiation [26]. Although the results showed that permanent quenching centres are produced by the irradiation, and that this can be used as a method of radiation dosimetry, the magnitude of the effect is much too small to account for the ionization quenching effect (1). Recent studies have confirmed that molecular ions are the entities responsible for the quenching of the singlet excited molecules which yield the

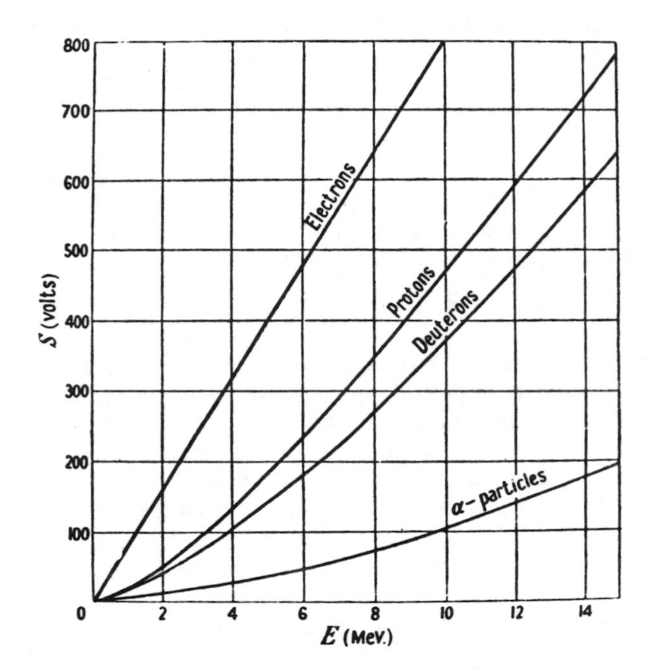

Figure 1. Predicted scintillation response S(≡ L) of anthracene crystal to particles of energy E [27] .

scintillation photons.

A deviation from relation (1) was observed for short-range (< a few μm) α-particles and low-energy external electrons [29]. An analysis showed that the scintillation efficiency is reduced in the surface layer of the scintillator, an effect attributed to surface quenching of the excitation [30]. As a test of this theory, observations were made of the scintillation response of an anthracene crystal to 6-30 keV photoelectrons generated internally by characteristic X-rays [34,36]. The scintillation response is consistent with relation (1), and it is significantly greatly than that with external electrons.

To complete these studies the absolute scintillation yield of crystalline anthracene for α-particles was determined, and the corresponding value of S for β-particles was evaluated [33]. The value of S = 0.04 compares satisfactorily with those of other observers [2], although later observations on other scintillators suggest that the true value may be as high as S = 0.06.

Fluorescence spectra and lifetimes. The later phase of these studies was undertaken during 1951-4 in the Department of Physics, Rhodes University, South Africa. By this time a distinction could be drawn between the primary scintillation processes, associated with ionization, excitation, ion recombination, ionization quenching and radiation damage, and the secondary scintillation processes, associated with internal conversion, energy migration, energy transfer, and fluorescence [8, 9, 12]. The secondary processes can be more readily observed using non-ionizing ultraviolet radiation for excitation, and a programme to study the photophysics of aromatic molecular systems was therefore initiated.

Studies were made of the fluorescence spectra of a wide range of aromatic crystals [48-50], and these were subsequently extended to include scintillator solvents [93] and solutes [19,52]. A direct experimental comparison of the β-particle-excited scintillation spectrum and the optically-excited fluorescence spectrum of several typical organic solutions showed the spectra to be independent of the means of excitation [38].

Similar studies were made of the fluorescence lifetimes of organic crystals [59,60,97] and of organic molecules in solution [61,63,86-7,96]. The techniques used for the determination of fluorescence lifetimes have included phase and modulation fluorometry [59,63,86,130], pulse-sampling fluorometry [60-1, 86-7, 96-7, 108] and the single-photon technique [17,43]. A theoretical and experimental study was made of the relations between the fluorescence lifetime and the absorption and fluorescence spectra of organic molecules [63]. These and other studies of the fluorescence lifetimes of aromatic molecules have been reviewed elsewhere [15].

The scintillation and fluorescence lifetimes of anthracene crystals were shown to be identical, provided that the relatively large effect of self-absorption is taken into account [37,62,65]. In a later paper [79] it has been proposed that the self-absorption in anthracene crystals is due to crystal defects, which act as exciton traps, and the influence of these traps on the fluorescence behaviour of pure and mixed crystals has been analysed.

Energy transfer. Electronic excitation energy transfer is a key process in all solution scintillators. One of the last studies at Rhodes University was of solvent-solute energy transfer from toluene to p-terphenyl by observations of the fluorescence excitation spectra [67]. It was found that the energy transfer rate exceeds that of a diffusion-controlled collisional process, a point to be discussed later.

This work was resumed in 1957 on my appointment at the University of Manchester after a brief spell in industry. A study was made of solvent-solute transfer in plastic solution scintillators [69]. This showed that there is a small radiative transfer component, limited by the fluorescence quantum yield of the polystyrene solvent, and a larger radiationless transfer component which dominates at high solute concentrations (Figure 2). The latter is due to dipole-dipole interaction, the theory of which was formulated by Th. Förster. The probability of energy transfer from a stationary excited donor molecule $^1M^*$ to a stationary acceptor molecule ^1Y is given by

$$k'_{YM} = \frac{1}{\tau_M}\left(\frac{R_o}{r}\right)^6 \qquad (2)$$

where τ_M is the $^1M^*$ lifetime in the absence of ^1Y, r is the intermolecular distance, and R_o is the Förster critical transfer distance, at which the probability of transfer is equal to the probability of $^1M^*$ decay by other means. R_o depends on the overlap of the $^1M^*$ fluorescence spectrum and the ^1Y absorption spectrum and on the relative orientation of the two molecules. The theory, which only strictly applies to stationary donor and acceptor molecules, agrees satisfactorily with the experimental data on plastic solution scintillators.

Further studies were made of energy transfer between a primary solute (p-terphenyl) and a secondary solute (tetraphenylbutadiene) in a low viscosity solvent (toluene) [70]. The results (Figure 3) show that for the particular specimen thickness of 13 mm, radiative transfer is dominant except at the highest concentrations of secondary solute.

The experimental value of R_o = 42.2 Å for TP-TPB radiationless transfer in toluene agrees closely with the theoretical value of R_o = 42.5 Å [70]. This was a surprising result, since the experiments were in a low viscosity solution, while the theory relates to stationary molecules. A series of experiments were therefore undertaken to investigate the influence of diffusion on solute-solute transfer. Energy transfer from anthracene to perylene in benzene solution was studied by measurements of the fluorescence response functions of the donor and acceptor [74]. The fluorescence response functions, defined as the fluorescence quantum intensities as a function of time t following δ-function excitation at t = 0 [86], depend only on the radiationless transfer component, and they are unaffected by any parallel radiative transfer. Due to the molecular diffusion in the anthracene-perylene-benzene system, the critical transfer distance is increased from R_o = 31 Å (for stationary molecules) to $(R_o)_{eff}$ = 43 Å (in benzene at room temperature).

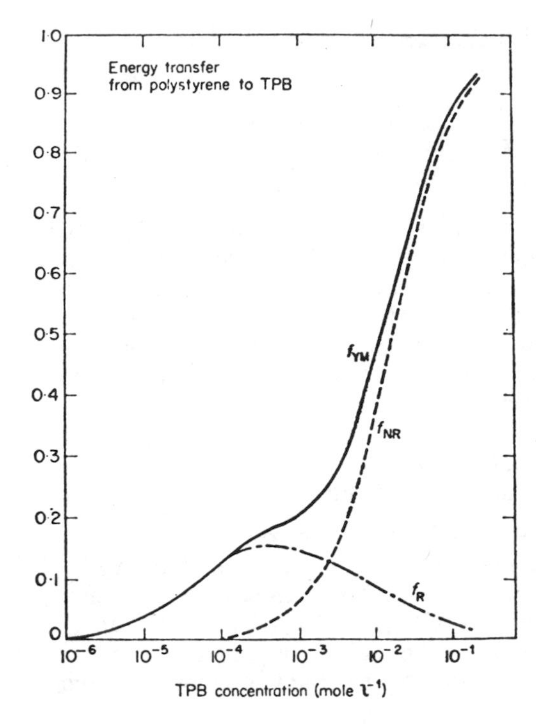

Figure 2. Solvent-solute energy transfer from polystyrene (^1M) to tetraphenylbutadiene, TPB (^1Y). Energy transfer quantum efficiency f_{YM}, its radiative component f_R and radiationless component f_{NR} against TPB concentration [3,69] .

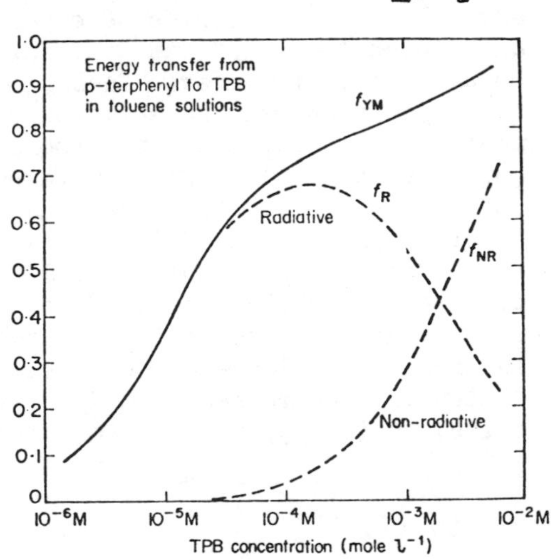

Figure 3. Solute-solute energy transfer from p-terphenyl (2.17×10^{-2}M) to TPB in toluene solution. Energy transfer quantum efficiency f_{YM}, its radiative component f_R and radiationless component f_{NR} against TPB concentration [70].

In a theoretical analysis [75] of fluorescence response functions and scintillation pulse shapes (the two functions are usually identical, since excitation by an ionizing particle is equivalent to δ-function optical excitation) three cases were distinguished, depending on the relative magnitudes of R_o and the molecular diffusion length $\sqrt{2 D \tau_M}$, where D is the diffusion coefficient.

(i) $R_o \gg \sqrt{2 D \tau_M}$. The molecules are effectively stationary during the energy transfer process. The energy transfer is described by Förster kinetics (time-dependent rate parameters) with non-exponential fluorescence response functions.

(ii) $R_o \simeq \sqrt{2 D \tau_M}$. There is some motion of the molecules during energy transfer. The system is described by intermediate kinetics with semi-exponential fluorescence response functions.

(iii) $R_o \ll \sqrt{2 D \tau_M}$. Complete molecular mixing occurs during energy transfer. The system obeys Stern-Volmer kinetics (time-independent rate parameters) with exponential fluorescence response functions.

The TP-TPB system in toluene (Figure 3) corresponds to case (i), because of the low value of τ_M = 1.2 ns and the high value of R_o = 42.2 Å.

The transition from case (i) to case (iii) is shown clearly in studies of the fluorescence response functions of phenanthrene ($^1M^*$) in the presence of acridine (^1Y) in solvents of different viscosities [64,76]. (Figure 4). In the absence of ^1Y the $^1M^*$ decay is exponential. In the presence of ^1Y, in a high viscosity (η = 64.6 cP) solvent the $^1M^*$ decay is non-exponential and agrees with that predicted by Förster kinetics (case i); in low viscosity ($\eta \leq$ 3.7 cP) solvents the $^1M^*$ decay is exponential, corresponding to Stern-Volmer kinetics (case iii); and in medium viscosity solvents the $^1M^*$ decay is intermediate between these limits (case ii). A related study of the influence of diffusion on the efficiency of solute-solute energy transfer for various solution systems [78] gave results consistent with the theoretical studies.

In fluid solutions molecular diffusion influences the rate of all bimolecular processes, including solvent-solute energy transfer, solute-solute energy transfer, impurity quenching, concentration quenching and excimer formation. Naphthalene and biacetyl were chosen as a suitable donor and acceptor system for the study of diffusion-controlled collisional quenching in different solvents, since the Förster radius R_0 for energy transfer from naphthalene to biacetyl is small and it approximates to the sum of the molecular radii [77]. The results of these and related studies have been recently analysed in a comprehensive review of diffusion-controlled rate processes [127].

Excimers. In 1954 Th. Förster and K. Kasper discovered that the concentration quenching of the molecular ($^1M^*$) fluorescence of pyrene in solution is accompanied by the appearance of a new structureless fluorescence band at longer wavelengths. Typical fluorescence spectra of pyrene in cyclohexane solution are shown in Figure 5 [84]. The absorption spectrum is independent of concentration, and the structureless emission is due to excimers ($^1D^*$), i.e. excited dimers which are dissociative in the ground state, formed by the collisional process

$$^1M^* + {}^1M \rightleftharpoons {}^1D^* \qquad (3)$$

For 8 years after the discovery of the pyrene excimer the subject lay dormant, as though the scientific world considered excimer formation to be restricted to the pyrene molecule. In 1962 we decided to look for other aromatic hydrocarbons which formed excimers, and suddenly the subject erupted [81]. Almost every compound that we investigated showed a structureless excimer fluorescence band in concentrated solution, and at one stage new "excimer-formers" were being found at the rate of 2 or 3 a week [81-5, 88]. Many aromatic liquids were found to exhibit dominant excimer fluorescence just above their melting points [89]. Some aromatic crystals also have an intense excimer fluorescence component [50,103]. Within a year or two excimer formation had been shown to occur in practically all aromatic hydrocarbons and their derivatives [11, 20].

Of the condensed polycyclic hydrocarbons with 4 or less rings excimer formation occurs in

(i) benzene and its alkyl derivatives [93],
(ii) naphthalene and its methyl and dimethyl derivatives [94],

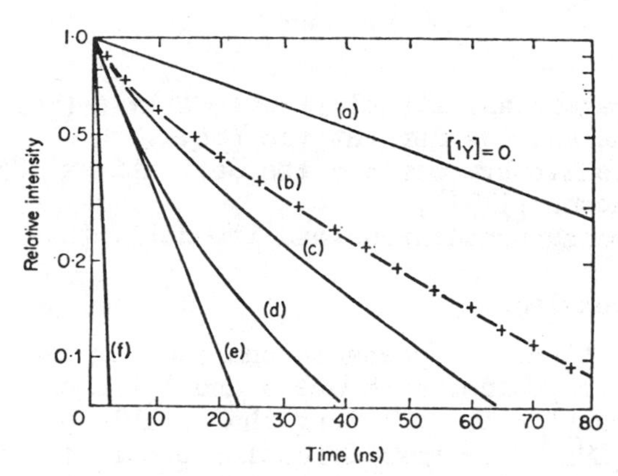

Figure 4. Influence of solvent viscosity η on solute-solute energy transfer from phenanthrene (^1M) to acridine (^1Y). Donor ($^1M^*$) fluorescence decay curves.

(b) η = 64.6 cP; + expt; solid curve, theory;

(c) η = 27 cP; (d) η = 10 cP; (e) η = 3.7 cP;

(f) η = 0.4 cP [76] .

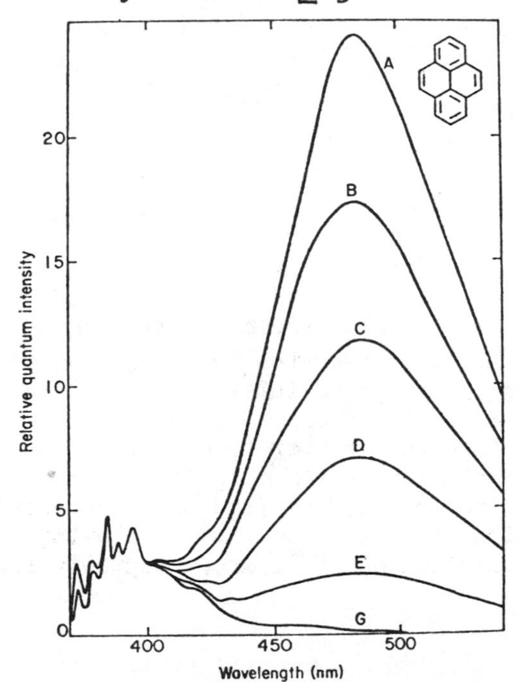

Figure 5. Fluorescence spectra of pyrene solutions in cyclohexane. Intensities normalized to a common molecular fluorescence quantum yield Φ_{FM}. A, 10^{-2} M; B, 7.75 x 10^{-3} M; C, 5.5 x 10^{-3} M; D, 3.25 x 10^{-3} M; E, 10^{-3} M; G, 10^{-4} M [84] .

(iii) anthracene and its alkyl derivatives [98,118] ,
(iv) pyrene and its derivatives [84] ,
(v) 1:2-benzanthracene and its derivatives [85] ,
(vi) tetracene [119] ,
(vii) 3:4-benzophenanthrene and its derivatives [50] , and
(viii) triphenylene.

The only exceptions are phenanthrene and chrysene. In two of the above compounds, anthracene and tetracene, the excimer interaction is so strong that stable photodimers are formed [119] . Excimer formation occurs in liquid scintillator solvents (benzene, toluene, p-xylene, mesitylene, etc.), in plastic scintillator solvents (polystyrene, polyvinyltoluene) and in several scintillator solutes (PPO, BPO, α-NPO, α-NPD) and other solutes exhibiting concentration quenching.

In view of the generality of the phenomenon we initiated a series of studies of excimer formation. The following reaction scheme is applicable [86] .

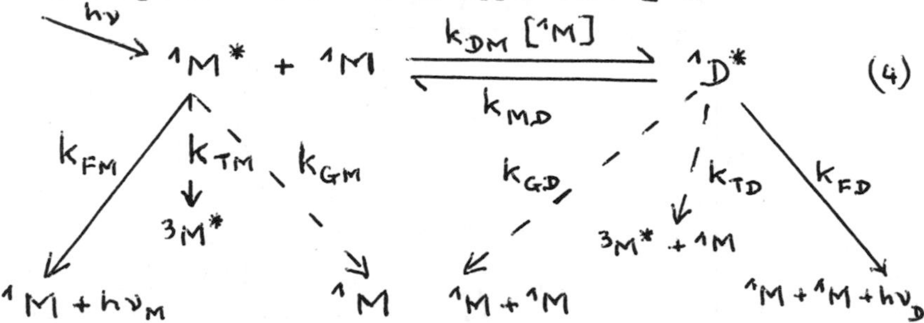

k_{FM} and k_{FD} are the fluorescence rates, k_{GM} and k_{GD} the rates of internal conversion to the ground state ^1M, and k_{TM} and k_{TD} the rates of intersystem crossing to the triplet state $^3M^*$, of $^1M^*$ and $^1D^*$, respectively. k_{DM} $[^1M]$ and k_{MD} are the excimer formation and dissociation eates, respectively, where $[^1M]$ is the molar concentration. From a kinetic analysis [86] the fluorescence response functions of $^1M^*$ and $^1D^*$, following δ-function excitation of $^1M^*$, are, respectively,

$$i_M(t) \propto e^{-\lambda_1 t} + A e^{-\lambda_2 t} \qquad (5)$$

$$i_D(t) \propto e^{-\lambda_1 t} - e^{-\lambda_2 t} \qquad (6)$$

λ_1, λ_2 and A are functions of k_M (= $k_{FM} + k_{GM} + k_{TM}$), k_D (= $k_{FD} + k_{GD} + k_{TM}$), k_{DM} $[^1M]$ and k_{MD}.

Observations were made of $i_M(t)$ and $i_D(t)$ at different values of $[^1M]$ to determine λ_1 and λ_2, and hence the rate parameters k_M, k_D, k_{DM} and k_{MD} [86-7]. Typical fluorescence response functions are shown in Figure 6. Observations of the $^1M^*$ and $^1D^*$ fluorescence quantum yields, Φ_{FM} and Φ_{FD}, then enabled k_M and k_D to be split into their radiative components, k_{FM} and k_{FD}, and their radiationless components k_{IM} (= $k_{GM} + k_{TM}$) and k_{ID} (= $k_{GD} + k_{TD}$). Recent observations of the triplet quantum yield as a function of $[^1M]$ have enabled k_{IM} and k_{ID} to be split into their internal conversion components k_{GM} and k_{GD} and their intersystem crossing components k_{TM} and k_{TD} [110]. In this manner all the rate parameters in (4) have been determined experimentally for pyrene and other aromatic molecules in solution.

Similar studies at different temperatures have enabled the frequency factors and activation energies of the temperature-dependent rate processes to be determined [86, 91-2]. The enthalpy ΔH and the entropy ΔS of the excimer formation process have also been evaluated. From an analysis of the fluorescence spectrum of the pyrene crystal excimer, we have determined the excimer interaction potential, and the force constants and zero-point vibrational energies of the ground and excited states of the dimer [101, 103-4, 106]. The potential diagram of the pyrene crystal excimer is shown in Figure 7. The results have been used to explain excimer formation in other aromatic hydrocarbons, and its absence in phenanthrene and chrysene [102].

Excimer formation in the liquid alkyl benzenes is of interest, in view of their use as liquid scintillator

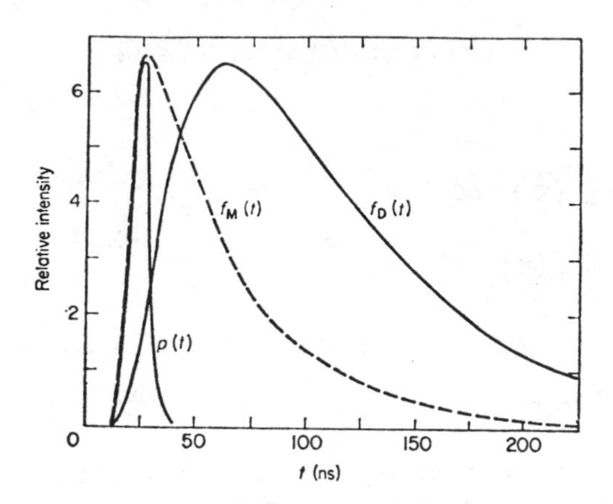

Figure 6. Pyrene (5×10^{-3} M) in cyclohexane solution. Molecular fluorescence response $f_M(t)$ and excimer fluorescence response $f_D(t)$ to excitation light pulse p(t) [86] .

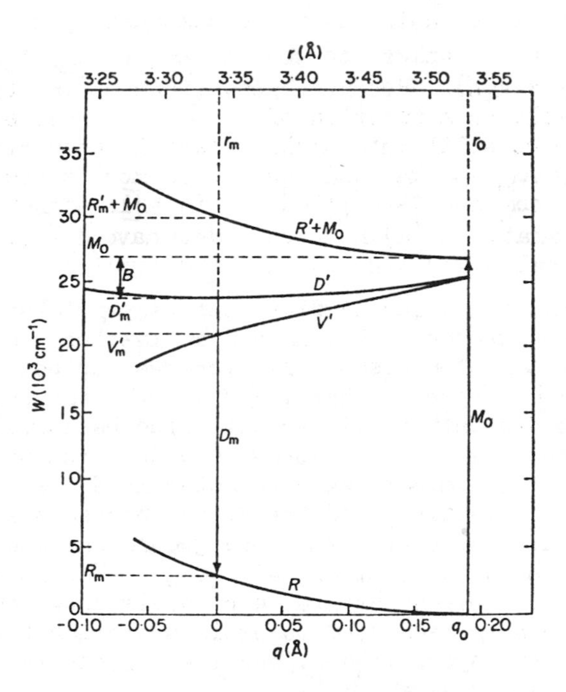

Figure 7. Pyrene crystal dimer and excimer. Potential energy diagram derived from experimental data. Energy (W) against intermolecular separation (r) and displacement (q) from excimer equilibrium [103] .

solvents. If liquid benzene, toluene and p-xylene at room temperature are excited optically or by low-intensity ionizing radiation, then 57%, 37% and 14%, respectively, of the excited species are in the excimer state [93, 21]. The excimer fraction d (= 1 - m) increases at lower temperatures [107-8]. The excimer formation and dissociation rates, k_{DM} $[^1M]$ and k_{MD}, are very high in the alkyl benzenes, and the $^1M^*$ and $^1D^*$ fluorescences decay at a common rate

$$\lambda_1 = m\ k_M + d\ k_D \quad (7)$$

The fluorescence spectrum F of the liquid alkyl benzenes excited optically or by low-intensity ionizing radiation consists of a mixture of the molecular fluorescence spectrum F_M and the excimer fluorescence spectrum F_D [93], so that

$$F = m F_M + d F_D \quad (8)$$

L. G. Christophorou and co-workers discovered that when an <u>intense</u> electron beam is used for excitation, only the excimer fluorescence emission F_D is observed. The effect has been observed in benzene, toluene, the three xylene isomers, mesitylene, and other alkyl benzenes and in alkyl derivatives of naphthalene. The present author [41] has interpreted the effect in terms of the formation and subsequent neutralization of dimer cations

$$^2M^+ + {^1M} \longrightarrow {^2M_2^+} \quad (9)$$

$$^2M^+ + {^2e^-} \longrightarrow {^1D^{**}} \longrightarrow {^1D^*} \quad (10)$$

yielding excimers, and the strong ionization quenching of the excited molecules $^1M^*$ which result from dissociation of $^1D^*$. It is concluded that excimers are much less prone to ionization quenching than excited molecules.

<u>Primary processes and internal conversion</u>. The primary processes are those by which the energy W dissipated by the ionizing particle in the solvent is converted into molecular ions ($^2M^+$, $^2M^-$), electrons ($^2e^-$), molecules in higher excited singlet states ($^1M^{**}$) and triplet states ($^3M^{**}$), and excimers in higher excited singlet states ($^1D^{**}$). The excited species $^1M^{**}$, $^3M^{**}$ and $^1D^{**}$ then

undergo radiationless internal conversion and vibrational relaxation into their lowest excited states $^1M^*$, $^3M^*$ and $^1D^*$ respectively.

From various studies [2-3, 9, 21-2] it is concluded that the most important primary and internal conversion processes are as follows.

$^1M + W \longrightarrow {}^2M^+ + {}^2e^-$ Ionization (11)

$^1M + W \longrightarrow {}^1M^{**}$ Singlet excitation (12)

$^2M^+ + {}^2e^- \longrightarrow {}^1M^{**}$ Excited singlet from ion recombination (13)

$^2M^+ + {}^2e^- \longrightarrow {}^3M^{**}$ Excited triplet from ion recombination (14)

$^1M + {}^2e^- \longrightarrow {}^2M^-$ Electron attachment (15)

$^1M^{**} + {}^2M^+ ({}^2M^-) \longrightarrow {}^1M + {}^2M^+ ({}^2M^-)$ Ionization quenching (16)

$^2M^+ + {}^1M \longrightarrow {}^2M_2^+$ Dimer cation formation (17)

$^2M_2^+ + {}^2e^- \longrightarrow {}^1D^{**}$ Excimer from dimer cation neutralization (18)

$^2M_2^+ + {}^2e^- \longrightarrow {}^3M^{**} + {}^1M$ Excited triplet from dimer cation neutralization (19)

$^1M^{**} + {}^1M \rightleftharpoons {}^1D^{**}$ Excimer formation and dissociation (20)

$^1M^{**} \longrightarrow {}^1M^*$ Internal conversion in singlet manifold (21)

$^3M^{**} \longrightarrow {}^3M^*$ Internal conversion in triplet manifold (22)

$^1D^{**} \longrightarrow {}^1D^*$ Internal conversion in excimer manifold (23)

In an anthracene crystal the $^1M^{**} \longrightarrow {}^1M^*$ internal conversion (21) quantum yield $\Phi_{IC} = 1.0$. In the alkylbenzenes and the polyvinyl alkylbenzenes $\Phi_{IC} \leq 1.0$ for the $^1M^{**} \longrightarrow {}^1M^*$ (21) and $^1D^{**} \longrightarrow {}^1D^*$ (23) internal conversion processes [126] . This is one reason why the scintillation yields S of liquid and plastic solution scintillators are less than that of an anthracene crystal. The values of Φ_{IC} are determined from observations of the fluorescence excitation spectra of liquid alkylbenzenes and scintillator solutions [67,126] . For liquid benzene, toluene and p-xylene Φ_{IC} = 0.45, 0.76 and 0.95, respectively.

A theory of the scintillation yields of liquid scintillators has been developed [22] , which takes account of the ionization yield (11) which is inversely proportional to the solvent ionization potential, the singlet excitation yield (12) which is inversely proportional to the mean solvent excitation energy, and the internal conversion yield Φ_{IC}. The theory predicts relative scintillation yields of 82, 100 and 116 for benzene, toluene and p-xylene solutions, respectively, which compare satisfactorily with the observed values of 85, 100 and 112 for 3 gl^{-1} PPO solutions in these three solvents.

The reduced internal conversion yield ($\Phi_{IC} < 1.0$) in the alkylbenzenes and their polyvinyl derivatives is due to a competing radiationless transition which is absent in the higher aromatic hydrocarbons like anthracene where $\Phi_{IC} = 1.0$. The competing process, which is known simply as "channel 3" since its exact nature is uncertain, occurs when the excitation energy is more than 2000 - 3000 cm^{-1} above the zero-point level of the first excited singlet state ($^1M^*$) of benzene [58] . It corresponds to a fast ($\sim 10^{12}$ s^{-1}) internal conversion to the ground state (^1M) of benzene, and it competes with the $^1M^{**} \longrightarrow {}^1M^*$ internal conversion which has a similar rate. It has been proposed that the "channel 3" transition occurs via a non-planar "physical isomer" of benzene, which reverts collisionally to the ground state of benzene [58] .

"Channel 3" has stimulated fresh interest in the spectroscopy and photophysics of benzene [55, 57, 125] and toluene [108] . From observations [128] of the linewidth of the electronic S_0 - S_2 absorption transition in crystal benzene, extrapolated to $0^{o}K$, it is concluded that S_2 has a radiationless decay rate of 9.4×10^{12} s^{-1}. This value agrees satisfactorily with that of 9.9×10^{12} s^{-1} obtained from the $S_2 - S_o$ fluorescence quantum yield ($= 8 \times 10^{-6}$) of liquid benzene at $25^{o}C$ recently observed by F. Hirayama and co-workers. This is the first time that the lifetime of a higher excited state of a complex molecule has been determined by both linewidth and fluorescence measurements. From comparison with the fluorescence excitation spectrum of crystal benzene, it is concluded [128] that the $S_2 - S_1$ internal conversion rate is 6.6×10^{12} s^{-1} and that the "channel 3" rate is 2.8×10^{12} s^{-1}.

Triplet-triplet interaction. F. D. Brooks and G. T.Wright, two of my colleagues at Rhodes University, observed that the scintillation pulse shape of an anthracene crystal excited by α -particles differs from that excited by electrons. The discovery that the scintillation pulse shape depends on the specific energy loss dE/dr of the particle led to the development of the pulse shape discrimination (P.S.D.) technique in which organic scintillators are used to distinguish neutrons, which give recoil protons, from γ -rays, which give Compton electrons.

The scintillation decay consists of two components, a prompt component I_P, which decays exponentially, and a delayed slow component I_S, which decays non-exponentially. I_P corresponds to the fluorescence decay of the $^1M^*$ and $^1D^*$ excited species, produced by internal conversion from the $^1M^{**}$ and $^1D^{**}$ excited species (21, 23) generated by the primary processes (12, 13, 18, 20). I_s is due to the bimolecular interaction of the $^3M^*$ excited species, produced by internal conversion from the $^3M^{**}$ excited species (22) generated by the primary processes (14, 19). The rate of the triplet-triplet interaction process

$$^3M^* + {}^3M^* \longrightarrow {}^1M^* + {}^1M \qquad (24)$$

is determined by the $^3M^*$ concentration and diffusion rate, and it produces the delayed $^1M^*$ fluorescence which constitutes the slow component I_s. I_p depends on dE/dr (1), because of the ionization quenching of $^1M^{**}$ (16) but I_s is approximately independent of dE/dr, indicating that the ionization quenching of $^3M^{**}$ is less, possibly

because the $^3M^{**}$ species are only produced by ion recombination. Moreover the process (24) which generates the delayed $^1M^*$ fluorescence is separated in space and time from the primary ionization column. This model [2, 8] accounts reasonably for the scintillation pulse shape and its dependence on dE/dr.

The delayed fluorescence due to triplet-triplet interaction in solution has been studied using optical excitation [113-4, 116-7, 109]. With solute molecules like pyrene, which form excimers, triplet-triplet interaction produces both excited molecules and excimers

$$^3M^* + {}^3M^* \xrightarrow{k_{MTT}[^3M^*]} {}^1M^* + {}^1M \quad (24)$$

$$^3M^* + {}^3M^* \xrightarrow{k_{DTT}[^3M^*]} {}^1D^* \quad (25)$$

The inclusion of these two bimolecular processes in the previous excimer formation and dissociation scheme (4) complicates the kinetic analysis, which has been discussed in detail in Chapter 8 of ref. [3]. Due to process (25) the ratio of the $^1D^*$ and $^1M^*$ fluorescence intensities is greater in the delayed emission than in the prompt emission. This has been used to show that the ruby-laser-excited fluorescence of pyrene solutions is due to two-photon absorption into $^1M^*$, giving the prompt fluorescence spectrum, and not due to one-photon absorption into $^3M^*$, which would give the delayed fluorescence spectrum [95].

Liquid solution scintillators, in the absence of dissolved oxygen, exhibit prompt and slow scintillation components, I_p and I_s, respectively. I_p results from solvent-solute singlet energy transfer

$$^1M^* + {}^1Y \longrightarrow {}^1M + {}^1Y^* \quad (26)$$

leading to solute ($^1Y^*$) fluorescence. R. Voltz proposed that I_s is due to solvent ($^3M^*$) triplet-triplet interaction (24) followed by solvent-solute singlet energy transfer (26). However the $^3M^*$ lifetime of the liquid alkylbenzenes is only 10-20 ns, while the lifetime of I_s is about 1 μs. As an alternative to Voltz' model it has been proposed [22, 42] that solvent-solute triplet energy transfer occurs

$$^3M^* + {}^1Y \longrightarrow {}^1M + {}^3Y^* \quad (27)$$

and that the delayed solute fluorescence then results from solute ($^3Y^*$) triplet-triplet interaction

$$^3Y^* + {}^3Y^* \longrightarrow {}^1Y^* + {}^1Y \quad (28)$$

This model is now generally accepted.

Energy transfer and quenching in liquid scintillators. In a liquid scintillator containing excited solvent ($^1M^*$), solute (^1Y) and impurity (^1Q) molecules, there are three bimolecular rate processes,

$$^1M^* + {}^1Y \xrightarrow{k_{YM}\,[^1Y]} {}^1M + {}^1Y^* \quad \text{Solvent-solute energy transfer} \quad (29)$$

$$^1M^* + {}^1Q \xrightarrow{k_{QM}\,[^1Q]} {}^1M + {}^1Q \quad \text{Solvent quenching} \quad (30)$$

$$^1Y^* + {}^1Q \xrightarrow{k_{QY}\,[^1Q]} {}^1Y + {}^1Q \quad \text{Solute quenching} \quad (31)$$

Over the last two decades there have been many studies of the solvent-solute transfer process (29) in the author's laboratory [40, 65, 67, 72-3, 80, 126] and elsewhere, and several reviews of the subject have been published [2, 3, 16, 22]. In typical liquid scintillators k_{YM} is independent of whether the solvent is excited by ionizing or ultraviolet radiation, showing that the energy transfer normally occurs from the lowest excited singlet state ($^1M^*$) of the solvent.

The rate parameter of a diffusion-controlled process is given by the Smoluchowski-Sveshnikov relation [3, 127]

$$k_{diff} = 4\pi N' \; D \; p \; R \quad (M^{-1}\,s^{-1}) \quad (32)$$

where N' is Avogadro's number x 10^{-3}, D is the sum of the diffusion coefficients of the reactant species, p (≤ 1) is the reaction probability per collision, and R is the molecular interaction distance. A transient term of the order of unity is omitted from (32) to simplify the discussion. Applying (32) to efficient (p = 1) solute quenching (31) we obtain

$$k_{QY} = 4\pi N' \; (D_Y + D_Q) \; (r_Y + r_Q) \quad (33)$$

where D_Y and D_Q are the diffusion coefficients, and r_Y and r_Q are the molecular radii of $^1Y^*$ and ^1Q, respectively. Relations (32) and (33) have been verified for pyrene excimer formation [86], quenching of naphthalene fluorescence by biacetyl in various solvents [77], and quenching of PPO fluorescence by carbon tetrabromide in toluene [80]. In each case R is the sum of the molecular radii and $p = 1$ in room-temperature solution.

The solvent-solute transfer parameter k_{YM} and the solvent quenching parameter k_{QM} each exceed the diffusion-controlled solute quenching parameter k_{QY} by a factor of 2 to 3 [73, 100]. This indicates that

(i) the effective $^1M^*$ diffusion coefficient $\bar{D}_M$ exceeds the molecular diffusion coefficient D_M, and/or

(ii) the effective $^1M^*$ interaction radius $\bar{r}_M$ exceeds the molecular radius r_M,

so that

$$k_{YM} = 4\pi N' (\bar{D}_M + D_Y)(\bar{r}_M + r_Y) \tag{34}$$

$$k_{QM} = 4\pi N' (\bar{D}_M + D_Q)(\bar{r}_M + r_Q) \tag{35}$$

with $\bar{D}_M \geqslant D_M$, $\bar{r}_M \geqslant r_M$. In solution systems in which $D_Y \simeq D_Q$, $r_Y \simeq r_Q$, it is observed that $k_{YM} \simeq k_{QM}$, and we have therefore concluded that $\bar{r}_M$ is the same for both processes [100]. The $^1M^*$ migration process responsible for $\bar{D}_M > D_M$ reduces the $^1M^*$ lifetime of a molecule in the bulk solvent to $\sim 10^{-3}$ that of an isolated molecule, and this reduces the Förster critical transfer distance R_o to molecular dimensions comparable with the quenching interaction radius [3].

R. Voltz proposed that

$$\bar{D}_M = D_M + \Lambda_M \tag{36}$$

where Λ_M is the coefficient of $^1M^*$ migration, due to intermolecular Coulombic and exchange interaction, analogous to exciton migration in an aromatic molecular crystal. With the discovery of excimers in the alkyl benzenes [93], relation (36) was modified to

$$\bar{D}_M = m(D_M + \Lambda_M) + d\ D_D \qquad (37)$$

where m and d are the fractions of solvent excited species in the $^1M^*$ and $^1D^*$ states, and D_D ($\sim 0.75\ D_M$) is the $^1D^*$ diffusion coefficient. Birks and Conte [40, 100] proposed that the $^1M^*$ migration is due to successive rapid excimer formation and dissociation

$$^1M_A^* + {}^1M_B \rightleftharpoons {}^1D_{AB}^* \longrightarrow {}^1M_A + {}^1M_B^* \text{ etc.} \qquad (38)$$

where suffixes A and B denote different solvent molecules. On this model [100]

$$\Lambda_M = d\ k_{MD}\ \bar{\alpha}^2 \qquad (39)$$

where $\bar{\alpha}$ is the r.m.s. displacement for the $^1M_A - {}^1M_B^*$ migration step.

To test these models, observations have been made of the three rate parameters k_{YM}, k_{QM} and k_{QY} for the toluene ($^1M^*$), PPO (^1Y), carbon tetrabromide (^1Q) system at temperatures from -20 to +50°C [80]. The results are shown in Figure 8. The solute quenching parameter k_{QY} agrees satisfactorily with (33). The unexpected result is that k_{YM} (over the full temperature range) and k_{QM} (at $t \leq 20°C$) are each proportional to k_{QY}. The behaviour of k_{QM} at $t > 20°C$ is attributed to the factor p, omitted from (35), decreasing below unity.

The observed temperature dependence of k_{YM} and k_{QM} is not consistent with either the Voltz model (36, 37) or the Birks-Conte model (38, 39) of $^1M^*$ excitation migration. It is, however, consistent with the relation

$$k_{YM} = 4\pi N' (\bar{D}_M + D_Q)(\bar{r}_M + r_Q) \qquad (34)$$

provided that either

(a) $\bar{D}_M = 3\ D_M$, and $\bar{r}_M = r_M$, or

(b) $\bar{D}_M = D_M$, and $\bar{r}_M = 3\ r_M$.

In case (a) the $^1M^*$ excitation migration coefficient $\Lambda_M = 2\ D_M$, and the $^1M^*$ interaction radius equals the molecular radius. In case (b) there is no $^1M^*$ excitation migration, only diffusion, but the $^1M^*$ interaction radius is 3 times the molecular radius, possibly due to excimer interaction (38).

Initially we favoured [80] interpretation (b) since it accounts for k_{YM} and k_{QM} (at $t \leq 20°C$) each being

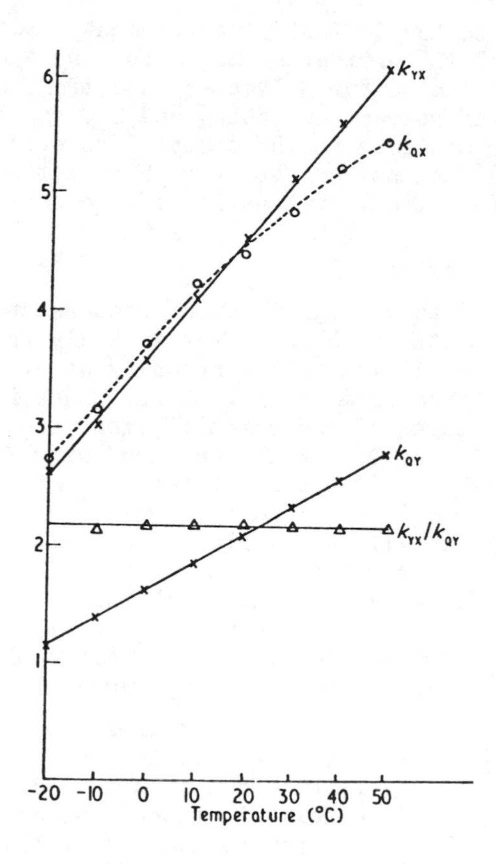

Figure 8. Toluene ($^1M^*$) solutions of PPO (^1Y) and carbon tetrabromide (^1Q). Rate parameters (in 10^{10} M^{-1} s^{-1}) of solvent-solute transfer (k_{YX}), solvent quenching (k_{QX}) and solute quenching (k_{QY}) and k_{YX}/k_{QY} as a function of temperature [80]. ($X \equiv M$)

proportional to the diffusion coefficient. However, in the absence of $^1M^*$ excitation migration ($\bar{D}_M = D_M$), the interaction distance for solvent-solute transfer should exceed that for solvent quenching and k_{YM} should exceed k_{QM}, which is contrary to the observed behaviour. Interpretation (a) does not suffer from these limitations, although a theoretical explanation of the relation

$$\Lambda_M = 2 \; D_M \qquad (40)$$

is currently lacking.

Liquid scintillators. Scintillation process studies are made under conditions which differ markedly from those in which liquid scintillators are normally used. Oxygen is usually eliminated from the solutions, since it interferes with several stages of the scintillation process. It captures electrons, it quenches excited singlet ($^1M^*$, $^1Y^*$) and triplet ($^3M^*$, $^3Y^*$) states of the solvent and solute, and it yields singlet-excited oxygen $^1O_2^*$ which is very chemically active [112]. A liquid scintillator at ambient temperature in contact with air contains about 2×10^{-3}M of dissolved oxygen, which reduces its scintillation yield by up to 20 per cent. In liquid scintillation counting it is usually considered impracticable or unnecessary to expel the dissolved oxygen.

In scintillation process studies techniques like vacuum distillation, recrystallization, zone-refining, microsublimation and chromatography are used to eliminate impurities. In liquid scintillation counting high purity chemicals may be used to prepare the scintillator "cocktail", but the high scintillation yield thus obtained is partially dissipated by exposure to air, and more drastically by the addition of the specimen and any solubilizing, dispersing, bleaching, diluting or other agents used for its incorporation in the solution [21].

In an attempt to bridge the gap between the theory and the practice of liquid scintillation counting, we have investigated a series of liquid scintillators under practical conditions [23]. The relative pulse heights (RPH) of air-equilibrated solutions at ambient temperature and normal atmospheric pressure were measured using a Philips automatic liquid scintillation counting analyzer. The solutions were contained in standard 20 ml glass vials, and a small quantity of ^{14}C-labelled hexadecane, which is a non-quencher, was added to give about 20,000 d.p.m. The RPH values were determined using a "channels-ratio" method.

We used 6 solvents (benzene, toluene, xylene, p-xylene, mesitylene and p-dioxan with 100 g/l naphthalene), 7 primary solutes (TP, PPO, BBOT, PBO, butyl-PBD, PBD and BIBUQ) and 7 secondary solutes (PBBO, POPOP, dimethyl-POPOP, BBO, bis-MSB, α-NPO and DPH). The RPH of the 42 possible binary solution systems were measured as a function of primary solute concentration $[^1Y]$. The RPH of 78 ternary solutions based on TP, PPO or PBD in benzene, toluene, xylene or p-xylene were measured as a function of the concentration $[^1Z]$ of each of the 7 secondary solutes. The results provide a practical guide to the relative performance of 120 liquid scintillator formulations under standard conditions in the absence of the specimen [23].

To simulate the effect of the specimen, we studied the impurity quenching of the RPH of each of the binary solutions due to the addition of carbon tetrachloride [23]. The magnitude of the quenching depends both on the solvent and on the solute. The quenching susceptibility of the solvents increases in the following order: benzene, p-dioxan + 100 gl^{-1} naphthalene, toluene, xylene, p-xylene, mesitylene. For the alkyl benzenes this corresponds to the inverse order of their ionization potentials, indicating that the quenching is due to exciplex formation [14, 51]. The quenching susceptibility of the solutes increases in the following order: PBD, butyl-PBD, PBO, BBOT, PPO, BIBUQ and TP. The solutions most resistant to quenching are those in which both the solvent and the solute lie in the early parts of the quenching susceptibility lists.

Table 1, prepared from the experimental data [23], compares the RPH of different binary solutions in the absence (V_o) and presence (V_Q) of 0.1 M carbon tetrachloride. The values of V_Q are expressed relative to $V_Q = 100$ for the PBD-toluene solution, and the values of V_o are expressed relative to $V_o = 100$ for the same solution. The values of V_o may be normalized to the same scale as V_Q by multiplication by 2.73, the quenching factor of the PBD-toluene solution.

Comparison of the values of V_Q and V_o shows dramatic changes in the relative merits of the solutions due to the quencher. The BIBUQ-toluene and BIBUQ-p-xylene solutions, which have the highest V_o values, have much inferior V_Q values. $V_Q < V_o$ for all the p-xylene solutions, while in contrast $V_Q > V_o$ for all the benzene solutions in the upper part of Table 1. The PBD-toluene solution maintains its superior position in the absence or presence of the quencher.

In liquid scintillation counting the magnitude of the quenching depends not only on the nature of the scintillator, but also on the nature and concentration of the specimen. For those engaged in regular radio-assay of the same type of specimen, the above procedures [23] provide a straightforward method of choosing the optimum scintillator composition under given quenching conditions. For those seeking simpler recipes for efficient liquid scintillators, the values of V_Q (Table 1) determined in the presence of a quencher should prove a more practical guide than those of V_o determined in its absence.

Conclusion. In keeping with the personal nature of this paper, it concludes with a bibliography of my publications on radiation physics and photophysics. This includes the names of my co-workers at Glasgow, Rhodes and Manchester Universities, to whom I wish to express my appreciation. They have come from all over the world, from Britain, Belgium, Germany, Czechoslovakia, Poland, Portugal, Greece, Cyprus, Iraq, Iran, Pakistan, India, Nigeria, South Africa and the United States, so that what started as a solitary journey has developed into a small international expedition.

Table 1.
Liquid scintillators. Relative pulse heights in presence (V_Q) and absence (V_o) of 0.1 M carbon tetrachloride.

Solute	Solvent	V_Q	V_o
10 g/l PBD	Toluene	100	100
10 g/l PBD	Benzene	95	77
10 g/l Butyl PBD	Toluene	86	99
10 g/l PBD	p-Xylene	84	96
10 g/l Butyl PBD	Benzene	84	80
8 g/l PBO	Toluene	82	98
8 g/l PBO	Benzene	81	77
10 g/l Butyl PBD	p-Xylene	79	97
8 g/l BBOT	p-Dioxan, 100 g/l naphthalene	77	67
8 g/l PBO	p-Dioxan, 100 g/l naphthalene	75	80
10 g/l Butyl PBD	p-Dioxan, 100 g/l naphthalene	74	72
10 g/l PBD	p-Dioxan, 100 g/l naphthalene	74	70
8 g/l PBO	p-Xylene	70	99
8 g/l BBOT	Toluene	60	79
8 g/l BBOT	Benzene	60	61
15 g/l BIBUQ	p-Xylene	56	103
8 g/l BBOT	p-Xylene	51	79
15 g/l BIBUQ	Toluene	51	103
15 g/l BIBUQ	Benzene	49	79
15 g/l BIBUQ	p-Dioxan, 100 g/l naphthalene	49	82
6 g/l PPO	Benzene	44	58
6 g/l PPO	p-Xylene	44	76
6 g/l PPO	Toluene	44	72
6 g/l PPO	p-Dioxan, 100 g/l naphthalene	35	59
5 g/l TP	Toluene	32	65
5 g/l TP	Benzene	30	52
5 g/l TP	p-Xylene	28	64

Bibliography

Books

1. Scintillation Counters
   (a) U.K. edition (London, Pergamon Press) 1953
   (b) U.S. edition (New York, McGraw-Hill) 1953
   (c) Russian edition (Moscow, Foreign Lit. Press) 1954
   (d) Polish edition (Warsaw, Panstwowe Wydawnictwa Techniczne) 1956
2. The Theory and Practice of Scintillation Counting (London, Pergamon Press) 1964
3. Photophysics of Aromatic Molecules (London, Wiley-Interscience) 1970
4. Organic Molecular Photophysics, Volume 1 (editor) (London, Wiley-Interscience) 1973
5. Organic Molecular Photophysics, Volume 2 (to be published 1974) (editor) (London, Wiley-Interscience)

Reviews

6. Nuclear scintillation counters. J.Brit.I.R.E. 11, 209-23 1951
7. Recent developments in scintillation counting. Science Progress 47, 484-94 1959
8. The scintillation process in organic systems. I.R.E. Trans.Nucl.Sci. NS-7, 2-11 1960
9. The efficiency of organic scintillators. Proc.Univ.New Mexico Conf. on "Organic Scintillation Detectors." August 1960 (Washington, A.E.C.) 1961
10. Improved organic scintillation detectors. Proc.Intern.Conf.(Belgrade 1961) on "Nuclear Electronics" (Vienna, I.A.E.A.) Vol. 1, 17-26 1962
11. Excimer fluorescence of aromatic hydrocarbons. Proc.Intern.Conf.Luminescence (Torun, 1963) Acta Physica Polonica 26, 367-78 1964
12. Scintillations in organic solids. In "Physics and chemistry of the organic solid state" ed. Fox, Labes and Weissberger (New York, Interscience) Vol. 2, 433-508, 909-24 1965
13. The history of organic scintillators. Organic Scintillator Symposium, Argonne National Laboratory, June 1966

14. Excimers and exciplexes. Nature 214, 1187-9 1967
15. The fluorescence lifetimes of aromatic molecules (with I. H. Munro)
Progress in Reaction Kinetics 4, 239-303 1967
16. Energy transfer, migration and diffusion in aromatic liquid solutions.
Proc.Intern.Symp.(Portmeirion 1967) on "Energetics and mechanisms in radiation biology" ed. Phillips (London, Academic Press) 203-21 1968
17. Energy transfer in organic scintillators.
Proc.Intern.Coll.Nuclear Electronics and Radioprotection, Toulouse 1968
18. An introduction to liquid scintillation counting.
(Philips, Eindhoven and Koch-Light Labs.) 24pp. 1968
19. Solvents and solutes for liquid scintillation counting. (Koch-Light Labs.) 40 pp. 1969
20. Excimer fluorescence of aromatic compounds.
Progress in Reaction Kinetics 5, 181-272 1970
21. Physics of the liquid scintillation process.
Proc.Intern.Symp.(M.I.T.)April 1969) on "The current status of liquid scintillation counting", ed.Bransome. (New York, Grune and Stratton) 1-12 1970
22. Liquid scintillator solvents.
Proc.Intern.Conf.(San Francisco, August 1970) on "Organic Scintillators and Liquid Scintillation Counting", ed.Horrocks and Peng (New York, Academic Press) 3-23 1971
23. Liquid scintillators.(with G.C. Poullis)
Proc.Intern.Symp.(Brighton 1971) "Liquid Scintillation Counting, Vol. 2" ed. Crook, Johnson and Scales (London,Heyden) 1-21 1972

Radiation physics

(a) Organic scintillators

24. Scintillations from naphthalene-anthracene crystals. Proc.Phys.Soc. A 63, 1044-6 1950
25. Scintillation efficiency of anthracene crystals. Proc.Phys.Soc. A 63, 1294-5 1950
26. Deterioration of anthracene under alpha-particle irradiation. (with F.A. Black)
Proc.Phys.Soc. A 64, 511-2 1951
27. Scintillation from organic crystals; specific fluorescence and relative response to different radiations. Proc.Phys.Soc. A 64, 874-7 1951

28. The specific fluorescence of anthracene and other organic materials. Phys.Rev. 84, 364-5 1951
29. Scintillation response of organic crystals to low energy alpha-particles. (with J.W.King) Phys.Rev. 86, 568-9 1952
30. Theory of the response of organic scintillation crystals to short-range particles. Phys.Rev. 86, 569 1952
erratum : Phys.Rev. 90, 1131 1953
31. Nuclear emulsion and scintillation crystal processes. Phys.Rev. 86, 791 1952
32. The radiation decomposition of organic molecules. J.Chem.Phys. 20, 1655-6 1952
33. The absolute scintillation efficiency of anthracene. (with M.E.Szendrei) Phys.Rev.91,197-8 1953
34. Scintillation response of anthracene to soft X-rays. (with F.D.Brooks) Phys.Rev.94,1800-1 1954
35. Temperature dependence in organic phosphors. Phys.Rev. 95, 277 1954
36. Scintillation response of anthracene to 6-30 keV photoelectrons. (with F.D.Brooks) Proc.Phys.Soc. B 69, 721-30 1956
37. The fluorescence and scintillation decay times of crystalline anthracene. Proc.Phys.Soc. 79, 494-6 1962
38. A comparison of the scintillation and photofluorescence spectra of organic solutions (with C.L. Braga and M.D. Lumb) Brit.J.Appl.Phys. 15, 399-404 1964
39. A modified theory of the organic scintillation process. Proc.Intern.Symp.Luminescence (Munich 1965) on "The physics and chemistry of scintillators" ed. Riehl and Kallmann (Munich, Verlag Karl Thiemig) 120-8 1966
40. The influence of excimer formation on solvent-solute energy transfer in organic liquid scintillators (with J.M.de C.Conte and G. Walker) I.E.E.E.Trans.Nucl.Sci. NS 13, No.3, 148-52 1966
41. The luminescence of liquid alkyl benzenes under high-intensity ionizing radiation. Chem.Phys.Lett. 4, 603-6 1970
42. Triplet-triplet interactions in organic scintillators. Chem.Phys.Lett. 7, 293-5 1970

43. Organic scintillators with improved timing characteristics (with R.W.Pringle) Proc.Roy.Soc. Edinburgh, (A) 70, 233-244 1972
(See also refs. 1, 2, 3, 8, 9, 10, 12, 16, 17, 18, 19, 21, 22, 23.)

(b) Inorganic materials.

44. The luminescence of air, glass and quartz under alpha-particle irradiation (with J.W.King) Proc.Phys.Soc. B 66, 81-4 1953
45. Cathodoluminescence of air, quartz and metal films (with W.A.Little) Nature 174, 82 1954
46. Some questions on inorganic scintillators. Proc.Intern.Conf.(New York 1961) on "Luminescence of organic and inorganic materials", ed.Kallmann and Spruch (New York,Wiley)645-6 1962
47. The scintillation process in alkali halide crystals. I.E.E.E.Trans.Nucl.Sci.NS-11, No.3, 4-10. (See also refs. 1, 2). 1964

Photophysics of organic molecular systems.

(a) Spectra

48. Fluorescence spectra of organic crystals (with G.T.Wright) Proc.Phys.Soc. B67, 657-63 1954
49. Fluorescence spectra of some organic crystals (with A.J.W.Cameron) S.A.Journal of Science, 53, 16-9 1956
50. Crystalline fluorescence of carcinogens and related organic compounds (with A.J.W.Cameron) Proc.Roy.Soc. A 249, 297-317 1959
51. The π-electronic excitation and ionization energies of aromatic hydrocarbons (with M.A.Slifkin) Nature, 191, 761-4 1961
52. The emission spectra of organic liquid scintillators (with J.E.Geake and M.D.Lumb) Brit.J.Appl.Phys.14, 141-3 1963
53. The dual fluorescence spectra of biphenylene (with J.M.de C.Conte and G. Walker) Phys.Letters 19, 125-6 1965
54. Fluorescence of the helicenes (with E. Vander Donckt, J. Nasielski and J.R.Greenleaf) Chem.Phys.Lett. 2, 409-10 1968
55. Assignment of two electronic states of benzene. Chem.Phys.Lett. 3, 567-8 1969

56. The photophysics of azulene
Chem.Phys.Lett. 17, 370-2 1972

57. The $^{1}E_{2g}^{-} \leftarrow {}^{1}A_{1g}^{-}$ transition in benzene
(with A.M.Taleb and I.H.Munro)
Chem.Phys.Lett. in press 1973

58. The spectroscopy of the π-electronic states of aromatic hydrocarbons. In "Organic molecular photophysics" ed. Birks (London, Wiley-Interscience) Vol. 1, 1-55 1973

(See also refs. 2, 3, 4, 5, 19, 20, 63, 84, 85, 88, 89, 93, 94, 98, 103, 106, 122, 123, 124, 125, 126, 128)

(b) Lifetimes.

59. Photofluorescence decay times of organic phosphors (with W.A.Little)
Proc.Phys.Soc. A 66, 921-8 1953

60. The photoluminescence decay of organic crystals (with T.A.King and I.H.Munro).
Proc.Phys.Soc. 80, 355-61 1962

61. Fluorescence lifetime studies of pyrene solutions (with I.H.Munro) Proc.Intern.Conf. (New York 1961) on "Luminescence of organic and inorganic materials" ed.Kallmann and Spruch (New York, Wiley) 230-4 1962

62. The decay times of anthracene crystals
ibid, 292 1962

63. The relations between the fluorescence and absorption properties of organic molecules (with D.J.Dyson) Proc.Roy.Soc. A 275,135-48 1963

64. Fluorescence decay and energy transfer in viscous solutions (with S. Georghiou)
Chem.Phys.Lett. 1, 355-6 1967

(See also refs. 2, 3, 4, 15, 37, 43, 74, 75, 76, 79, 86, 87, 96, 97, 110, 130)

(c) Energy transfer.

65. Energy transfer in organic phosphors
Phys.Rev. 94, 1567-73 1954

66. Transfer of molecular energy between solutes in liquids. J.Chem.Phys. 28, 730-1 1958

67. Energy transfer in organic systems. I. Photofluorescence of terphenyl-toluene solutions (with A.J.W.Cameron) Proc.Phys.Soc. 72, 53-64 1958

68. Transfer of excitation energy and radiationless transitions. J.Chem.Phys. 31, 1135-6 1959
69. Energy transfer in fluorescent plastic solutions (with K.N.Kuchela). Disc.Faraday Soc.27,57-63 1959
70. Energy transfer in organic systems
II. Solute-solute transfer in liquid solutions (with K.N.Kuchela) Proc.Phys.Soc. 77, 1083-94 1961
71. Energy transfer in organic systems
III. Spectral effects in solutions (with K.N.Kuchela and F.H.Read) Proc.Phys.Soc. 77, 1095-6 1961
72. Energy transfer and oxygen quenching in solution of 2,5-diphenyloxazole in benzene, toluene, p-xylene and mesitylene (with C.L.Braga and M.D.Lumb) Trans.Faraday Soc. 62, 1830-7 1966
73. Energy transfer in organic systems IV. Solvent diffusion, excitation migration and quenching (with J.Nafisi-Movaghar and K.Razi Naqvi) Proc.Phys.Soc. 91, 449-58 1967
74. Energy transfer in organic systems V. Lifetime studies of anthracene-perylene transfer in benzene solutions (with S.Georghiou and I.H.Munro) J.Phys.B (At.Molec.Phys.) 1, 266-73 1968
75. Energy transfer in organic systems VI. Fluorescence response functions and scintillation pulse shapes. J.Phys.B (At.Molec.Phys.) 1, 946-57 1968
76. Energy transfer in organic systems
VII. Effect of diffusion on fluorescence decay (with S.Georghiou)
J.Phys.B (At.Molec.Phys.) 1, 958-65 1968
77. Energy transfer in organic systems
VIII. Quenching of naphthalene fluorescence by biacetyl (with M.S.S.C.P.Leite)
J.Phys.B. (At.Molec.Phys.) 3, 417-24 1970
78. Energy transfer in organic systems
IX. Effect of diffusion on transfer efficiency (with M.S.S.C.P.Leite). J.Phys.B. (At.Molec. Phys.) 3, 513-25 1970
79. Energy transfer in organic systems
X. Pure and mixed crystals.
J.Phys.B.(At.Molec.Phys.) 3, 1704-14 1970

80. Energy transfer in organic systems
    XI. Temperature dependence of singlet excitation transfer and quenching in toluene solutions (with H.Y.Najjar and M.D.Lumb)
    J.Phys.B. (At.Molec.Phys.) 4, 1516-22 — 1971

    (See also refs. 2, 3, 16, 17, 21, 24, 40, 64, 100).

(d) Excimers

81. Excimer fluorescence of aromatic hydrocarbons in solution (with L.G.Christophorou)
    Nature 194, 442-4 — 1962
82. Resonance interactions of fluorescent organic molecules in solution (with L.G.Christophorou)
    Nature 196, 33-5 — 1962
83. Excimer formation in polycyclic hydrocarbons and their derivatives (with L.G.Christophorou)
    Nature 197, 1064-5 — 1963
84. Excimer fluorescence spectra of pyrene derivatives (with L.G.Christophorou) Spectrochim. Acta 19, 401-10 — 1963
85. Excimer fluorescence I. Solution spectra of 1:2-benzanthracene derivatives (with L.G. Christophorou) Proc.Roy.Soc. A274, 552-64 — 1963
86. Excimer fluorescence II. Lifetime studies of pyrene solutions (with D.J.Dyson and I.H.Munro)
    Proc.Roy.Soc. A 275, 575-88 — 1963
87. Excimer fluorescence III. Lifetime studies of 1:2-benzanthracene derivatives (with D.J.Dyson and T.A.King) Proc.Roy.Soc. A 277, 270-8 — 1964
88. Excimer fluorescence IV. Solution spectra of polycyclic hydrocarbons (with L.G.Christophorou)
    Proc.Roy.Soc. A 277, 571-82 — 1964
89. Excimer fluorescence of aromatic liquids (with J.B.Aladekomo) Spectrochim.Acta 20,15-21 — 1964
90. Fluorescence of organic mixed excimers
    Nature 203, 1062-3 — 1964
91. Temperature studies of the fluorescence of pyrene solutions (with M.D.Lumb and I.H.Munro)
    Acta Physica Polonica 26, 379-84 — 1964
92. Excimer fluorescence V. Influence of solvent viscosity and temperature (with M.D.Lumb and I.H.Munro) Proc.Roy.Soc. A 280, 289-97 — 1964

93. Excimer fluorescence VI. Benzene, toluene, p-xylene and mesitylene (with C.L.Braga and M.D.Lumb) Proc.Roy.Soc. A 283, 83-99 1965
94. Excimer fluorescence VII. Spectral studies of naphthalene and its derivatives (with J.B. Aladekomo) Proc.Roy.Soc. A 284, 551-65 1965
95. Double-photon excitation of excimer fluorescence of pyrene solutions (with H.G. Seifert) Phys.Letters 18, 127-8 1965
96. Excimer fluorescence VIII. Lifetime studies of 1,6-dimethylnaphthalene (with T.A. King) Proc.Roy.Soc. A 291, 244-56 1966
97. Excimer fluorescence IX. Lifetime studies of pyrene crystals (with A.A. Kazzaz and T.A.King) Proc.Roy.Soc. A 291, 556-69 1966
98. Excimer fluorescence X. Spectral studies of 9-methyl and 9,10-dimethyl anthracene (with R.L.Barnes) Proc.Roy.Soc. A 291,570-82 1966
99. The quintet state of the pyrene excimer Phys.Letters 24A, 479-80 1967
100. Excimer fluorescence XI. Solvent-solute energy transfer (with J.C.Conte) Proc.Roy.Soc. A 303, 85-95 1968
101. Experimental determination of the pyrene crystal excimer interaction potential (with A.A.Kazzaz) Chem.Phys.Lett. 1, 307-8 1967
102. Exciton resonance states of aromatic excimers. Chem.Phys.Lett. 1, 304-6 1967
103. Excimer fluorescence XII. The pyrene crystal excimer interaction. (with A.A.Kazzaz) Proc.Roy.Soc. A 304, 291-301 1968
104. The pyrene excimer. Acta Physica Polonica, 34, 603-17 1968
105. Higher excited states of benzene and toluene excimers Chem.Phys.Lett. 1, 625-6 1968
106. The ground and excited states of the pyrene crystal dimer. Molecular Crystal Symposium, Enschede, July 1968 1968
107. Some photophysical processes in toluene. Proc.Intern.Conf. (Loyola 1968) on "Molecular Luminescence," ed. Lim (New York, W.A.Benjamin, Inc.) 219-36 1969
108. Molecular and excimer fluorescence of toluene (with J.R.Greenleaf and M.D.Lumb) J.Phys.B. (At.Molec.Phys.) 1, 1157-9 1968

109. The luminescence of pyrene in viscous solutions (with B.N.Srinavasan and S.P.McGlynn) J.Mol.Spectrosc. 27, 266-84 1968
110. Influence of environment on the radiative and radiationless transition rates of the pyrene excimer (with A.J.H.Alwattar and M.D.Lumb) Chem.Phys.Lett. 11, 89-92 1971
(See also refs. 2, 3, 11, 14, 20, 21, 22, 40, 41, 50, 72, 73, 80, 113, 114, 116, 118, 121)

(e) Molecular complexes and quenching

111. Interaction of amino-acids, proteins and amines with chloranil (with M.A.Slifkin) Nature 197, 42-5 1963
112. Quenching of excited singlet and triplet states of aromatic hydrocarbons by oxygen and nitric oxide. Proc.Intern.Conf.Luminescence (Delaware, 1969) J.Luminescence, 1-2, 154-65 1970
(See also refs. 3, 14, 23, 72, 73, 77, 80, 90, 129)

(f) Delayed fluorescence and triplet-triplet interaction

113. On the delayed fluorescence of pyrene solutions J.Phys.Chem. 67, 2199-2200 1963
erratum. J.Phys.Chem. 68, 439 1964
114. Delayed excimer fluorescence (with G.F.Moore and I.H.Munro) Spectrochim.Acta 22, 323-31 1966
115. Prompt and delayed fluorescence of dyes in solid solutions (with J. Grzywacz) Chem.Phys.Lett. 1, 187-8 1967
116. Delayed luminescence and triplet quantum yields of pyrene solutions (with G.F.Moore) Proc. Intern.Conf. (Beirut 1967) on "The Triplet State" ed. Zahlan (Cambridge University Press) 407-14 1967
117. Triplet-triplet interaction of aromatic molecules in solution Chem.Phys.Lett. 2, 417-9 1968
(See also refs. 2, 3, 21, 22, 42)

(g) Photodimers

118. The photodimerization and excimer fluorescence of 9-methylanthracene (with J.B.Aladekomo) Photochem.Photobiol. 2, 415-8 1963
119. The photodimers of anthracene, tetracene and pentacene (with J.H.Appleyard and R. Pope) Photochem.Photobiol. 2, 493-5 1963

(h) Miscellaneous

120. A physical theory of carcinogenesis by aromatic hydrocarbons. Nature, 190, 232-5 1961
121. Singlet-triplet intersystem crossing in organic monomers and excimers (with T.A.King) Phys.Letters 18, 128-9 1965
122. Symmetry selection rule for π-electronic transitions in cata-condensed hydrocarbons Phys.Letters 19, 25-6 1965
123. π^*-duality in aromatic molecules Phys.Letters 19, 214-6 1965
124. Excitation of triplet states of organic molecules. Proc.Intern.Conf.(Beirut 1967) on "The Triplet State", ed. Zahlan (Cambridge University Press) 403-5 1967
125. Excited states of benzene and naphthalene (with L.G.Christophorou and R.H.Huebner) Nature 217, 809-12 1968
126. Fluorescence excitation spectra of aromatic liquids and solutions (with J.C.Conte and G.Walker). J.Phys.B. (At.Molec.Phys.) 1, 934-45 1968
127. Diffusion-controlled rate processes (with A.H.Alwattar and M.D.Lumb) in "Organic Molecular Photophysics", ed. Birks (London, Wiley-Interscience) Vol. 1, 403-56 1973
128. The $^1B_{1u} \leftarrow {}^1A_{1g}$ 0-0 transition in crystal benzene (with E. Pantos and T.D.S. Hamilton). Chem.Phys.Letters, in press 1973
129. The benzene-oxygen complex in the vapour phase (with E. Pantos and T.D.S. Hamilton). Chem. Phys.Letters, in press 1973

<u>Experimental methods and instruments</u>

130. Phase and modulation fluorometer (with D.J.Dyson) J.Sci.Instr. 38, 282-5 1961

131. Spectralresponse of antimony-caesium photocathodes (with I.H.Munro) Brit.J.Appl. Phys. 12, 519-22 1961

132. Improved spectrofluorimetry of organic crystals from -170 to +160°C (with A.A.Kazzaz) J.Sci.Instr. 43, 172-5 1966

133. Directional characteristics of cylindrical scintillators (with I. Petr and A.Adams) Nucl.Instr.Meth. 95, 253-7 1971

134. The composite directional gamma-ray scintillation detector (with I. Petr and A. Adams) Nucl.Instr.Meth. 99, 295-93 1972

(See also refs. 1, 2, 3, 6, 7, 10, 15, 18, 23, 43, 95).

A TRIPLE SAMPLE LIQUID SCINTILLATION COUNTER

Barton H. Laney
Searle Analytic Inc.
Des Plaines, Illinois (USA)

ABSTRACT

One of the major limitations of present liquid scintillation counters is their inability to count more than one sample at a time. Sample throughput would be greatly enhanced if two or more samples could be counted at the same time in one instrument. Functional parts of the liquid system could be shared, rather than duplicated, resulting in significant economies.

A liquid scintillation counter is described which measures three liquid scintillation samples simultaneously. Three phototube detectors are used to measure the three samples in a shared coincidence counter. Detection efficiency and background count rate are comparable to single sample counters. Isolation between the three sample chambers is better than 100 parts per million.

Four potential sources of error particular to analyzing events produced in a shared multiple detector system are examined: 1) pulse pile up, 2) dead time, 3) accidental coincidence, and 4) cross coupling between detectors.

INTRODUCTION

An important measure of liquid scintillation counting performance is the rapidity with which an assay can be performed. Most applications in the clinical and biological research laboratory require rapid throughput of relatively high activity samples. Improvements in detection efficiency and sample transport time would reduce counter time little. Except for tritium, most isotopes count at nearly 100% efficiency and transport time is typically less than

10 percent. In the environmental, health physics, and geological applications low activity samples are prevalent. E^2/B is an important figure of merit when the activity of the sample being assayed is equal to, or less than, the background rate (1). However, E^2/B is also a measure of minimum time to assay an unknown.

In either case an instrument which measures three samples simultaneously has counting efficiency and background rate essentially equal to that of all three samples added together. Whether measured by counting time, efficiency, or E^2/B, performance is improved by nearly a factor of three over the single sample counter. This paper describes a liquid scintillation counter which conserves time by counting three samples simultaneously.

MATERIALS AND METHODS

Detector

A new detector, using three 75mm phototubes to measure three samples, is shown top view in figure 1. The three samples are separated by a "Y" partition in the prism-shaped optical cell. Scintillations produced in chamber A are visible to multiplier phototube (MPT) 2 and MPT 3, but not MPT 1. Similarly, scintillations produced in chamber B are visible to MPT 1 and MPT 3, but not MPT 2. Three coincidence detectors, one for each sample, are used to separate the scintillations from each chamber. Similarly three summed amplifiers are used to analyze the pulse height spectrum from each chamber.

The detector head is constructed of 48mm steel plates. Each phototube is mounted in a steel housing (not shown) which slips into one of three ports on the side. Lead bricks, 5 cm thick, surround the detector (excluding phototubes).

The prism shaped cell is constructed of 1.5mm steel plates welded and machined for dimensional stability. Over all dimensions are 9.5cm on each side by 7.7cm high, sufficient to accomodate three 20 ml counting vials. Each sample chamber is lined with specular aluminum to minimize optical losses due to absorption. The "D" shaped opening from each chamber to the phototube has an area of $12cm^2$ (78% of a 50mm tube).

Phototubes are EMI type D247B, 75mm diameter with 10

stages of SbCs venetian blind dynodes. Each bialkali (K-Cs) photocathode has an effective area of $33cm^2$, twice that of the 50mm tube. Since the photocathode is shared by two counting chambers, less than half is available after partitioning. The cathode to dynode 1 inside surface of each phototube is coated with an anti-reflection coating to reduce the optical coupling between chambers. The smooth flat faceplates are made of 4mm Pyrex glass.

The interface between the "Y" shaped optical barrier and the phototube faceplate is designed to absorb light reflecting off the cathode from one chamber to another (figure 2). The black rubber seal is optically coupled to the faceplate. Because of the 4mm thickness, T, of the faceplate, the barrier seal is 7mm wide (D) losing over $3cm^2$ of cathode area.

Detector Logic

Coincidence gating has been an accepted method of reducing background in a liquid scintillation counter for over 20 years (2). Thermionic emissions from the photocathods of the phototube detector are virtually eliminated by imposing the requirement of coincidence. Although each phototube may produce several thousand thermionicly generated output pulses per minute, the chance of a simultaneous occurrance in both is rare. On the other hand, each scintillation from a beta decay produces many photons within several nanoseconds, some of which are detected, almost simultaneously, by each phototube. The coincidence detector is arranged to have a time window, τ, sufficiently wide (30 nanoseconds) to accept pulse pairs which are not exactly simultaneous.

Three coincidence detectors A, B and C, one for each chamber, are required to properly locate the origin of the scintillation in the three sample system (figure 5). Scintillations in chamber A excite MPT 2 and MPT 3. Each event will produce, almost simultaneously, pulses at both inputs to coincidence detector A and therefore produce an output from coincidence detector A. Coincidence detectors B & C will not produce an output because neither will be energized at the #1 input. All three coincidence detectors are identical. Each has a detection threshold of approximately ten microamperes and a resolving time, τ, of about 30 nanoseconds.

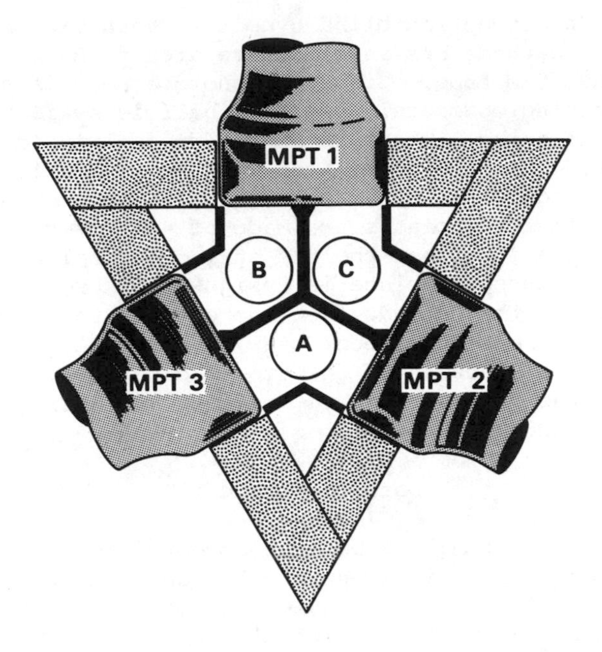

Figure 1 Detector partitioning.

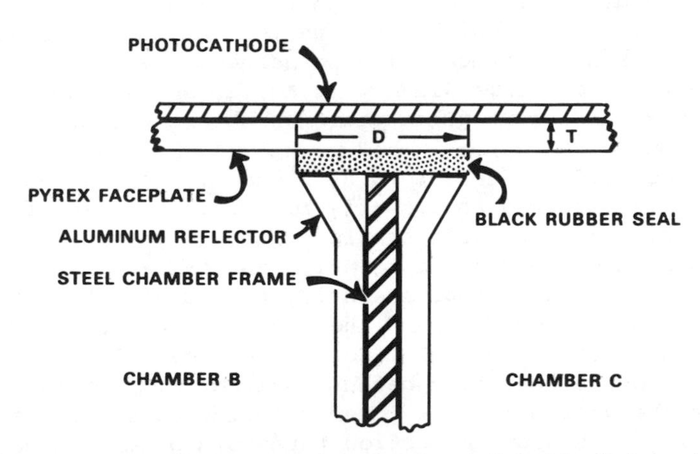

Figure 2 Cross-sectional view of optical barrier between counting chambers.

A first differential pulse height discriminator (3) was employed to discriminate between light emmissions from the phototubes (4) and scintillations produced by the sample. This circuit imposes a requirement of symmetry on the detector. Light generated in the sample tends to produce equal output from both phototubes and is accepted, whereas light generated asymmetrically, as within the phototube, is rejected because it exceeds the threshold of the Differential Discriminator. The differential threshold increases linearly with pulse height to accommodate any pulse height pair.

The Differential Discriminator was redesigned for the three sample system to operate upon three inputs. Rather than triplicate the Differential Discriminator function (one for each sample), it was modified to always respond to the largest and median pulse height from the three phototubes, independent of the phototube pair. Thus, the same differential discriminator criteria is automatically applied to the two phototubes with the largest pulse amplitude. Signals are also derived from this circuit to identify which phototube pair is the largest.

When two of the samples are high in activity the probability of chance simultaneous disintegrations from both is increased. If the scintillations from both samples are sufficient to excite all three phototubes simultaneously, the origin of the events is uncertain. Furthermore, a pulse would be added to all three sample channels. Fortunately, these occurrences are rare and represent a small fraction of the total count rate. They, therefore, can be rejected and neglected to be accounted for with little loss in system accuracy. The Triple Coincidence Detector shown in Figure 5 is used to inhibit all three outputs at "AND" gates A, B and C.

Optical isolation between chambers is imperfect. Intense scintillations from a sample in chamber A have increasingly good probability of exciting MPT 1 because of reflections from the inner surfaces of the phototubes. In an experiment to determine the amount of optical coupling, a ^{36}Cl liquid scintillation reference source was positioned in chamber A, and the spectrum from chamber C was recorded (figure 3). Chamber B was filled with opaque material to prevent coupling via MPT 3, chamber C contained a blank sample, and the Triple Coincidence and Differential Discriminator were disabled.

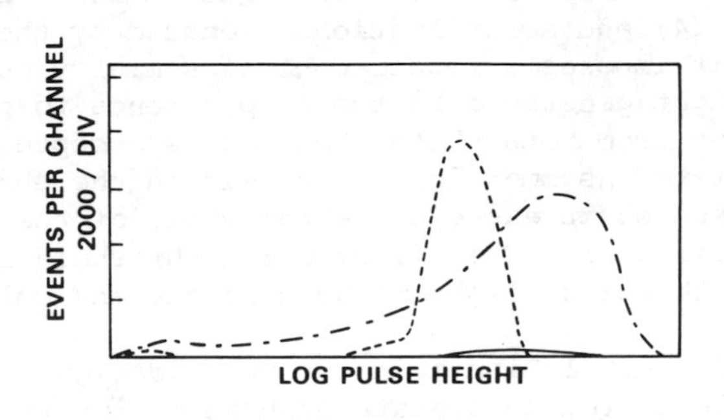

Figure 3 ^{36}Cl Cross coupling spectra without triple coincidence rejection.

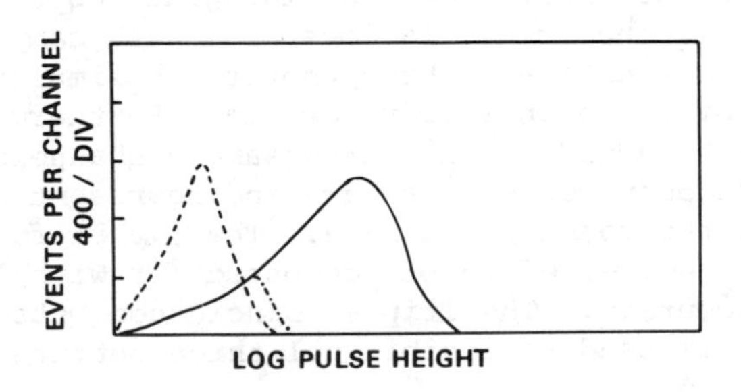

Figure 4 ^{14}C Cross coupling spectra without triple coincidence rejection.

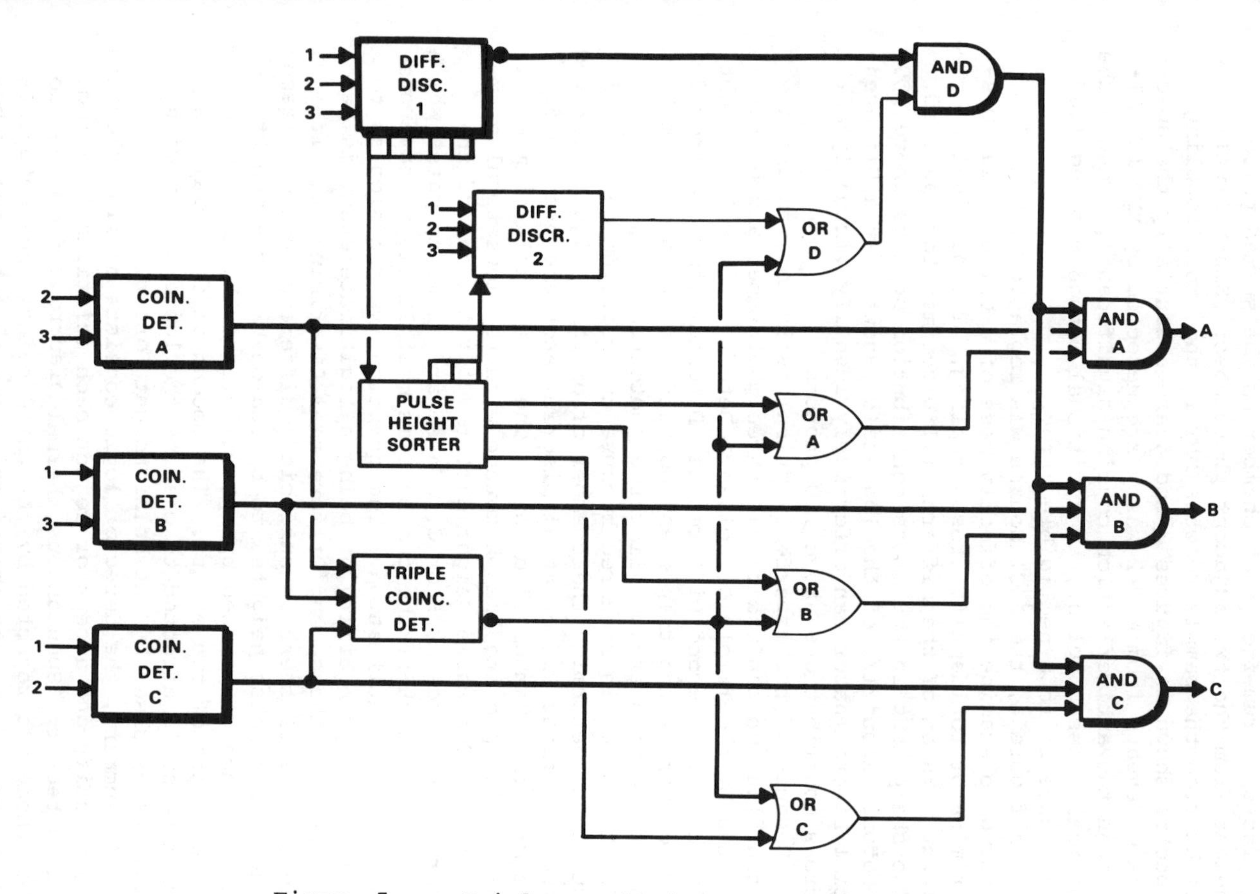

Figure 5 Triple sample detector logic diagram.

The spectrum from each chamber is the summed pulse height from the two adjacent phototubes. Since little light from the sample reaches MPT 1, the cross coupling spectra shown in figures 3 and 4 are essentially the spectra of events from MPT 2 which are detected by MPT 1. Although the antireflection coating significantly reduced the optical cross coupling, an additional method was required to reduce it to a negligible error.

In figure 4, the ^{36}Cl source was replaced with a ^{14}C source to examine the effectiveness of a Differential Discriminator to identify cross coupling. The original 2 input version of the Differential Discriminator was used. The sharp drop in the cross contribution spectra above 100 photons (about 1/3 of the log scale) shows that a Differential Discriminator can effectively identify virtually all cross contributions above 100 photons.

Because of the imperfect optical isolation, an intense scintillation can excite all three phototubes and be rejected by the Triple Coincidence Detector. A Pulse Height Sorter and a second Differential Discriminator are provided to overide the triple anticoincidence, when legitimate scintillations excite all three phototubes. The Pulse Height Sorter is required because the coincidence logic is unable to determine the chamber of origin. Signals derived from the first Differential Discriminator are used to determine the chamber of origin. The most likely chamber of origin is opposite the phototube with the lowest pulse amplitude. The Pulse Height Sorter activates one of the three "OR" gates A, B or C. These are steering gates which determine which of the chambers is most likely to have produced the scintillation. The quantitative decision as to sufficient probability of being a legitimate event is determined by the second Differential Discriminator. It determines if there is sufficient difference between lesser and median pulse heights, to be reasonably sure the triple coincidence was caused by optical crosstalk, not by random coincidence of two events. The second Differential Discriminator is directed to examine the lesser and median amplitude pulses by the Pulse Height Sorter.

In summary, the detector logic consists of five elements to identify and accept or reject each scintillation from the detector. Spontaneous thermal electrons from the photocathode are rejected by the three coincidence detectors. Spontaneous luminous emissions from within the phototubes are rejected by Differential Discriminator 1. Occasionally random coincidences of events from different samples are rejected by the triple anti-coincidence circuit. Scintillations of undeterminable origin which excite all three photocathodes are also rejected by the triple anti-coincidence circuit. Scintillations with sufficient intensity to excite all three phototubes are located by the Pulse Height Sorter. Differential Discriminator 2 determines if there is sufficient probability that the location determined by the Pulse Height Sorter is sufficiently correct to accept the event. Very few triple coincidences are indeterminable.

RESULTS

Tritium detection efficiency averaged 51.3% in the three chambers, about 13% (absolute) lower than in a single sample detector. The Carbon 14 and chlorine 36 efficiencies measured 95.4 and 100% respectively. The difference is attributable to chamber optics. The optical cell efficiency can be improved by the addition of a reflector around the top of the sample vial and by minimizing the area of black seal which is exposed. A reflecting top surface with a round hole for the sample, as is used in conventional detectors, would add about 10% more reflector surface to the chamber. Elimination of the black rubber seal exposed around the circumference of each window in the chamber would add another 15% to the reflecting surface.

Background count rate measured about 20 cpm greater than in a conventional detector. Total integral background averaged 62.9 cpm; 23.8 cpm in the ^{3}H energy range (unquenched) and 17.6 cpm between the ^{3}H and ^{14}C endpoints. About 8 cpm can be removed in the ^{3}H energy range by replacing the Pyrex face plates with low background material, such as quartz (5).

Chamber isolation was determined by placing an active sample in one chamber and measuring the cross-contribution detection efficiency in each of the other two chambers. Blank samples were placed in the other two chambers and background was subtracted from the gross count rate to obtain the net cross-contribution. Tritium, Carbon 14 and chlorine 36 cross-contribution efficiency measured 0.005%, 0.0015% and 0.0012% respectively. Cross-contribution is less at higher energies because both the triple anticoincidence and differential discrimination functions are more effective at higher energies.

DISCUSSION

Pile Up and Dead Time

Potential errors in pulse height analysis, due to pulse pile up and dead time losses, are slightly more than in a single sample counter. The essential difference is that the activity of all three samples contribute to the error. Conventional pulse amplifier and linear gating techniques can virtually eliminate pulse pile up as a source of error. Similarly, conventional live timing can eliminate potential errors caused by analyzer dead time. Because of the additional dead time, it takes a little more time to count three samples simultaneously than one at a time. The difference becomes significant only at high count rates.

Assume each sample has 10^6 cpm and the average dead time per event is 2 microseconds. The triple sample counter is busy analyzing events 10% of the time whereas the single sample counter is busy only 3% of the time. Therefore, the triple sample detector will take 7% longer to count three 10^6 cpm samples than to count one.

Accidental coincidences can either increase or decrease the count rate in a triple sample detector. A decrease occurs when events are discarded by the triple coincidence detector. Let us distinguish between double and triple accidental coincidences.

Double Accidentals

The probability of an accidental double coincidence event for any phototube pair is the same as in a single sample detector:

$$A_2 = 2\tau S_1 S_2$$

where τ is the coincidence resolving time for each input to the coincidence detector (30 nanoseconds). S_1 and S_2 are the single phototube count rates. The singles rates include chemiluminescence and phosphorescence excitation in addition to thermionic emission.

Present day bialkali phototubes have thermal emissions typically from 1 to 20 thousand counts per minute and produce negligible accidental coincidence. However, chemiluminescence and phosphorescence samples may easily produce hundreds of thousands of counts per minute at each phototube. Decreasing the resolving time, τ, only linearly diminishes the accidental count rate; whereas the accidental count rate increases with the square of the luminescent rate.

In a triple sample detector a second type of double accidental coincidences can occur. High monophotonic event rates can cause not only an erroneous increase in count rate but can indicate the wrong sample. Assume chamber A is empty and there are two samples in chambers B and C with intense monophotonic event rates. Each random coincidence between a pulse at MPT 3 and MPT 2 caused by samples B and C respectively, will appear as a coincidence from the empty chamber A. These events will not be rejected by the Triple Coincidence Detector because MPT 1 is unexcited during these chance coincidences. It is best to avoid high chemiluminescence or phosphorescent samples or to wait until the luminescent energy has decayed to an emission rate which produces negligible accidental coincidences.

Low energy scintillations can produce the same kind of counting error when the intensity of the scintillations are low enough to excite one but not both phototubes. The maximum probability of an accidental coincidence in the wrong detector is less than 0.01% even when both samples have an activity of 10^6 disintigrations per minute at the worst intensity, 3-5 photons. Less intense scintillations are less likely to produce any detectable output. More intense scintillations are more likely to be rejected as triple coincidences.

Triple Accidentals

Triple coincidences may occur due to optical crosstalk between chambers or from the accidental coincidence of events in two different samples. Triple coincidences caused by optical crosstalk are identified by the Pulse Height Sorter and Differential Discriminator 2 and correctly counted. Accidental triple coincidences cannot be properly analyzed and are rejected. Although two legitimate events are discarded the fractional loss is small. Even if all three samples have count rates of 10^6 cpm the loss amounts to only 0.2% assuming a coincidence resolving time of 30 ns.

The probability of an accidental triple coincidence is:

$$A_3 = 2\tau(R_1R_2 + R_2R_3 + R_1R_3)$$

where A_3 = accidental triple coincidence rate

R_1, R_2, R_3 = count rates in each chamber

However, each triple coincidence rejected is a loss of two events for all three samples. Therefore the loss per sample is 2/3 the accidental coincidences (assuming equal count rates).

Gamma Emitters

Samples containing gamma emitting radionuclides can produce cross coupling between chambers. Scintillations from Compton electrons may be produced in one sample by a gamma emitted from another. Cross coupling from a low energy gamma emitter like ^{125}I (35keV) can be virtually eliminated with 1.5mm of lead shielding. However, shielding becomes unpractical for gammas above 100keV. Because of the required close proximity of the samples little shielding can be added to protect against cross coupling from high energy gammas.

A triple chamber detector is a practical method of improving sample throughput. Improvements in detector optics are required to achieve performance equal to that of current single sample detectors. Data obtained from any sample is virtually independent of the activity or beta energy of adjacent samples. Gamma emitters below 100keV may also be assayed with negligible cross-contribution.

ACKNOWLEDGMENT

The author wishes to thank Mr. H. Engberg and Mr. J. M. Dudley for their assistance particularly in analyzing the sources of cross-contribution

REFERENCES

1. R. Loevinger and M. Berman, Nucleonics 9, No. 1, 26 (1951)
2. R. D. Hiebert and R. J. Watts, Nucleonics 11, No. 12, 38 (1953)
3. B. H. Laney *in* Organic Scintillators and Liquid Scintillation Counting, p. 991 (D. L. Horrocks and C. T. Peng Ed.). New York and London: Academic Press (1971)
4. H. R. Krall, IEEE Transactions on Nuclear Science, NS 14, No. 1, 455, (1967)
5. I. S. Boyce and J. F. Cameron *in* Tritium in the Physical and Biological Sciences, p. 231, IAEA, Vienna (1962)

A NEW LIQUID SCINTILLATION COUNTER FOR MEASUREMENT OF TRACE AMOUNTS OF ^3H and ^{14}C

John E. Noakes
Michael P. Neary
James D. Spaulding
Geochronology Laboratory
The University of Georgia, Athens, Georgia U.S.A.

Abstract

A newly designed liquid scintillation counter has been built specifically for the measurement of low specific activity samples of ^3H and ^{14}C. The instrument's electronics measure fast-pulse time intervals rather than performing pulse height analysis. Massive graded shielding and a background guard shield are utilized to minimize external radiation for low background counting. Figure of merit (E^2/B) of the counter is an order of magnitude superior to present commercial liquid counters. The external background guard shield was evaluated in the absence of a massive graded lead shield. The use of such a background guard shield in commercial liquid scintillation systems is considered.

Introduction

In an earlier paper (1) we described a new kind of low-level liquid scintillation counter that was designed, built, and tested in our laboratory. The importance of tritium measurements were stressed and performance data for the counter were given there in terms of ^3H.

It is certainly true that the quantative detection of other β emitting nuclides in low specific activity is of interest. The counter previously mentioned is not limited to the analysis of ^3H, but rather can be used to detect and quantify other β emitters.

^{14}C is a nuclide which is widely used and counted by both the physical (2,3,4) and life scientist. More specifically, carbon dating has become a widely used tool of the geologist, archaeologist, etc. By converting the carbonaceous sample to benzene (5) and counting in a liquid scintillation counter of the conventional type, age measurements can be estimated from the counts collected during the counting interval, t, with some precision . However, with the instrument described here and the above sample, either its' age could be estimated on the basis of a shorter counting interval, t' and the same precision or its' age determined with greater precision where the count is taken during the same interval t; and, in principle, older samples can be analyzed with the new system. This advantage is based on the larger sized sample that can be counted with a lower background. Moreover, such an advantage can be applied to other experimental systems.

Modern, commercially available, liquid scintillation counters employ pulse height discrimination (where pulse height is a much convolved energy function) as a means of utilizing the coincidence counting technique. In the low-level liquid scintillation counter described here, we employ time discrimination to implement the coincidence counting technique. The use of an annular NaI(Tl) crystal scintillator shield which served to reduce the counter's background by 82% was another difference between the low-level liquid scintillation counter and a commercial liquid scintillation counter. The high cost of the crystal annulus precludes its' general use. Therefore, a lower cost scintillating plastic annulus was tested with the low-level liquid scintillation counter. The results were compared with those obtained under the same circumstances, but with the NaI(Tl) crystal in place in an effort to determine the efficacy of such an approach to low level counting.

Experimental Procedure

A) COUNTER: Conventional liquid scintillation counters use the coincidence counting technique which places the sample between two photomultiplier tubes which share a common axis. For a data pulse to be generated at the counter's scaler, two pulses per beta event must arrive at the coincidence gate within a specified time interval, usually 20 nsec. Differential counting is accomplished by the use of pulse height discriminators which select or reject data pulses on the basis of their beta energy analog, pulse height. Background radiation is excluded through the use of lead shielding. No method is employed to discriminate between external and internal background noise.

The counter reported in this paper differs considerably from commercially available liquid scintillation counters. Particularly in the way in which raw data pulses are tested and selected and/or rejected; as well as, the manner in which shielding is employed. Fig. 1 shows the instruments' control panel and shielded counting chamber. A detailed discussion of these features follows.

B) ELECTRONICS: Fig. 2 shows a block diagram of the counter. With the exception of the photomultiplier tubes, all of the electronics were made up from state of the art ORTEC NIM bin components. A pair of RCA 4501/V4, 2" diameter, head-on "Quantacon" photomultiplier tubes (PMT) were obtained from the Industrial Tube Division of RCA Inc. as well as, the four RCA 6199 photomultipliers used in conjunction with both the crystal and plastic shield. The 4501/V4 tubes were selected because of their high quantum efficiency (31% at 385$\pm$50 nm), high gain (secondary emission ratio of 50 for the first dynode with 1000 V between the cathode and the first dynode) and low dark current (less than 10,000 c/m at 22°C). The high quantum efficiency is achieved by using a bialkali photocathode (potassium-cesium-antimony) and high gain by using a secondary emitting surface of galliumphosphide.

Electrical pulses, whose area is proportional to the number of photons sensed at the photocathode of the two photomultiplier tubes, are amplified by two preamplifiers (PA), one for each PMT. Only pulse shaping due to variation in the fall-time of the PA is done by the PA.

The output of the two PA's is directed to two delay line amplifiers (DLA) which provide shaping for all output pulses and expand the pulse amplitude by a selected gain factor of from 3 to 1000. Since no two PMTs can be expected to exhibit exactly equal gain for a given number of photons, equal sensitivity in each DLA is effected with the DLA gain adjustments. While both unipolar and bipolar output pulses are generated by the DLA, the bipolar output pulse was selected in our case since it is double delay-line shaped and thus provides a precision measure of time at the baseline crossover which is independent of pulse amplitude. The fastest integration time was selected to give the best pulse shapes. The height of an output pulse can range from 0 to 10 V and is dependent only on the area of the input pulse.

The output pulses of both DLAs are collected at a dual sum and invert amplifier (DSI). The DSI serves only to collect and make available for displaying and/or recording all of the raw data pulses produced by the two PMTs. This includes not only pulses dependent on beta

Figure 1. Over-all view of the low-level liquid scintillation counter - includes massive graded lead shield and electronics.

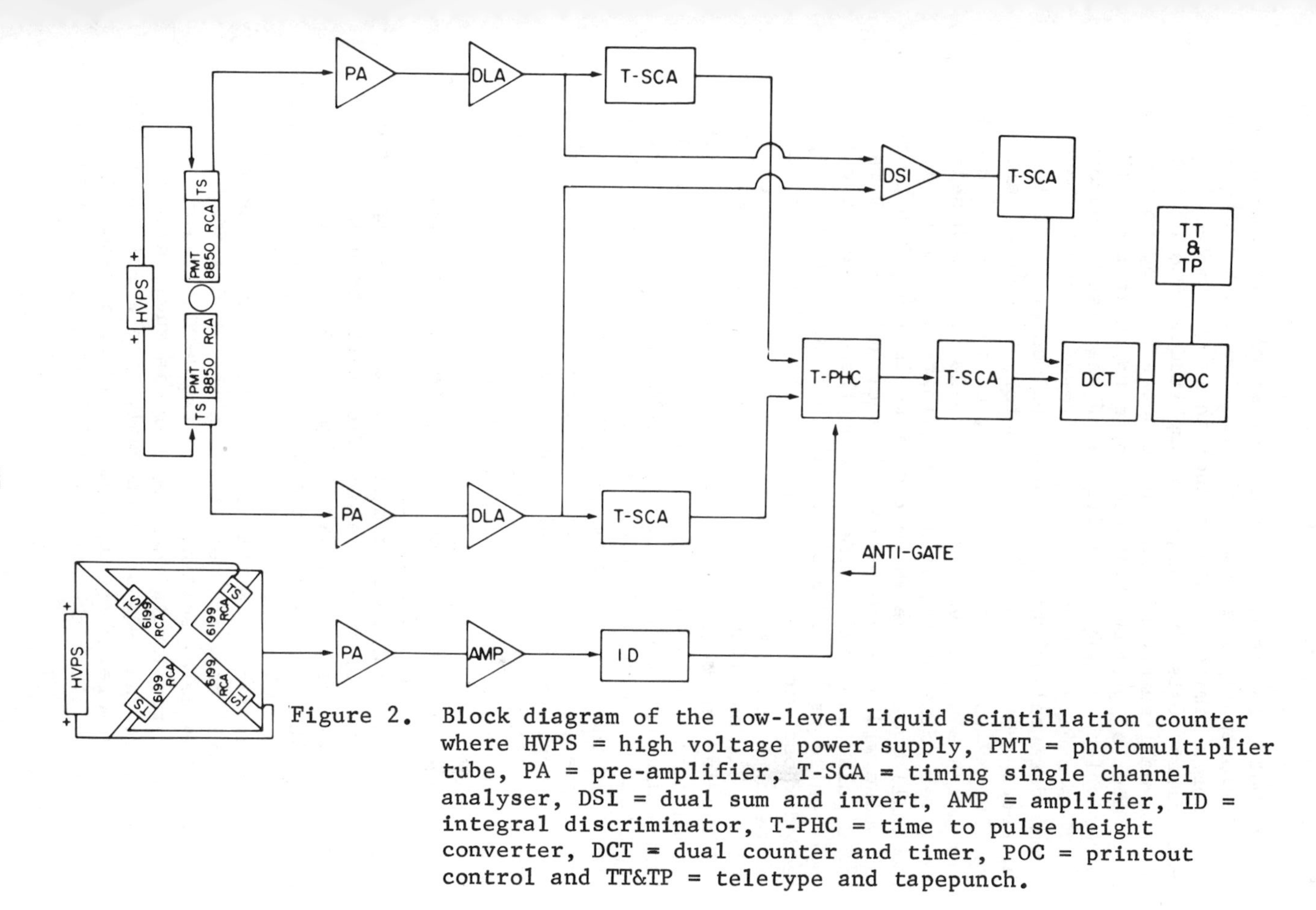

Figure 2. Block diagram of the low-level liquid scintillation counter where HVPS = high voltage power supply, PMT = photomultiplier tube, PA = pre-amplifier, T-SCA = timing single channel analyser, DSI = dual sum and invert, AMP = amplifier, ID = integral discriminator, T-PHC = time to pulse height converter, DCT = dual counter and timer, POC = printout control and TT&TP = teletype and tapepunch.

decay, but also the dark current or singles from each PMT.

The output pulse from each DLA also becomes the input pulse for a timing single channel analyzer (T-SCA). The T-SCA first tests the pulse for amplitude. The amplitude of this well-shaped bipolar input pulse is proportional to the number of photons that give rise to it at the PMT. Pulses which satisfy the criteria of the discriminators and thus occupy the pulse height interval delta E above E, become data pulses while the other pulses are rejected. The setting of the lower discriminator, E may be varied from 100 mV to 10 V and the setting of the interval above E or delta E may be varied from) to 10 V. The baseline crossover point of the data pulses initiates both a +5 V (nominal) square wave whose rise time is less than 20 nsec. and whose width is nominally 500 nsec. and a -0.6 V fast logic pulse whose rise time is less than 5 nsec. The fast logic pulse is selected since only a fast trigger is needed whose width is less than 20 nsec. A delay of 1.1 microsec to 100 nsec. can be introduced, as required, between the baseline crossover point of the data pulse and the leading edge of the fast logic pulse. It is worth noting that regardless of the input pulse amplitude, the fast logic output pulses of the T-SCAs are all of equal amplitude and differ from one another only in time. Note also that the fast logic output pulses from the T-SCA are initiated not only by scintillations but also by PMT dark current or singles. The contribution to the output of the T-SCAs by the PMT singles is eliminated by the coincidence requirement imposed by the next electronic operation.

The time to pulse height converter (T-PHC) receives three inputs, two of which are the output pulses of the two T-SCAs and one anti-gate input pulse from either the crystal or the plastic shield within which both the sample and the PMTs are placed. The latter signal's source will be dealt with in detail in the section of this paper concerned with background shielding. Briefly, however, this signal is used to eliminate a large fraction of the external radiation which gets through the lead graded shield. The delay setting of the two T-SCAs, which provide two of the inputs for the T-PHC is such that if two pulses leave the DLAs simultaneously, the resulting T-SCAs, and usually more due to differences in intrinsic electronic delay along the two paths followed by the data pulses. So that the output pulse of a given T-SCA will always occur prior to that of the other T-SCA, sufficient delay is introduced into the latter. For the experiment described here a minimum difference of 35 nsec. is used, the reasons for which will be discussed below.

The T-PHC, under the influence of the T-SCA's input pulses, generates a bipolar output pulse whose amplitude is proportional to the time difference between the two fast logic input pulses, (note that two such fast logic pulses are produced per beta event). Of the two fast logic pulses, the one from the T-SCA with the least delay always arrives at the T-PHC first and serves as a start pulse for the timing gate; the other fast logic pulse serves as it's stop pulse. If the time interval separating the start and the stop pulses is less than the selected gate range, an output pulse is produced. If, however, the stop pulse arrives after a time interval which is longer than the selected gate range, no output data pulse is generated. Moreover, if an anticoincidence pulse arrives at the T-PHC within the selected gate range along with a start and stop pulse, the generation of a data pulse is disabled. The logic being that if all three input pulses exist within the gate range, then an external event has given rise to them and is thus determined to be false data or background. For our purposes the 50 nsec. range was selected so that the probability of selecting false data would be minimized. Other gate ranges that are switch selectable included: 250 and 500 nsec. The selected gate range defines the range of the pulse height distribution of the T-PHC output pulses. Recall that the amplitude of the output pulses of the T-PHC is proportional to the time interval separating the start and stop fast logic pulses. The output pulse of the T-PHC is bipolar and has a constant pulse shape which is independent of it's amplitude. These pulses provide the input for yet another T-SCA whose task is to impose time discrimination on them and generate a square wave output pulse whose amplitude is +5 V (nominal) and 500 nsec. wide. For this T-SCA the E or lower discriminator is properly termed the T (for time) discriminator and the delta E termed delta T.

One scaler of a dual counter and time (DCT) receives the output pulses of the T-SCA discussed above. The other scaler of the DCT may receive as an input either the output of the DSI or that of it's timer. In this study the timer output was selected as this scaler's input so that count rates could be measured. Besides creating the time base with which any or all of the data are scaled, the DCT can be set up to display the data in either scaler. By feeding the output of the DCT to a print-out control POC, shown in Fig. 2 , the contents of both scalers can be permanently preserved on either a punched paper tape or a teletype printout or both.

C) NaI(Tl) CRYSTAL: A Bicron Inc., Mod. 7.5 HWS annular crystal, whose dimentions are 8" O.D., 5" in length, and 3½" I.D. was used in conjunction with four RCA 6199 PMTs and associated electronics as a background shield and is shown in Fig. 2 & 3 . With the sample and the two RCA 4501/V4 nested within the annulus of the crystal, all but a small fraction of the background radiation reaching the crystal and sample is detected, giving rise to an anti-coincidence pulse at the T-PHC. Such a pulse identifies the two fast logic pulses from the preceeding T-SCA's as false data and aborts a data pulse produced from them.

As before, the PA provides amplification, but virtually no shaping. The amplifier (Amp) is used to provide further amplification and Gaussian shaping of the unipolar output pulse. The integral discriminator (ID) requires an imput pulse of at least a 50 nsec. width and up to +10 V amplitude. Such input pulses are provided by the Amp when the proper gain of X4 is selected. The output of the ID is composed of those pulses initiated by Amp pulses whose amplitude exceeds or equals the discriminator setting. The output signals are 500 ns wide square pulse of +5 amplitude. It is necessary to introduce considerable delay in the anticoin pulse so that its time of arrival at the T-PHC corresponds to within 50 nsec. of that of the associated start and stop fast logic input pulses.

The effect of the NaI(Tl) crystal, an anticoincidence guard, was measured with a 25 ml background sample in a quartz vial of high optical quality. When the crystal was disabled, a count rate of 10.40 ± 0.09 c/m was observed in an optimized window, but with the crystal in operation, the background dropped to a count rate of 2.10 ± 0.02 c/m. The difference in the two count rates shows a 81.5% reduction in background when the NaI(Tl) crystal shield is used.

D) PLASTIC SHIELD: The plastic (NE-102) background shield was obtained from Bicron Inc. It was an annulus

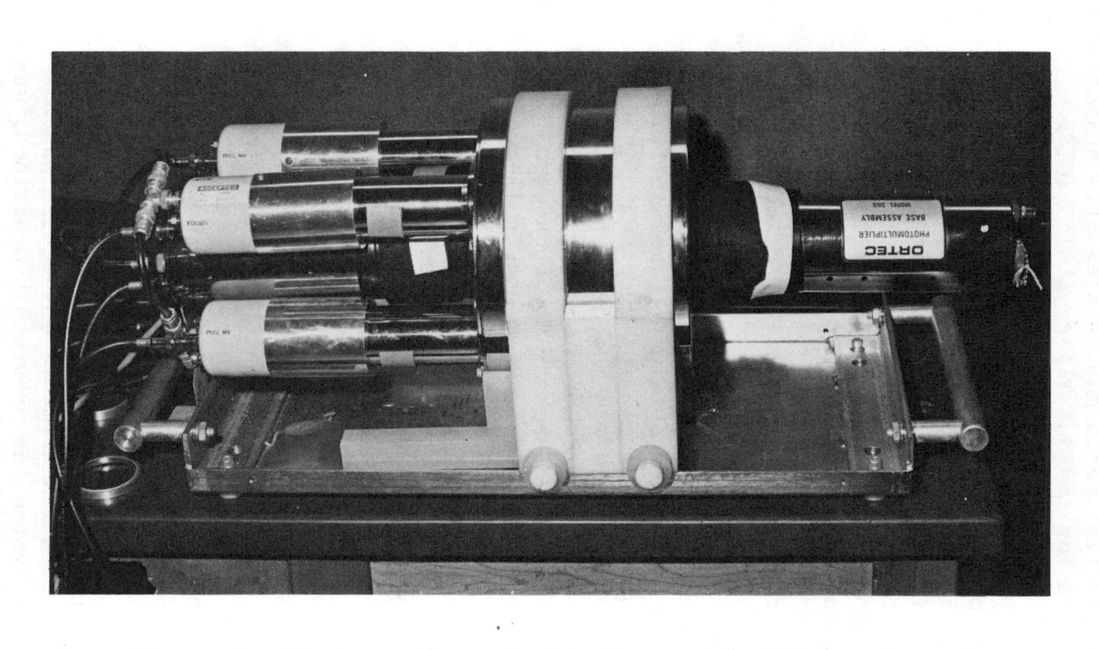

Figure 3. Background guard shield exposed with sample and PMT's mounted. (This particular one is the NaI(Tl) crystal.

whose dimensions are 5" in length, 9.5" O.D. and 3.5" I.D. Fig. 4 shows the above shield alone and Fig. 2 shows it diagramatically as a part of the system. Mechanically and electronically its' performance is exactly analogous to that described for the NaI(Tl) crystal shield.

Results and Discussion

It was our objective with this study to compare the performance of the NaI(Tl) crystal shield with that of the plastic (NE-102) shield, where, in both cases, no graded lead shield was present.

The composition of the scintillation solution used for all measurements was: 6.25 g PPO (2, 5-diphenyl-oxazole) and 0.4 g POPOP (1, 4-di-2-(5-phenyl-oxazole) benzene) thoroughly mixed into 1000 ml of benzene. In all cases the final concentration of the fluors in the sample was adjusted to that given above so that constant scintillator efficiency was insured. All chemicals used were of scintillation grade and obtained commercially.

Each background sample was composed of the scintillator described above and was measured into the vials with class "A" pipets. One hundred minutes was selected as the counting interval, and each of the background count values reported is the average of the counts collected during many counting intervals. The error associated with each background was computed to a 2σ confidence level and is based on the sum of all of the counts collected during each of the counting intervals of a given measurement.

The electronics previously described were routinely calibrated. A precision pulser operating at either 60 or 220 Hz and producing 1 nsec. rise up to +10 V tail pulse was used in conjunction with a fast oscilloscope. The test BNC connector on each PA was used to accept the pulser's input and was subsequently treated, as any data pulse would be, by the system's electronics. Pulse shapes and timing were measured with the scope.

Initially both long and short term stability was studied. Long term stability is illustrated by Fig. 5. Short term stability was checked with an external γ source (^{137}Cs). In both measurements Chi Square was obeyed.

When the system, as previously described, was used to study ^{3}H, ^{14}C, and background samples, the results shown in Table I were obtained. It is clear that an E^2/B maximum is exhibited at 64.7% relative efficiency. It

Figure 4. The plastic background guard shield exposed and without the PMT in place.

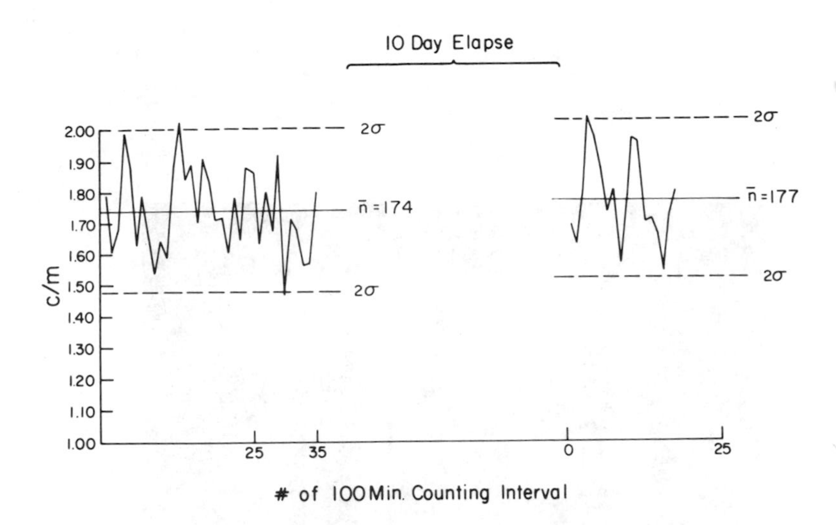

Figure 5. Long term system stability test count data.

TABLE I

Performance data of NaI(Tl) crystal background guard shield when inside the massive graded lead shield. Both ^{14}C and ^{3}H where considered.

Window (%)[a]	Bkgd in c/m[b]	Eff in %[c]	E^2/B[c]	E^2/B per ml[c]
100.0	4.20±0.5	96.0 (64)	2194 (975)	100 (44)
95.0	3.55±0.41	91.2 (61)	2332 (1041)	106 (47)
88.8	2.11±0.40	85.2 (57)	3440 (1540)	156 (70)
75.9	1.22±0.26	72.9 (48)	4356 (1889)	198 (86)
64.7	0.79±0.24	62.1 (41)	4882 (2170)	222 (98)
55.3	0.71±0.24	53.1 (29)	3971 (1764)	180 (80)
Typical commercial counter (d)	37 (29)[c]	94 (59)	237 (118)	15 (7)

a) relative % efficiency.

b) all errors expressed in c/m at 2σ confidence level.

c) numbers in parenthesis are for ^{3}H; those without parentheses are for ^{14}C.

d) Nuclear Chicago Mark II bench-top LS counter without the crosstalk level feature active.

should be born in mind that these results were obtained with the use of both the graded and NaI(Tl) crystal shield.

As shown in Fig. 3 , the background guard shield was outside of the graded lead shield, but placed on a bench-top, physically near to the electronics. The sample and PMT's were nested within the annulus, such that they all had a common axis. The PMTs were light-sealed by means of a felt tape of our own construction. Whether the NaI(Tl) crystal or the plastic served as the background guard shield, the above experimental configuration was used. The data shown in Table II illustrates the effectiveness of the background guard shield.

It is clear from an examination of Table II that the NaI(Tl) is a very good passive filter for background radiation and the plastic is not (passive is meant to denote a state of electronic inactivity). Since the average "Z" of the NaI(Tl) crystal is much higher than the plastic these results are reasonable. When the background guard shield is made active the NaI(Tl) again bests the plastic guard shield, by 30% (relative). In view of the high efficiency of the NaI(Tl) crystal, its' superior performance is expected.

Type	Mode	Bkgd in $^{C}/_{M}$ (a)	$^{E^2}/_{B}$ for ^3H (b)	$^{E^2}/_{B}$ for ^{14}C (b)
NaI(Tl)	on	17 ± 0.24	99	227
	off	58 ± 0.48	29	66
Plastic	on	70 ± 0.75	24	55
NE102	off	192 ± 0.88	9	20

(a) errors expressed to 2σ confidence interval.
(b) based on best E^2/B window as shown in Table I.

Table II Performance data of background guard shields in the absence of the massive graded lead shield.

From these results it can be seen that the counting system which employs the NaI(Tl) crystal and no lead gives results comparable with those of typical low cost commercial liquid scintillation counters. The performance of the plastic guard without lead was not as good. With an optimized background guard shield which makes use of a new scintillation material, Bicroguard^{TM}, counter performance equivalent to that of a commercial counter is to be expected. It is worth noting that such performance is obtained without massive graded lead shielding, and thus, such an arrangement may have future commercial application.

References

1. J.E. Noakes, M.P. Neary and J.D. Spaulding, Nuc. Insts. and Meth., 109, 177, (1973).
2. D.G. Jacobs, Trans. Am. Nucl. Soc., 14, 160, (1972).
3. P. Goldsmith and F. Brown, Nature, 191, 1033, (1961).
4. T. Wyerman, Private communication (U.S.G.S. Tritium Lab., Washington, D.C., 1972).
5. J.E. Noakes, S.M. Kim, and J.J. Stipp, "Proceedings of Sixth International Conference on Radiocarbon and Tritium Dating", Pullman, Washington, 68 (1965).

TM Bicroguard, Bicron Inc.

A LOW BACKGROUND LIQUID SCINTILLATION COUNTER FOR ^{14}C

P.E. Hartley and V.E. Church

Australian Atomic Energy Commission, Research Establishment, Lucas Heights, N.S.W.

Abstract

A liquid scintillation counter for ^{14}C with an E^2/B of over 1800 has been built and operated on a routine basis for over a year. It has a background of 1.96 ± 0.03 c.p.m. and an efficiency of 60.3% ± 0.5% has been measured using 5 ml of ^{14}C benzene prepared from N.B.S. Contemporary Standard Oxalic Acid (14.24 ± .07 d.p.m./g of C).

Signals from the last dynodes of the two EMI 9635QB photomultiplier tubes are fed through amplifiers, discriminators and a coincidence system which are entirely contained on three small printed circuit boards mounted near the tube bases. The anode signals of the two tubes are fed into a single N.I.M. preamplifier, amplifier, linear gate (controlled by the coincidence unit) and to a pulse height analyser. As the system gain is high, an EHT of less than 1000 volts can be used.

The sources of background are discussed. Optical and electronic methods of substantially reducing crosstalk, the major component of background, without reducing counting efficiency are described.

Introduction

A high figure of merit (% efficiency^2/background c.p.m.) is necessary for a counter used for ^{14}C dating. Commercial liquid scintillation counters have a typical figure of merit of 600, but this may be easily increased to 1200 by simple modifications to reduce the background (1). To reduce the background further special counters with anti-coincidence shields have been built (2).

The counter described in this paper achieves a figure of merit of 1800 without the need for anticoincidence shielding, by reducing the crosstalk between the two photomultipliers by optical and electronic means.

Sources of Background

The background of a coincidence type liquid scintillation counter may be divided into three components:

(a) Accidental coincidences between the two channels.

(b) Coincidences from light pulses in the sample itself, i.e. in the part of the system which is common to both channels.

(c) Coincidences from light pulses which originate in one channel and which are seen by the other channel.

Component (a) arises mainly from thermionic emission from the photocathodes and may be reduced to well below 0.1 c.p.m. by using a short resolving time. Component (b), light pulses in the sample, is caused by external radiation. It may be reduced by shielding, preferably with materials selected for low radioactivity. Little improvement will be obtained beyond 10 cm of lead or equivalent. A window discriminator almost eliminates the high energy cosmic ray components and further reduces the γ background. Component (c) is usually the major one. It originates from two sources, namely, light pulses associated with the operation of the photomultipliers, which may be minimised by using a low operating voltage (3), and those caused by cosmic rays and radioactivity of the photocathode window. Some of the light from these pulses will reach the other photomultiplier and may trigger the coincidence circuit causing a background count to register. Background arising from this process will be referred to as cross talk background, and the fraction of light leaving one photomultiplier which reaches the other as the cross talk ratio. Cross talk background is directly but not linearly related to cross talk ratio. If the ratio is minimised without affecting the light collecting efficiency the figure of merit will be improved. Even if light collecting efficiency is reduced along with the cross talk ratio an improved figure of merit may still result, especially with higher energy β emitters. Cross talk background may also be reduced by increasing the coincidence thresholds. Most instruments have fixed thresholds, but the same effect may be achieved by lowering the tube voltage and increasing the main (sum) amplifier gain.

Reduction of Cross Talk Ratio

The cross talk ratio may be reduced without reducing light collecting efficiency by using a vial made from a material which reflects back light from the photomultipliers but does not absorb any light from the scintillator. Consider a material which diffusely transmits a small fraction μ of the light incident on it and diffusely reflects all the remainder. A vial of this material is mounted between two photomultiplier tubes so that a fraction δ of its surface is coupled to the cathode of each photomultiplier with negligible loss. (This can be done with a light guide or mirror system which has constant or increasing cross section from vial to photocathode (4).) If absorption in the scintillator is negligible and μ is small enough for each photon to be reflected several times before leaving the vial, then the fraction of the total light emitted by the scintillator reaching each photocathode is δ and the light collecting efficiency is 2δ (5). Light originating from the photomultiplier will be reflected back by the vial and absorbed, and only a fraction μ will be transmitted to the scintillator. A fraction δ of this will then be transmitted to the other photomultiplier. The cross talk ratio is therefore $\mu\delta$ if there is no light leakage around the vial. It can be seen that a low μ and low light absorption are desirable properties for a vial material. μ for 1 mm thick Teflon is about 0.17 and for 2 mm or thicker about 0.1. Teflon has a very low light absorption and is unaffected by scintillator solutions, making it an ideal vial material. It is possible to fabricate a vial from Teflon with a light collection efficiency close to 100% and a cross talk ratio of less than .1.

Description of Counter

The counter is based on an early model Tracerlab liquid scintillation counter of which only the freezer and manual sample changer have been retained. The sample changer has been automated and a new lead shield with the photomultipliers mounted coaxially has been fitted. The shield which has an average thickness of about 5 cm encloses the photomultipliers as well as the sample. The

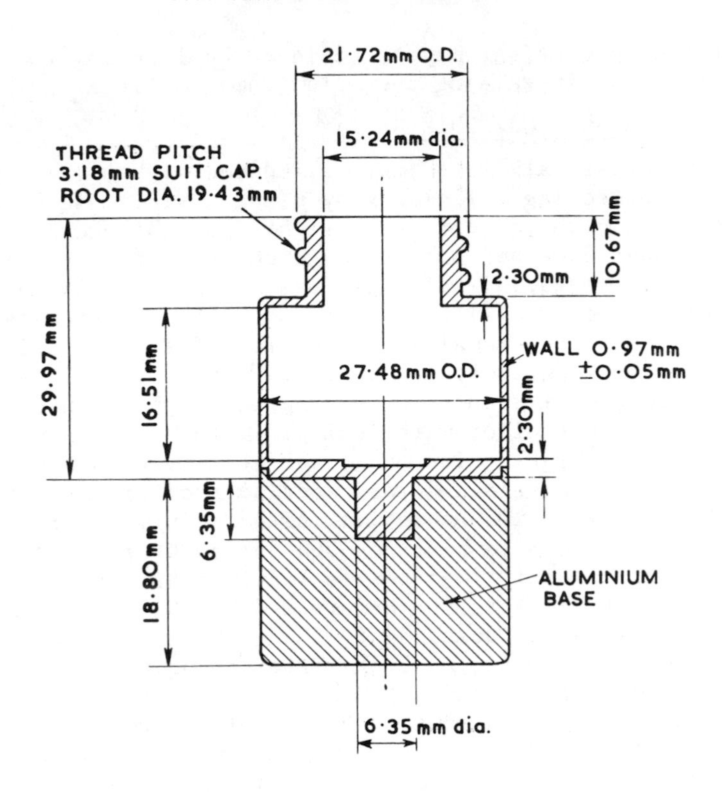

TEFLON VIAL

Figure 1 Teflon counting vial

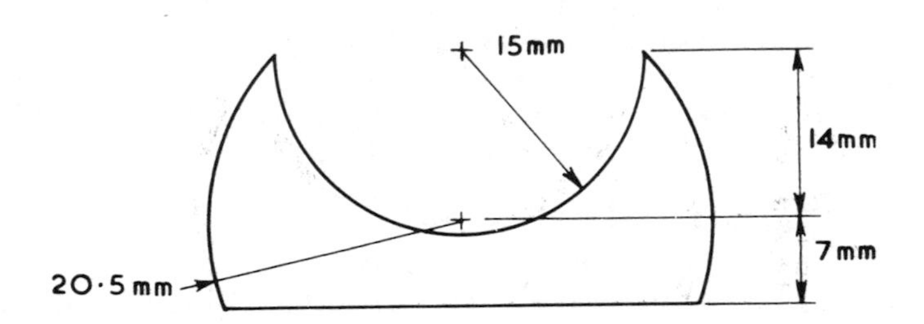

Figure 2 Plan view of one light guide. The light guide is made from 13 mm thick perspex. The outer curved surface is coated with evaporated aluminium.

sample vials which are shown in figure 1 are made of Teflon and have a volume of 5 ml. The vial is coupled to two EMI type 9635QB photomultipliers by two perspex light guides (figure 2) supported by a copper block which provides some additional shielding. Black paint and tape eliminate any stray light travelling from one tube to the other. μ for the optical system is .17, and δ is 0.274. The calculated light guide efficiency is .83, giving a light collecting efficiency of .46 and a cross talk ratio of .039. Figure 3 shows a block diagram of the electronics. The photomultipliers are operated with 150 V from cathode to first dynode and equal voltage distribution over the remaining stages. Signals from the last dynodes of the two photomultipliers are fed through specially designed amplifiers, discriminators and a coincidence gate with a 30 ns resolving time. The anode signals of the two tubes are fed through a single N.I.M. preamplier, amplifier and linear gate (controlled by the coincidence unit) and then to a single channel pulse height analyser and print out scaler, or to a multichannel analyser (figure 3). Alternatively, the linear gate may be omitted and the output of the single channel analyser gated by the coincidence unit.

Coincidence System Circuit Description (Figure 4, 5)

Amplifier

A charge sensitive amplifier with a field effect transistor (FET) Q1 in a folded cascode circuit is used as the input stage. The load resistor of Q2 is bootstrapped by the emitter followers Q3 and Q5 to increase the preamplifier open loop gain to approximately 2,000. With the feedback resistor R2 and the capacitor C3 connected, the preamplifier has an output of 1.6×10^{-7}V per electron applied to the input. This gives a 1V output at the emitter of Q6 for a 10^{-12} coulomb input. With a 7 cm input lead of 50 ohm coaxial cable the output rise time at Q6 is less than 30 ns and the fall time, determined by R2 and C3, is 100 ns.

Discriminator

The output of the preamplifier is buffered by Q7 which drives the monostable multivibrator Q8, Q9 and Q10. The transistor Q8 is normally held cut off by the positive voltage (1-3 V) set on RV1; when the negative pulse from

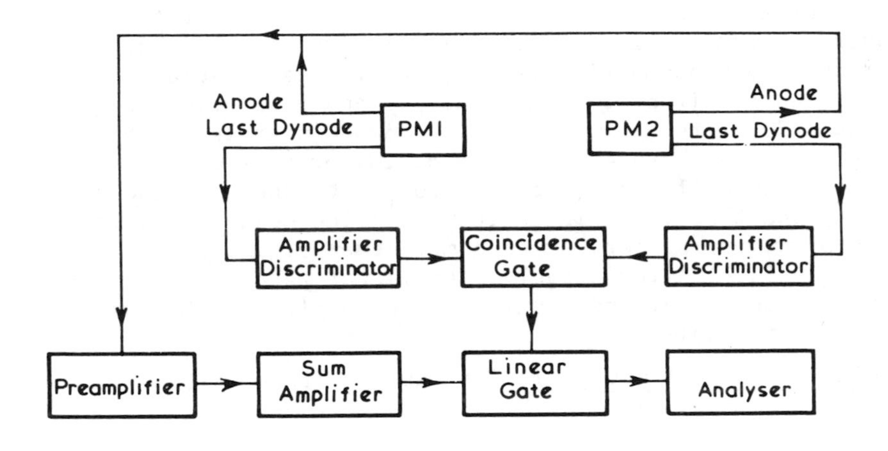

Figure 3   Block diagram of the counting system

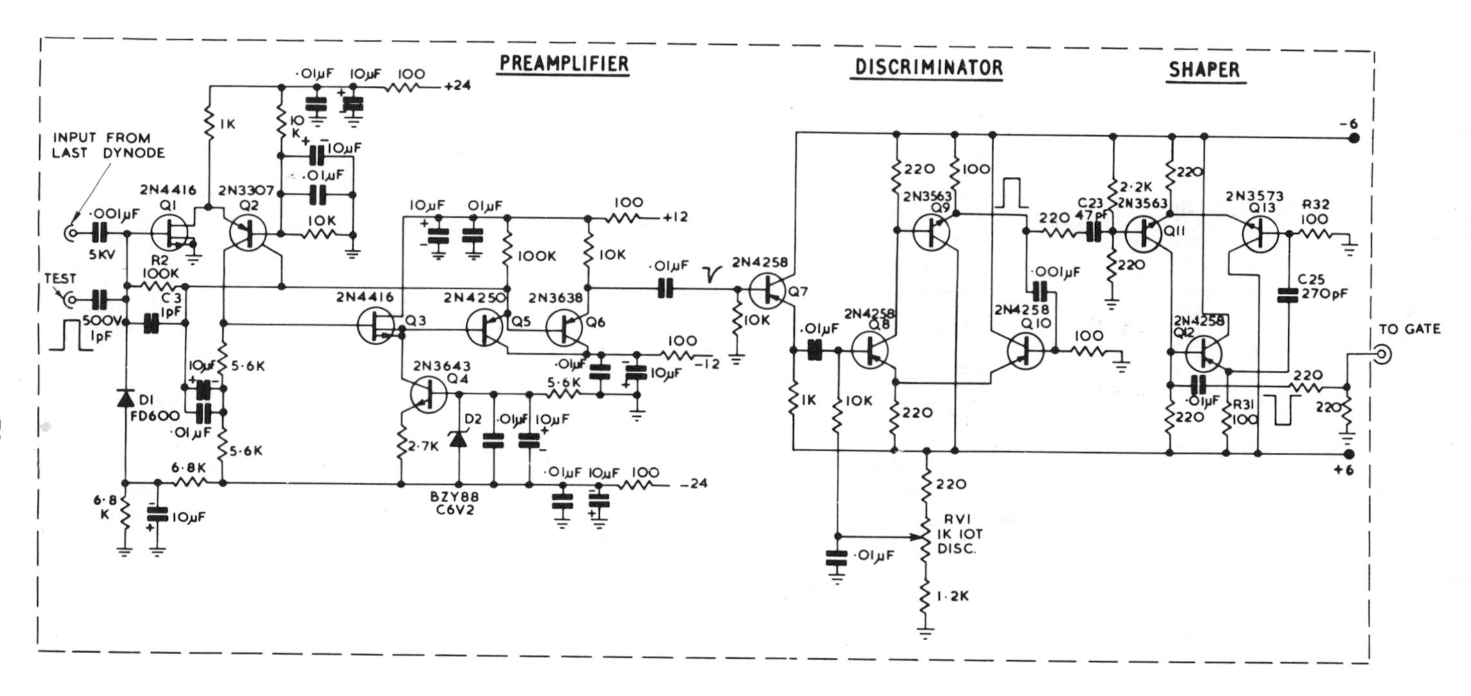

Figure 4   Amplifier - Discriminator circuit diagram

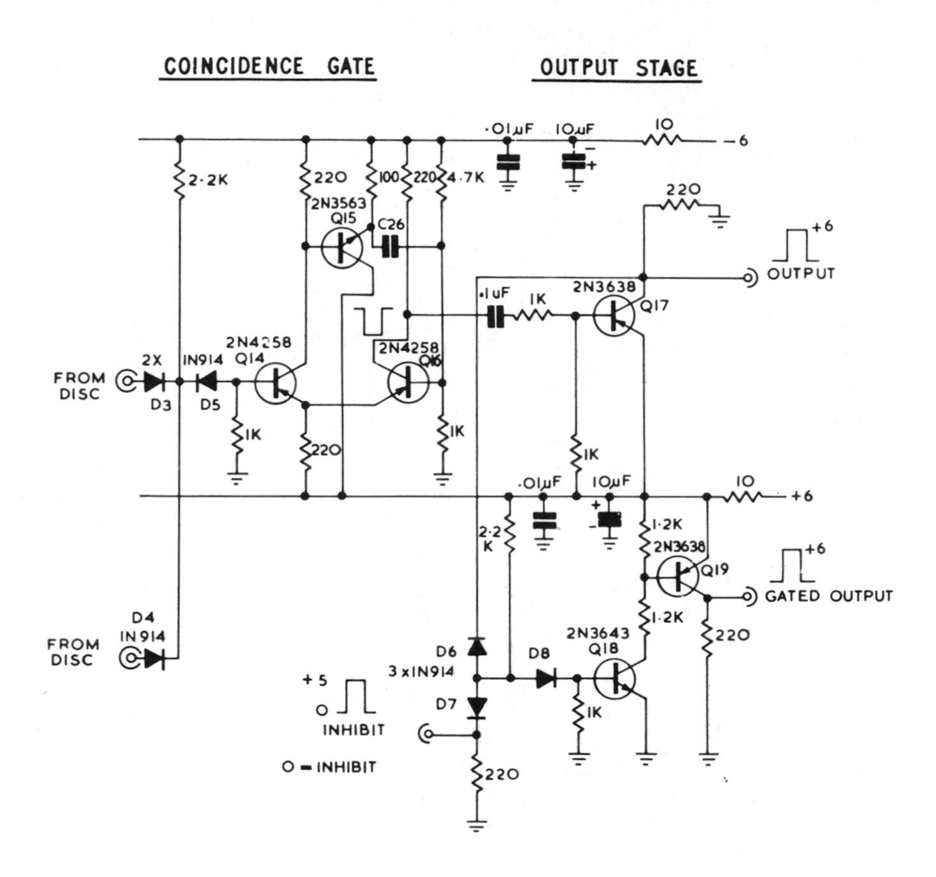

Figure 5   Coincidence gate circuit diagram

Q7 exceeds this preset voltage the monstable triggers giving a positive 100 ns wide pulse at the emitter of Q9.

Shaper

The output of Q9 is differentiated by C23 and the positive edge triggers the monstable Q11, Q12 and Q13. The width of the negative output pulse on the emitter of Q12 is set to 30 ns by C25 and R32.

Coincidence Gate

A diode AND gate is formed by D3, D4 and D5; when the outputs from Q12 in each channel are in coincidence D3 and D4 cut off and D5 conducts. This gives a positive trigger signal to the monstable Q14, Q15 and Q16, the width of the output at the collector of Q16 can be varied by selecting C26. If a coincidence does not occur only D3 or D4 will cut off and the monstable will not trigger.

Output Stage

The negative output pulse from the collector of Q16 saturates Q17 giving a positive 6V pulse capable of driving a 50 ohm cable. When Q17 saturates it cuts off D6 and if a positive strobe pulse of greater than 2V is present D8 will conduct allowing Q18 to saturate which in turn saturates Q19 giving a 6V output pulse to the gated output. Holding the INHIBIT input to 0 V or open circuit will prevent an output at the gated output. Both output pulses have less than 0.1 microsecond rise times.

Setting up Procedure and Results

A ^{14}C sample is loaded and the high voltage on each photomultiplier is set separately (with the linear gate held open) to give a maximum pulse height of about 5 volts. The linear gate is then returned to normal and a sample and background spectrum taken using the multichannel analyser. The cut off channels to obtain maximum E^2/B are then determined. The main amplifier gain is then changed (by a factor of 2) and the procedure repeated until the maximum E^2/B is obtained. The single channel analyser is then set to correspond to the optimum cut off channels using a mercury pulser. With the optimum settings (HV 945V and 1120V, gain 12, window .5 - 8.5 volts) a background of 1.96 $\pm$ 0.5 c.p.m. and an efficiency of 60.3 $\pm$ 0.5% has been measured using 5 ml of ^{14}C benzene prepared from N.B.S.

Contemporary Standard Oxalic Acid (14.24 ± .07 d.p.m./g of C).

Further Improvements

The background could probably be further reduced by using thicker shielding. Replacement of the light guides by a mirror system would reduce the amount of material in which cosmic ray induced Cerenkov radiation could occur. The optical efficiency could be increased by using a near spherical vial with a δ of close to .5, while the cross talk ratio could be kept about the same by using 2mm thick Teflon. If each of these vials were built into their own light guide assembly an automatic sample changer could still be used and light leakage around the vial would be completely eliminated.

Acknowledgements

The authors are indebted to Dr. G.E. Calf for his collaboration in many aspects of this work and to Mr. B.W. Seatonberry and Mr. L.W. Smith for their very considerable help in building the counter.

References

1. G.E. Calf and H.A. Polach, Teflon Vials for Liquid Scintillation Counting of Carbon-14 Samples, in Liquid Scintillation Counting: Recent Developments, (P.E. Stanley & B.A. Scoggins, eds.). New York and London: Academic Press Inc. (in press).

2. J.E. Noakes, M.P. Neary and J.D. Spaulding, Nucl. Instr. and Meth. 109, 177 (1973).

3. J. Sharpe and V.A. Stanley in Tritium in the Physical and Biological Sciences, Vol. 1, p.211, IAEA Vienna (1962).

4. R.L. Garvin, Rev. Sci. Instr. 23 (12), 755 (1952).

5. M. Mando, Nuovo Cim. 12 (1), 5 (1954): Translated by Technical Information Bureau of Chief Scientist, Ministry of Supply, London. Translation No. T.I.B. T4395 (1955).

A STOCHASTIC MODEL OF THE LIQUID SCINTILLATION COUNTING PROCESS

Philip J. Malcolm* and Philip E. Stanley†

**Biometry Section & †Dept. of Agricultural Biochemistry,*
Waite Agricultural Research Institute,
The University of Adelaide, Glen Osmond, South Australia

ABSTRACT

A stochastic multidimensional model for colour and chemical quenching is being developed for a coincidence liquid scintillation spectrometer. The programme, written in FORTRAN, is run on a CDC 6400 computer and requires a modest amount of memory but a considerable amount of central processor time.

The model produces a pulse height spectrum appropriate to the degree and type of quenching. The simulation is initiated within the computer by generating a β-event of random energy (following the energy probability spectrum of that radioisotope) in a random position within the vial. An appropriate number of photons of random wavelength (following the fluorescence spectrum of the phosphor solution) are emitted in random directions. The model also incorporates chemical and/or colour quenching, total internal reflection of light at the vial-air interface, quantum efficiency of the photomultipliers at the wavelengths of the photons, and the coincidence gate. Currently pulse height spectra are produced for a few thousand scintillations.

INTRODUCTION

Although mathematical modelling is well known to many scientists, it is worthwhile to restate briefly the basic concepts involved. The technique abstracts an observable closed system from the real world and describes, analyses and predicts its behaviour by developing and analysing a mathematical model of the system. In simulating the behaviour of the liquid scintillation counting (LSC) process on the computer, a 'pseudo-vial' containing a 'pseudo-sample' is counted by a 'pseudo-liquid scintillation counter'.

This work has evolved from an earlier attempt to produce a more comprehensive program package for LSC data reduction and processing. It became apparent that a deeper appreciation of the dynamics of the LSC process was needed, and the model has been developed with this aim in view.

The known appearance of the β-spectrum and the pulse height spectrum of an unquenched sample are compared with each other, and also with the pulse height spectra of chemical- (or impurity-) quenched and colour-quenched samples. For the case of ^{14}C it is known that the spectrum is compressed towards the left, or low-pulse-height end, as chemical quenching increases (see Fig. 1). An increase in colour quenching causes a similar compression towards the low-pulse-height end of the spectrum, but with the accompaniment of a greater spread in the spectrum than with chemical quenching. These two transformations of the pulse height spectrum are described *quantitatively* rather than *qualitatively*, and thus the dynamics of colour or chemical quenching can be considered.

A deeper understanding of the transformation of pulse height spectra for combined colour and chemical quenching can be applied in the following way. LSC measurements can be standardized for both colour and chemical quenching, and thus each type of quenching can be corrected for individually in DPM calculations; hopefully, more accurate and reliable results will be obtained.

Ways of improving the efficiency of the optics of the LSC process are also under investigation. Preliminary results indicate that transmitted light losses can be reduced as much as 50% when using counting vials of square rather than circular cross-section. This should improve counting efficiency, especially when counting heavily quenched or low-energy samples.

Several workers have proposed fairly complete models of the LSC process, but none appear to have developed a comprehensive stochastic model. ten Haaf (1) has presented a one-dimensional model which, although very simple, shows broader energy distributions and larger coincidence losses for colour-quenched samples compared with equivalent chemical-quenched samples. These results were in accordance with the earlier ones of Neary and Budd (2), who suggested that the wide divergence in behaviour between heavily colour- and chemical-quenched samples was due to the more probabilistic nature of colour quenching. Kaczmarczyk (3) has presented a detailed model of chemical quenching but has not extended it to include colour quenching, optical considerations, or details of photomultiplier response.

The work presented here is an attempt to combine the generality of ten Haaf's approach with the detail of Kaczmarczyk's work.

MATERIALS AND METHODS

Experimental. Counting was performed in a Packard Liquid Scintillation Spectrometer Model 3390-544. The instrument was fitted with RCA 4501 V4 bialkali photomultipliers which were operated at 20° ± 0.5°. Pulse height spectra were obtained with a 200-channel Packard Spectrazoom Multichannel Analyzer (MCA) which was connected to the test point at the rear of the RED pulse height analyzer. Here the pulses have been amplified according to the setting on the front panel. Discriminators were set at 3-1000. Coincidence was established by a gating pulse from the RED channel rate-meter output.

The phosphor solution contained 8g PPO (2,5-diphenyloxazole) in 1 litre toluene. PPO was obtained from Ajax Chemicals Ltd., Sydney, Australia. Its absorption spectrum was similar in toluene to that published (4). Toluene was "Proalys" grade of May and Baker (Aust.) Pty. Ltd. 15 ml samples of this phosphor were used in Packard Low Background vials, the glass of which had a refractive index close to 1.50. Vials were closed with Polyseal cones to minimize evaporation.

The phosphor was labelled with [Me-^{14}C]toluene obtained from the Radiochemical Centre, Amersham, U.K. It was diluted with unlabelled toluene so that 300 µl at 20° contained about 5×10^5 d.p.m. It was then calibrated against a standard [^{14}C]toluene sample. Each vial contained 300 µl of the diluted [^{14}C]toluene and this was checked by weighing. Two similar samples were quenched incrementally with either carbon tetrachloride as a chemical quencher, or a

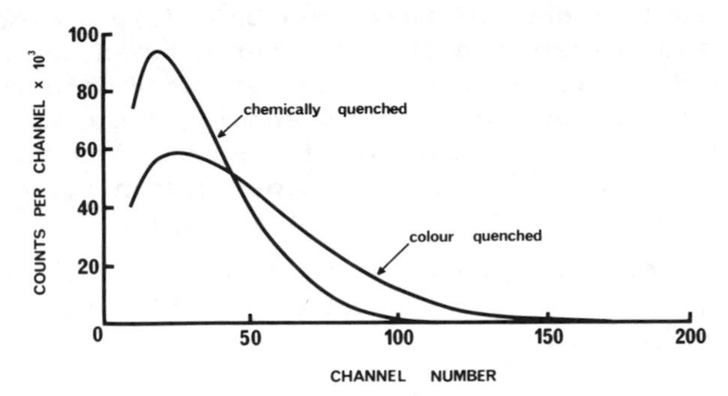

Fig. 1. OBSERVED PULSE HEIGHT SPECTRA OF COLOUR AND CHEMICALLY QUENCHED ^{14}C. Counting efficiency is 65%.

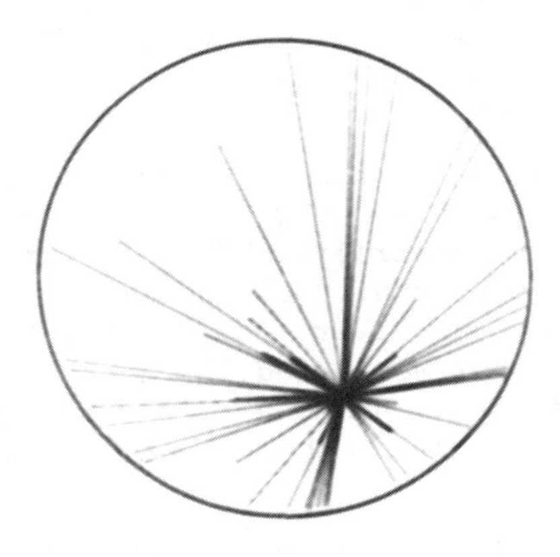

Fig. 2. A PHOTOGRAPH OF A SIMULATED SCINTILLATION OBTAINED AS SHOWN IN FIG. 4. Note the random positions of the 'scintillation' and the escape angles of the 'photons' and also the shorter distances travelled by the 'colour-quenched photons'.

saturated solution of methyl orange in ethanol as a colour quencher. Each sample was measured at 5% decrements in efficiency. Counting efficiencies of the two sets of samples were determined using spectrometer settings of 3-∞ at 100% amplification. For multichannel analysis the amplification was set so that the spectrum fitted easily into 200 channels. Counting time on the spectrometer was 2 min or 5×10^5 counts. A period of 600 seconds live time was used on the MCA.

The Stochastic Model. In this section the modelling process and the term 'stochastic' will be discussed, and then the model will be outlined in non-mathematical terms.

In developing a simulation of the behaviour of the LSC process an attempt has been made to isolate the fundamental components of this process, which have been described quantitatively rather than qualitatively. Programs following these quantitative descriptions have been prepared and have been assembled to produce a stochastic model of the LSC process.

A *stochastic* model is one in which the components are described (or modelled) probabilistically. The β-spectrum of the particular radioisotope is treated as a probability distribution, and 'β-events' with random energies are generated following that distribution. This means that, while the energies of any two 'β-events' are mutually independent, the spectrum produced mimics the β-spectrum.

The technique of stochastic modelling is particularly applicable to the LSC process because many of its components are random events. Energies of real β-particles, for example, are mutually independent, yet follow the β-spectrum of the emitter.

Fig. 2 is a photograph of the computer output showing the random trajectories of the simulated photons; the circle represents the vial surface and a trajectory ending short of it indicates that the 'photon' was colour quenched. The model is very simple; a point is chosen at random in a circle representing a cross-section of the counting vial. From a β-event at this point it is assumed that a fixed number, say 50, 'photons' would be emitted at random angles. Any consideration of energy forms, energy spectra, or energy transformations is ignored.

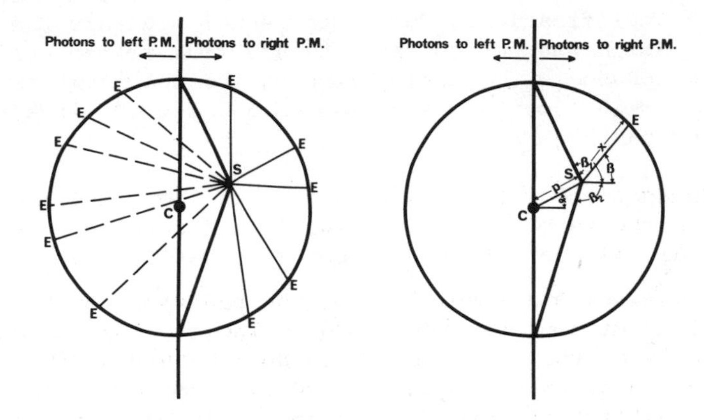

Fig. 3. THE GEOMETRY OF THE SIMPLE MODEL

C is the centre of the vial.
S is a random position of a 'scintillation' with polar coordinates (p,α) relative to C.

If a 'photon' is emitted from S at a random escape angle β, it will travel a distance x before reaching the vial surface at point E.

β_1 and β_2 are the maximum and minimum escape angles at which a 'photon' can be emitted and still leave the right-hand side of the vial.

For a vial of unit radius -

$$x = \sqrt{1-p^2\sin^2(\beta-\alpha)} - p\cos(\beta-\alpha)$$

$$\beta_1 = \tan^{-1}\left(\frac{p\sin\alpha-1}{p\cos\alpha}\right) \qquad \beta_2 = \tan^{-1}\left(\frac{-p\sin\alpha-1}{p\cos\alpha}\right)$$

Thus, 50 'photons' are emitted at random angles from a random point within a circle. Chemical quenching is simulated by discarding a proportion of these 'photons' immediately; for each 'photon', the computer generates a random number r between 0 and 1 and compares it with the expected proportion p of 'photons' lost by chemical quenching, p being set as a model parameter. If r is greater than p, then the 'photon' is not chemically quenched; if r is less than p, then it is.

For each 'photon' that avoids chemical quenching, the computer calculates the escape distance x (see Fig. 3) that it must travel to reach the vial surface, and also whether it would leave the vial on the right- or left-hand side (i.e. which photomultiplier). Colour quenching is simulated by calculating a probability of escape to the vial surface. The escape distance x and another model parameter q (to calibrate the amount of colour quenching) are used to calculated $u = 1\text{-exp}(-x \cdot q)$. u therefore gives the probability that the 'photon' is unquenched; as either i) the excape distance x, or ii) the optical density modelling parameter q, increase, then $\exp(-x \cdot q)$ decreases and therefore u increases exponentially. The computer now generates a random probability v and compares it with u. If v exceeds u, then the 'photon' is unquenched, and when v is less than u, the 'photon' is colour quenched. Whereas the unquenched 'photon' has now reached the vial surface, a colour-quenched 'photon' was absorbed somewhere between the point of scintillation and the surface. The computer simulates the actual distance travelled by a quenched 'photon' by taking a random proportion of the escape distance x; thus, the unquenched 'photon' would travel a distance x to reach the vial surface and, if it is quenched, it travels some random smaller distance before being absorbed.

A short ciné film* showing the behaviour of this very simple model was prepared as follows (see Fig. 4): The program for the model was written in FORTRAN and run on a CDC 6400 computer, which then punched data on a paper tape; this was read into a PDP 8 computer; displays of successive time slices of a 'photon burst' (or scintillation) were drawn on a Tektronix 611 storage oscilloscope under the control of a FOCAL program in the PDP 8. As the display of

*The film was viewed at the Symposium and each scintillation lasted about 5 seconds.

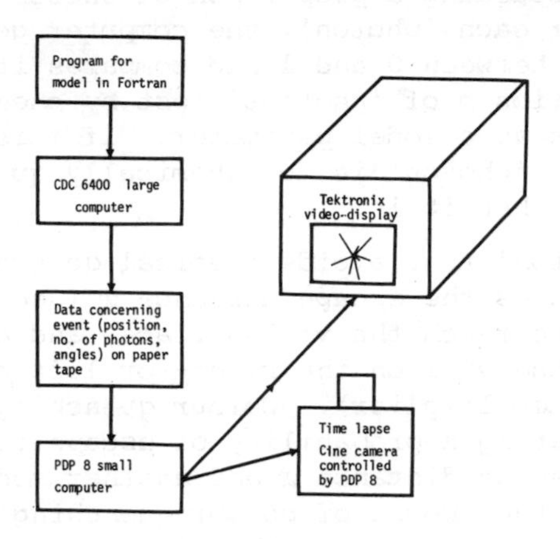

Fig. 4. STAGES IN PREPARING THE FILM

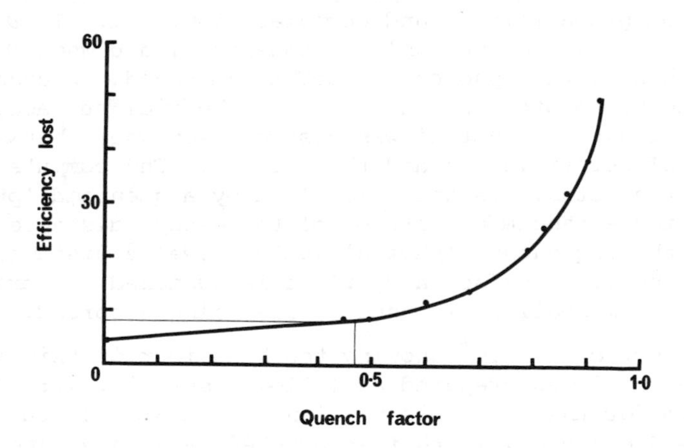

Fig. 5. A CHEMICAL QUENCH CALIBRATION CURVE

each time slice for each 'photon burst' was completed, the PDP 8 triggered a single half-second exposure by a ciné camera. The PDP 8 thus drew a picture, then photographed it, and then drew the next picture and so on.

The structure of a more refined model which includes the energetics of the LSC process is shown in Table I. Individual 'β-events' are modelled in a 3-dimensional 'vial' and, by repeating the simulation for a large number of these 'events', a pulse height spectrum is generated. The model simulates the behaviour of the LSC process using the probabilistic approach; a 'β-event' of a radioisotope is generated at some random position (3-dimensionally) within the 'vial' giving the emission of a 'β-particle' whose energy is randomly selected from a Fermi probability distribution following the β-spectrum of that isotope. About 4% of the energy is transferred to the fluor (PPO) *via* the solvent (toluene), and any losses due to chemical quenching are simulated randomly. Emission of 'photons' by the fluor [about 11 photons/keV of β-particle energy (4,5)] is the next step in the simulation; the random wavelengths of the emitted 'photons' are constrained to follow a probability distribution given by the fluorescence spectrum of PPO (4,6).

Colour quenching is simulated using an experimentally determined absorption spectrum for the vial and its contents. The computer calculates the absorbance of the solution for each 'photon' according to its wavelength and escape distance. The probability of transmittance is then calculated using the Beer-Bouguer Law, and transmittance or absorption of the photon is then simulated using an algorithm similar to the chemical-quenching one.

Any 'photons' left have by now reached the outer wall of the vial and are therefore susceptible to total internal reflection at the vial-air interface. This is simulated by calculating the angle of incidence of the 'photon' in a 3-dimensional vial; any 'photons' with an incident angle greater than the critical angle are lost.

Currently it is assumed that once a 'photon' leaves the vial it reaches the appropriate photomultiplier either directly or *via* a perfectly reflecting surface, and thus the next step in the simulation is to model the photocathodes of the photomultipliers. This is done using the quantum efficiency spectrum (7) of the photocathodes, and

TABLE I. THE LIQUID SCINTILLATION PROCESS AND ITS MODEL

Energy form	Stage in LSC process	Energy loss: Type	Energy loss: % Loss from previous stage	Included in model of LSC process
	β-disintegration			Yes. Random β-energies following β-spectrum are produced.
β-particle		heat	95	Yes. This loss is varied to calibrate counting efficiency.
	solvent (toluene)			
excited molecules		chemical quenching	40-95	Yes. Losses depend on sample.
	fluor (PPO)			
photons		colour quenching	38-85	Yes. Losses depend on sample.
	outer vial wall			
photons		total internal reflection	50-60	Yes. Partial internal reflection not modelled yet.
	detection chamber			
photons		reflection losses	?5-10	No. Second-order effect.
	photocathode			
photo electrons		quantum efficiency of photomultiplier is < 100%	75-90	Yes.
	dynode chain			
voltage pulse		pulse spread and losses	?	No. Lack of information.
	photomultiplier anode			
voltage pulse		assume zero losses		
	coincidence gate			
summed voltage pulse		coincidence losses (pulse from either anode below coincidence threshold)	various	Yes. Important with small numbers of photons.
	pulse height analyser and the remainder of the system			

randomly producing (or failing to produce) photoelectrons according to that spectrum.

The next stage in the model is to simulate coincidence by considering the pulses from each photomultiplier. Coincidence is enabled only when there is a pulse from each photomultiplier; the pulses are then summed as in a summation amplifier.

Many such β-events produce a pulse height spectrum. This spectrum is the main output of the model.

RESULTS

One particularly striking feature of the model is its requirement for computer time, since at least 5 hours per day are essential for work to proceed at a reasonable pace on the CDC 6400 (each event taking up to 1 second to simulate). This means that calculations equivalent to 10 man-years of work are required daily of the computer.

General trends in the behaviour of the model which are now apparent are presented here. Thus, the first-order effects (see Table I) modelled to date show close agreement with the experimental data. The current full-scale model was first calibrated for chemical quenching by plotting a graph of counting efficiency against the chemical quench factor p (see Fig. 5). A counting efficiency of 96-97% was found in the absence of chemical quenching (i.e. p=0), and as p increased, counting efficiency decreased according to an exponential-type relationship.

The accelerating rate of efficiency lost when p exceeds 0.7 corresponds to a moderately quenched sample. Note that for large p, even a small change in p will cause a marked change in counting efficiency. However, when p is small, there is only a small change in efficiency for a small change in p. An increase in p caused a compression of the pulse height spectrum towards the low-pulse-height end. These results are in accordance with experimental evidence.

Calibration for colour quenching in the model was achieved using an absorption spectrum determined for that sample. Colour quenching was found to occur even in the pure PPO-toluene phosphor, as the fluorescence and absorption spectra for PPO in toluene overlap. This is shown in Fig. 6.

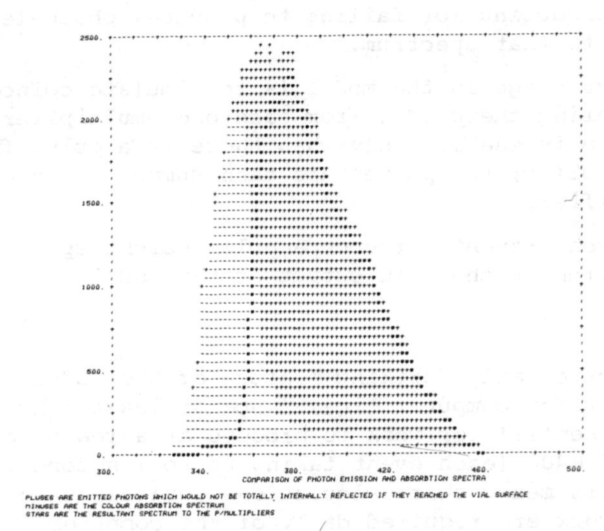

Fig. 6. A COMPUTER OUTPUT SHOWING COMPARISON OF THE PPO FLUORESCENCE (+ histogram) AND ABSORPTION (- histogram) SPECTRA INCLUDING THE RESULTANT SPECTRUM (* plot) LEAVING THE VIAL.

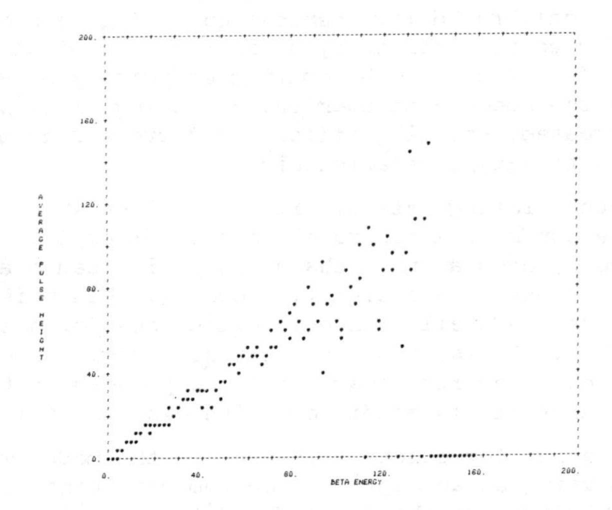

Fig. 7. A COMPUTER OUTPUT SHOWING PLOT OF THE AVERAGE PULSE HEIGHT AGAINST BETA ENERGY. Note the linear relationship and the greater scatter at higher energies.

An increase in colour quenching was simulated using absorption spectra determined for individual colour-quenched samples. Again the expected movement of the pulse height spectrum was found; as colour quenching increased, there was a compression of the spectrum toward the low-pulse-height end. As predicted, there was a greater spectral spread than found in the corresponding chemical-quenched sample. When quenching was due to the overlap of the PPO absorption and fluorescence spectra, then the model gave a counting efficiency of 96-97%. In practice, however, the nominally unquenched sample had a counting efficiency of only 92% and the difference can be ascribed to oxygen quenching. Purging the 'unquenched sample' with argon caused the counting efficiency to rise to 96%.

Another output of the model is a plot of mean pulse height against β-energy (see Fig. 7), which was expected to follow a linear relationship. Most (80%) β-events produced by ^{14}C fall in the energy range of 0-80 keV, and in this range the linear relationship between mean pulse height and β-energy is evident since there is little spread in the data points. At higher energies, however, there are relatively few β-events, and thus there are fewer recordings in each mean pulse height. This is reflected as an increased spread in the values of the mean pulse height, which also explains why the high end of a pulse height spectrum falls off more slowly than the corresponding β-spectrum. The few β-particles of such high energy are subject to a series of probabilistic processes, thus producing marked fluctuations in the resultant pulse height.

One weakness in the model is the energy transformation from the β-event to the fluor. Published values (5,8) indicate that 10 or 12.5 photons are produced for each keV of β-energy. The model is relatively insensitive to changes in this value for ^{14}C. In the case of an unquenched ^{3}H sample it is found that 13.5 photons/keV were needed to obtain a 67% counting efficiency, whereas an efficiency of only 63% was obtained with 12.5 photons/keV.

While relatively complete results are to hand for samples subjected to either colour or chemical quenching, it is clear that more experience with the model is required before reliable results can be obtained for their combined effects.

ACKNOWLEDGEMENTS

We are deeply indebted to Mr. R.J. Wilson of the Department of Psychology, University of Adelaide, for his assistance and for the use of his equipment for making the ciné film. The staff of the Computing Centre of the University of Adelaide is greatfully acknowledged for servicing the not inconsiderable computational demands of the model. One of us (P.E.S.) wishes to thank the Australian Wheat Industry Research Council for generous financial assistance.

REFERENCES

1. F.E.L. ten Haaf *in* Liquid Scintillation Counting, Vol. 2, pp. 39-48 (M.A. Crook, P. Johnson and B. Scales, Eds.). London : P. Heyden & Son (1972)
2. M.P. Neary and A.L. Budd *in* The Current Status of Liquid Scintillation Counting, pp. 273-282. (Edwin D. Bransome Jr., Ed.). New York : Grune & Stratton (1970)
3. N. Kaczmarczyk *in* Organic Scintillators and Liquid Scintillation Counting, pp. 977-990. New York : Academic Press (1971)
4. J.B. Birks *in* Solutes and Solvents for Liquid Scintillation Counting. Colnbrook, Bucks., U.K. : Koch-Light Laboratories Ltd. (1969)
5. D.L. Horrocks *in* Liquid Scintillation, Chapter 3. (M.P. Neary, Ed.). Beckman Instruments, Inc. (April 1972)
6. E. Langenscheidt *in* Liquid Scintillation Counting, Vol. 1, pp. 23-36. (A. Dyer, Ed.). London : P. Heyden & Son (1971)
7. RCA specification sheet for 4501 V4 photomultipliers
8. J.B. Birks *in* The Current Status of Liquid Scintillation Counting, pp. 3-12. (Edwin D. Bransome Jr., Ed.). New York : Grune & Stratton (1970)

LESSER PULSE HEIGHT ANALYSIS IN LIQUID SCINTILLATION COUNTING

C. Ediss, A.A. Noujaim and L.I. Wiebe
Division of Bionucleonics and Radiopharmacy
University of Alberta
Edmonton, Alberta, Canada

Abstract

"Lesser pulse height analysis", an instrumental development in liquid scintillation counting is described and compared with the "summed method" used by most manufacturers. A MKII® liquid scintillation counter (Nuclear Chicago Corporation) with lesser pulse height analysis installed was modified to allow direct comparison of the "lesser" and "summed" methods.

The calculation of DPMS in coloured samples using external standard ratio quench curves prepared from chemically quenched standards has been shown to yield appreciable errors. The performance of the "lesser" and "summed" methods with particular regard to this source of error is investigated. The "lesser" method is found to yield considerably smaller errors than the "summed" method. For tritium, the benefit in using the "lesser" method is clearly apparent for all counting efficiencies down to 10%. The "lesser" method is also superior for Carbon-14, but below a counting efficiency of 60% errors become more noticeable.

The unquenched counting efficiencies using the "lesser" method are only marginally smaller than those observed for the "summed" method.

Introduction

A steady progress in the development of improved liquid scintillation counting instruments has been maintained for several years (1,2). Most current commercially available machines employ "summed coincidence" circuitry to improve the signal to noise ratio of the detection system, and thus enhance the sensitivity of the liquid scintillation method (3,4,5). The external standard ratio method of quench correction has been introduced in order to facilitate the rapid determination of sample counting efficiencies (6).

The calculation of coloured sample activities using external standard ratio quench curves prepared from chemically quenched standards has been shown to yield appreciable errors (7,8). Lang (8) devised a technique to distinguish whether a sample was predominantly colour or predominantly chemically quenched and to make the appropriate corrections. This technique however cannot accomodate those samples that are subject to simultaneous colour and chemical quenching. One way of overcoming this problem is to devise a method so that both coloured and chemically quenched standards give rise to a single quench correction curve. This paper describes the performance of an instrumental development, "lesser pulse height analysis", which is shown to achieve this objective for a wide range of counting efficiencies.

The "summed coincidence" method processes the signals from two horizontally opposed photomultiplier tubes in the following way: If pulses are detected from both photomultiplier tubes within such a short resolving time that they might be regarded as being coincident, then those signals are summed and used for subsequent analysis (3,4,5). The "lesser" method selects only the smallest of the two coincident photomultiplier tube signals for subsequent analysis. The benefit of using the "lesser" method is most clearly demonstrated by comparing the results obtained for both the "lesser" and "summed" methods on the same machine.

Experimental

A Nuclear Chicago Mark II ® liquid scintillation counter with the "lesser" analysing system installed was modified so that a front panel switch would allow selection of either "lesser" or "summed" modes of operation. The simple modification required is shown in figure 1a. In order to confirm the proper operation of both "lesser" and "summed" circuitry, photomultiplier tube signals were simulated using a test pulse generator (fig. 1b), and the result measured with the single channel

Figure 1a LESSER - SUMMED SWITCH

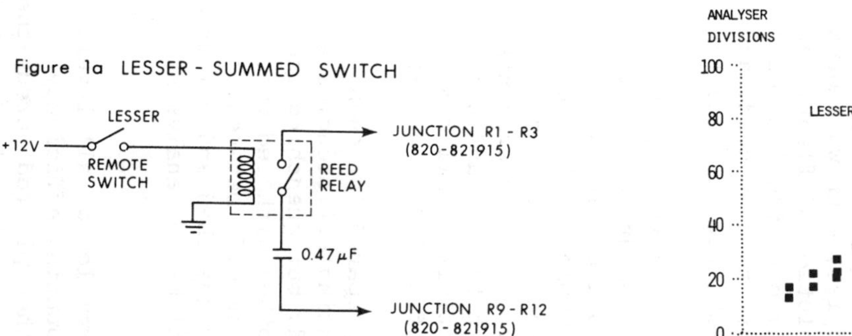

Figure 1b X and Y TEST PULSE INPUTS

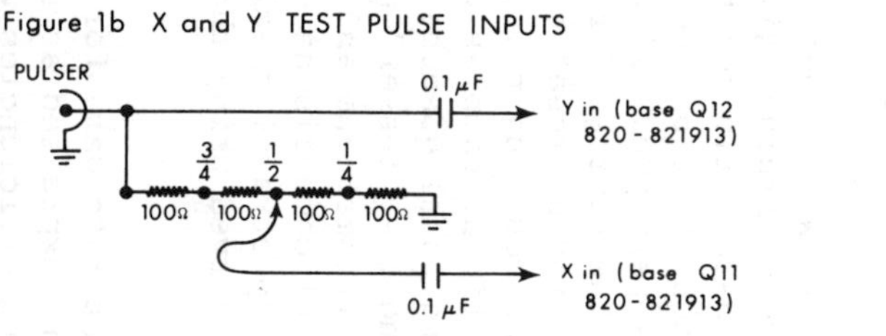

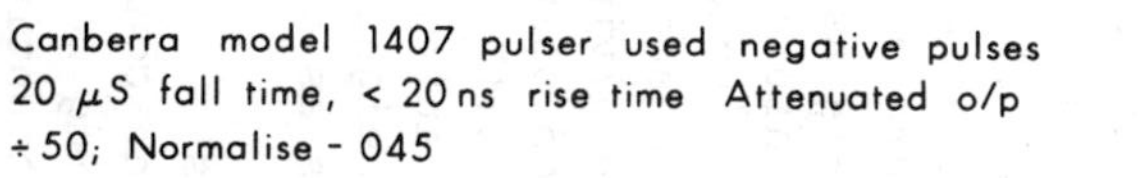

Canberra model 1407 pulser used negative pulses 20 μS fall time, < 20 ns rise time Attenuated o/p ÷ 50; Normalise - 045

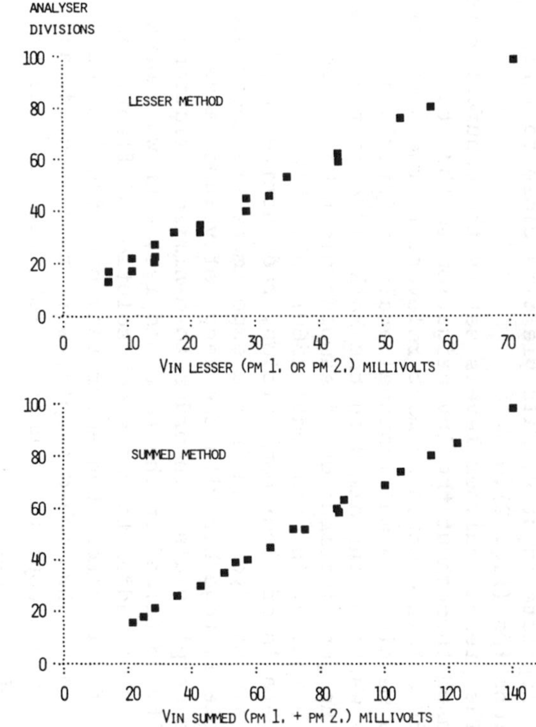

FIGURE 2. PERFORMANCE TEST OF LESSER AND SUMMED MODES OF OPERATION USING A PULSE GENERATOR

analysers built into the liquid scintillation counter. These single channel analyser levels were plotted versus the appropriate input signals yielding close to linear relationships (fig. 2).

The preset analyser levels set by the manufacturers were used throughout the investigation so that the results obtained would be most relevant to the users of such a machine. A multichannel analyser interface module was also installed in the Mark II® so that the pulse height distribution of samples might be viewed using a Nuclear Chicago model 25601 multichannel analyser.

Details of fluor composition, radioactive standards, chemicals and dyes used in the experiment are described in Table I, together with the range of volumes added to the sample vials. In order to minimize variations in activity contained in the test samples, a known amount of activity was added to the stock solution of fluor, and 15 ml of this fluor then accurately pipetted into each vial. These unquenched samples were then counted, and sets of vials containing uniform activity selected for use as test samples.

The test samples were counted repetitively and between each counting period small amounts of the various quenching agents were added each to a particular single vial. Data were obtained giving sample count rate and external standard ratio for both the "lesser" and "summed" methods at each level of quenching. This experimental procedure was performed using first tritium and then carbon-14 as the source of radioactivity. All samples were measured for a preset count of 200000 or a maximum of ten minutes counting time.

Two additional samples were prepared for each isotope giving equal quenched counting efficiencies as measured using the "lesser" method, but quenching one sample with nitromethane and the other with red dye. Thus meaningful comparisons might be made between the pulse height spectra of chemically quenched and coloured samples as observed using the multichannel analyser.

Results

The count rate data for all samples at all levels of quenching were expressed as a % counting efficiency. These counting efficiencies were then plotted versus the corresponding external standard ratio values in figures

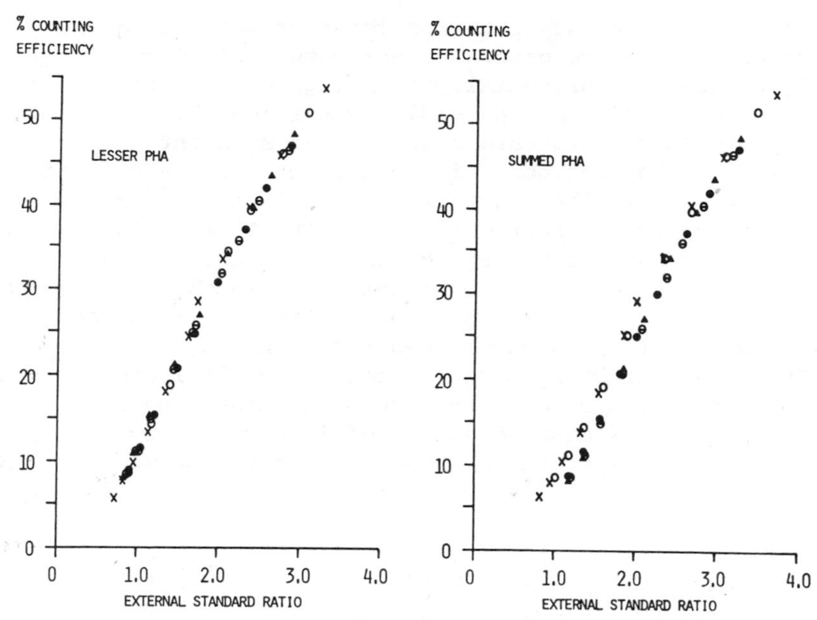

FIGURE 3. LESSER AND SUMMED TRITIUM COUNTING EFFICIENCY CORRECTION CURVES
QUENCHING AGENTS: - NITROMETHANE (X), CARBON TETRACHLORIDE (O), YELLOW (Θ), RED (▲) AND BLUE (●) DYES

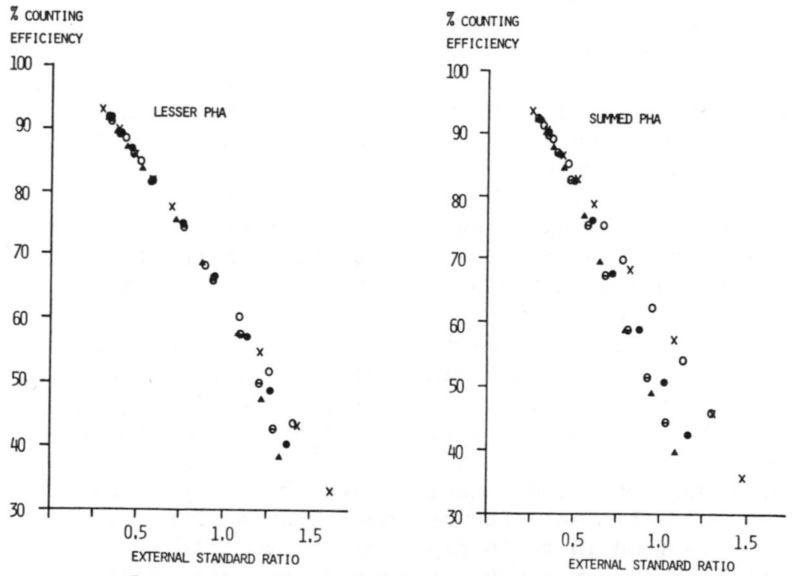

FIGURE 4. LESSER AND SUMMED CARBON 14 COUNTING EFFICIENCY CORRECTION CURVES
QUENCHING AGENTS: - NITROMETHANE (X), CARBON TETRACHLORIDE (O), YELLOE (Θ), RED (▲) AND BLUE (●) DYES

3 and 4. Using the data obtained for nitromethane, quench correction curves were generated using two methods; firstly by fitting polynomials according to the method of Carol and Houser (9), and secondly by straight line fitting between the standard points. The apparent activities in the remainder of the samples were calculated using counting efficiencies interpolated from the appropriate quench correction curves. The % error in calculating the activities using polynomial quench correction curves are shown in figures 5 through 8. Very similar results were obtained when counting efficiencies were linearly interpolated between the standard points.

The pulse height spectra for the carbon-14 isotope are shown in figure 9 for nitromethane quenched and red coloured samples of equal counting efficiency. The external standard spectra in these samples are shown in figure 10.

Discussion

The counting efficiency data for tritium shown in figure 3 indicate visually that all the quenching agents used give rise to a single quench curve for the "lesser" method. For the "summed" method however, chemically quenched and coloured samples give rise to distinctly separate curves.

Similar results are apparent for carbon-14 (fig. 4) for counting efficiencies above 60%. Below this counting efficiency differences between coloured and chemically quenched samples also become apparent for the "lesser" method. It is interesting to note that even though the "lesser" method always rejects information from one photomultiplier tube and thus might be expected to lose sensitivity, the unquenched counting efficiencies for the "lesser" method are very close to those obtained for "summed" (fig. 3,4).

Figures 6 and 8 show clearly that the "lesser" method reduces the error in determining coloured sample activities. This effect is particularly useful for samples that are quenched partly by chemicals and partly by colour.

The errors observed cannot be attributed to a volume dependancy effect, since it was found that a change in sample volume from 15 to 16 mls caused less than 1% change in both observed count rate and external standard

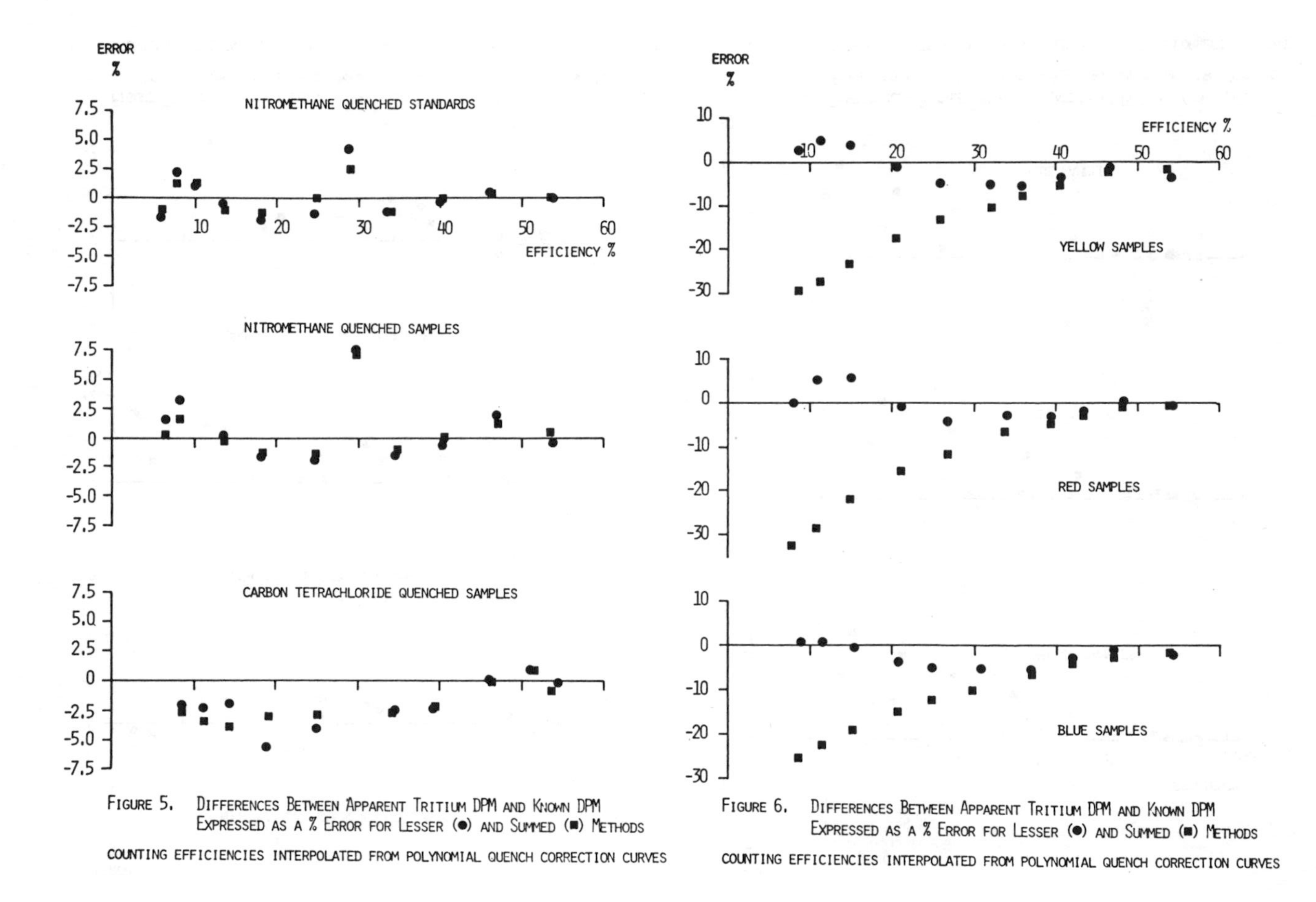

FIGURE 5. DIFFERENCES BETWEEN APPARENT TRITIUM DPM AND KNOWN DPM EXPRESSED AS A % ERROR FOR LESSER (●) AND SUMMED (■) METHODS

COUNTING EFFICIENCIES INTERPOLATED FROM POLYNOMIAL QUENCH CORRECTION CURVES

FIGURE 6. DIFFERENCES BETWEEN APPARENT TRITIUM DPM AND KNOWN DPM EXPRESSED AS A % ERROR FOR LESSER (●) AND SUMMED (■) METHODS

COUNTING EFFICIENCIES INTERPOLATED FROM POLYNOMIAL QUENCH CORRECTION CURVES

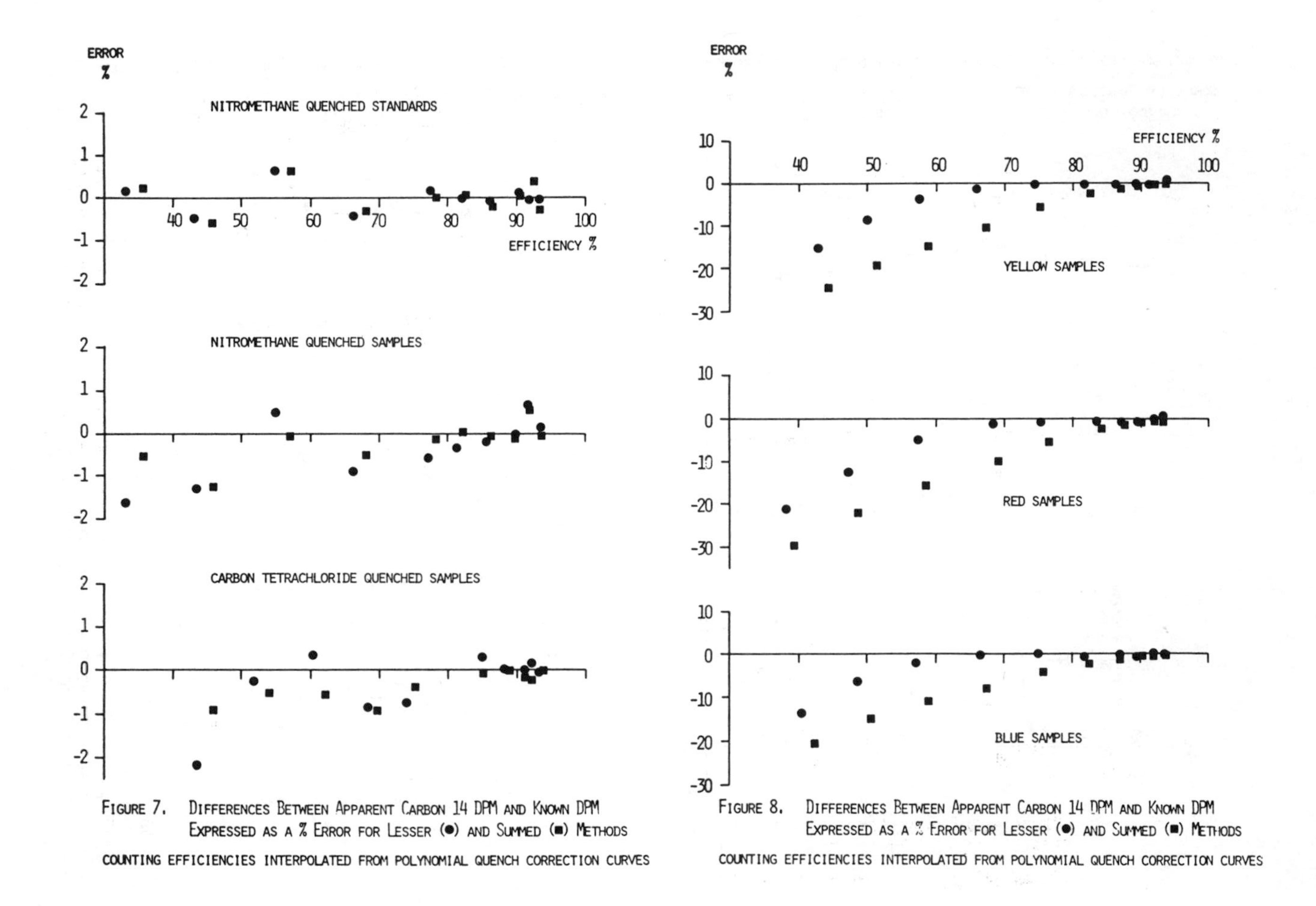

FIGURE 7. DIFFERENCES BETWEEN APPARENT CARBON 14 DPM AND KNOWN DPM EXPRESSED AS A % ERROR FOR LESSER (●) AND SUMMED (■) METHODS

COUNTING EFFICIENCIES INTERPOLATED FROM POLYNOMIAL QUENCH CORRECTION CURVES

FIGURE 8. DIFFERENCES BETWEEN APPARENT CARBON 14 DPM AND KNOWN DPM EXPRESSED AS A % ERROR FOR LESSER (●) AND SUMMED (■) METHODS

COUNTING EFFICIENCIES INTERPOLATED FROM POLYNOMIAL QUENCH CORRECTION CURVES

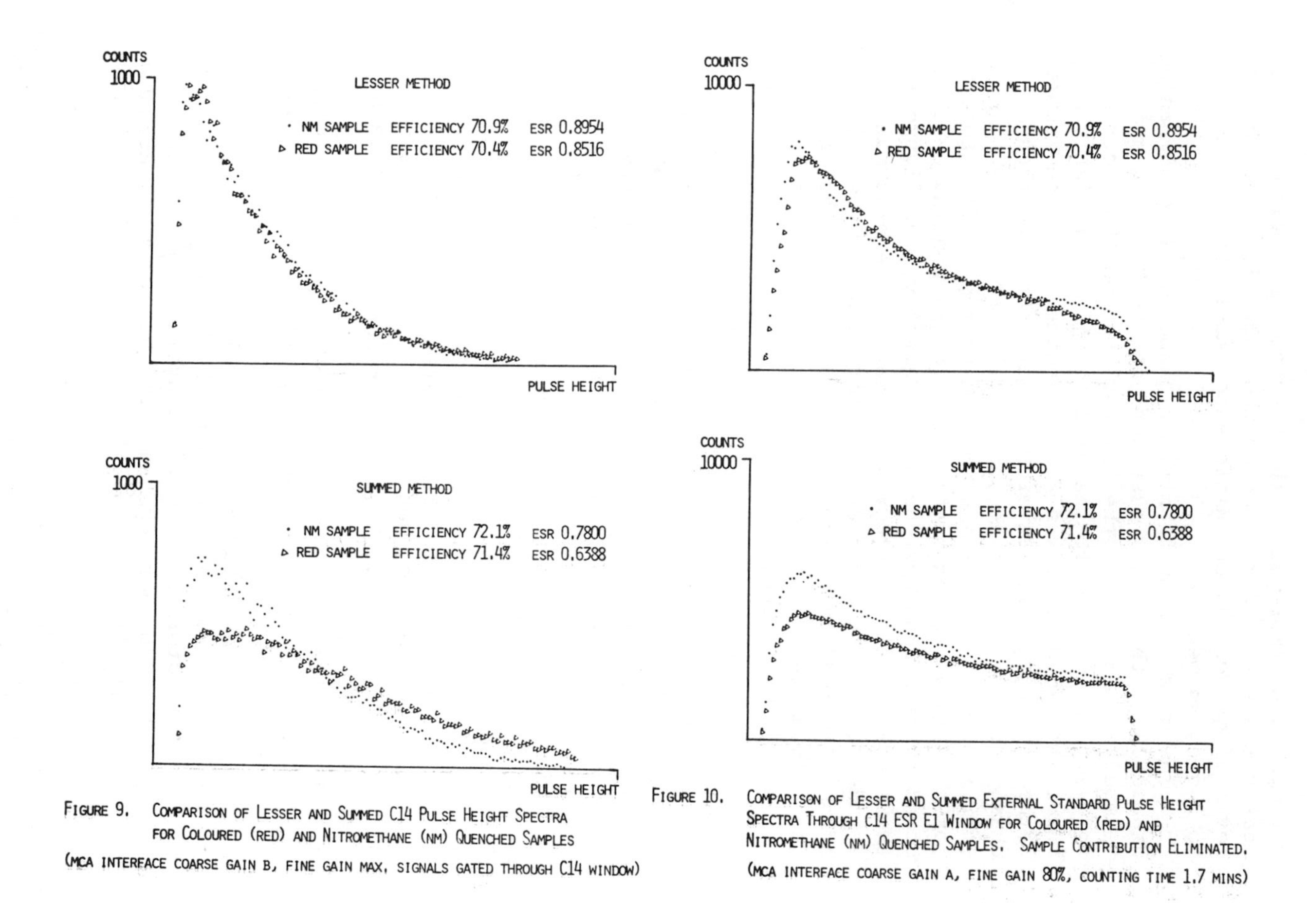

FIGURE 9. COMPARISON OF LESSER AND SUMMED C14 PULSE HEIGHT SPECTRA FOR COLOURED (RED) AND NITROMETHANE (NM) QUENCHED SAMPLES

(MCA INTERFACE COARSE GAIN B, FINE GAIN MAX, SIGNALS GATED THROUGH C14 WINDOW)

FIGURE 10. COMPARISON OF LESSER AND SUMMED EXTERNAL STANDARD PULSE HEIGHT SPECTRA THROUGH C14 ESR E1 WINDOW FOR COLOURED (RED) AND NITROMETHANE (NM) QUENCHED SAMPLES. SAMPLE CONTRIBUTION ELIMINATED.

(MCA INTERFACE COARSE GAIN A, FINE GAIN 80%, COUNTING TIME 1.7 MINS)

ratio. Furthermore since the "lesser" and "summed" measurements were performed on the same vial at each level of quenching, any differences must be attributable to those two methods.

The pulse height spectra of carbon-14 (fig. 9) shows dramatically the differences between the "lesser" and "summed" methods. For the "lesser" method the spectra for samples quenched to the same counting efficiency by chemicals and dyes very nearly coincide. For the "summed" method large differences are evident. Despite these changes in spectral shape, the integral counting efficiencies for the summed method do not differ greatly for the two quenching agents. The corresponding external standard ratios however do differ greatly, and figure 10 shows the associated spectral shapes of the external standard. It would appear therefore that the deviations of the "summed" data from a homogenous quench curve are due predominantly to displacements in the external standard ratio values.

Conclusions

The authors conclude from this investigation that the "lesser" method is a useful instrumental development in liquid scintillation counting. The reliability in determining the activities in samples that are both coloured and chemically quenched is improved by the application of this method.

References

1. E. Rapkin, Int. J. Appl. Rad. Isotopes 15, 69 (1964).
2. J.H. Parmentier and F.E.L. TenHaaf, Int. J. Appl. Rad. Isotopes 20, 305 (1969).
3. L.W. Price, World Medical Electronics, p. 282,310, 350 (1966).
4. L.W. Price, World Medical Electronics, p. 3 (1967).
5. D.R. White and J.L. Offerman, Bio-Medical Engineering 4 (8), 362 (1969).
6. D.L. Horrocks, Nature 202, 78 (1964).
7. A.A. Noujaim, C. Ediss and L.I. Wiebe *in* Organic Scintillators and Liquid Scintillation Counting, p. 705 (D.L. Horrocks and C.T. Peng, Eds.), Academic Press (1971).

8. J.F. Lang *in* Organic Scintillators and Liquid Scintillation Counting, p. 823 (D.L. Horrocks and C.T. Peng, Eds.), Academic Press (1971).
9. C.O. Carrol and T.J. Housar, Int. J. Appl. Rad. Isotopes *21*, 261 (1970).

Table I. Materials Used

A. Fluor Composition:
4 g/l PPO, 50 mg/l POPOP in toluene

B. Radioactive Standards:
(i) Tritiated Toluene NES 004 lot 552-220
3.12×10^6 DPM/ml - 25th April 1972
(ii) Carbon-14 Toluene NES 006 lot 552-285
4.15×10^5 DPM/ml - 8th March 1972

C. Chemical Quenching Agents:
(i) Nitromethane - maximum of 0.12 ml added to vials
(ii) Carbon Tetrachloride - maximum of 0.28 ml added to vials

D. Dyes - colour quenching agents:
(i) Carotene Yellow - 47.2 mg/100 ml fluor
maximum of 0.27 ml of solution added to vials
(ii) Scarlet Red - 115.9 mg/100 ml fluor
maximum of 0.31 ml of solution added to vials
(iii) Oracet Blue - 89 mg/10 ml fluor
maximum of 0.27 ml of solution added to vials

MEASUREMENT OF ^{125}I, ^{131}I and OTHER γ-EMITTING NUCLIDES BY LIQUID SCINTILLATION COUNTING

E. D. Bransome, Jr. and S. E. Sharpe, III
Division of Metabolic and Endocrine Disease
Medical College of Georgia, Augusta, Georgia U.S.A.

Other papers in this volume attest to the increasing interest in the biomedical sector in radioimmunoassays and radio-ligand assays of an ever-increasing list of polypeptides, glycopolypeptides, steroids, and pharmaceuticals. In a few procedures the substance of interest is labelled with ^{3}H but in most, iodination with high specific activity ^{125}I and occasionally ^{131}I, is carried out. It is still customary to count such samples in "gamma" counters where the scintillations (photofluorescence) emanate from a NaI crystal surrounding a shielded well.

We have recently summarized the evidence dating from 1956 that ^{125}I and ^{131}I can be counted at high efficiency and with little trouble by liquid scintillation counting. Calculation of quenchcorrection curves, of spill-over factors, and selection of optimum channels on any modern liquid scintillation counter are essentially the same as in the usual LS counting of βs(1). This is not surprising since, as Birks has pointed out, the principal effect of γ rays is a yield of Compton electrons in a broad spectrum of 0 to 75 percent of the γ ray energy (2). If energetic enough, X-rays may also excite dissolved scintillators. The k-X-ray (E_{max} 27 kEV) resulting from the decay of ^{125}I by electron capture thus contributes a second peak (see the Figures below) discernible from the low energy Compton peak of ^{125}I. We have made an effort to examine a number of γ emitters for their detection efficiency in counting solutions of toluene - 10% solubilizer - Biosolv BBS-3 with or without a standard primary scintillator PPO 7g/L (3).

Isotope	Principal Emission On Decay	E_{max}	Absolute Counting Efficiency With PPO	Without PPO
^{125}I	γ	0.035	73	6.4
	k-X-ray	0.027		
^{57}Co	γ	0.12	66	12.1
	γ	0.14		
^{85}SR	γ	0.51	68	14
^{131}I	γ	0.36		
	β	0.61	100	97

Table 1: Maximum efficiencies γ-emitting isotopes in a wide channel (0-1000) of a Beckman LS-150 liquid scintillation system.
Air equilibrated, in toluene - 10% BBS-3 with and without PPO 7g/L. Calibrated standards were generously provided by the Amersham-Searle Corporation.

Effects of Sample Geometry on Counting Efficiency

The results shown in Table 1 do not allow the systematic analysis of γs in LS counting we had hoped for. Unfortunately these are the only γ-emitters for which we have been able to obtain accurately calibrated standards.

The results do suggest that with more energetic γs, much of the radiation escapes the counting vial and therefore escapes detection. Addition of high electron-density materials to the scintillation solvent, eg. tetrabutyltin or tetrabutyl lead as suggested by Ashcroft (4) for counting ^{125}I (oddly enough, a procedure offering little advantage to ^{125}I counting efficiency) does not seem to be suitable for higher energy γs, however (5). It is obvious that an analysis of γ counting by liquid scintillation, sufficiently systematic to indicate optimum conditions for counting any particular nuclide is still needed.

We have previously emphasized (6) the converse problem that the absorbtion of low energy βs emitted by samples counted on solid supports may result in erroneously low estimates of sample radioactivity. We have carried out several experiments designed to assess the magnitude of either problem in ^{125}I and ^{131}I counting. For the sake of comparison we made up samples (without the primary scintillator PPO) and counted them in a well (NaI crystal) counter in solution and on two solid supports.

Sample	Absolute Efficiency (%) ^{125}I	^{131}I
Aqueous solution	80	20
Toluene	70	22
Cellulose nitrate	78	19
Filter Paper (Whatman #1)	64	19

Table 2. γ counting of ^{125}I and ^{131}I: effect of sample geometry.
Duplicate samples of ^{125}I-insulin (11,100 DPM) and ^{131}I-NaI (45,600 DPM) were dissolved in toluene-solubilizer or 15 mm spots were air dried on 20 mm discs. Each sample was placed in a perspex tube 18 x 100 mm and counted in the well of a Nuclear Chicago Model 1085 scintillation detector with a 2 in. NaI (Li) crystal for 10 min. Absolute activity was determined by reference to a standard quench correction curve after counting the same samples on a wide (0-1000) Channel of a Beckman LS-150 liquid scintillation system.

Table 2 shows that in the type of standard "γ-counter" the first problem: inadequate attenuation of γs, is very significant: ^{131}I efficiency was only a fifth of that obtained in LS counting. This loss of γs through penetration of the crystal detector is well known (7). The variation in efficiencies of ^{125}I detection could be attributed to the other problems: Compton electrons from ^{125}I were absorbed by the toluene-BBS-3 solution and by the solid supports, particularly the filter paper.

	^{125}I L.S. Counter, Settings		
A. ^{125}I	LS-150 0-1000	ISOCAP $^{131}I/^{125}I$*	LS-150 $(^{3}H)/(^{14}C)$*
Toluene - 10% BBS-3	73	77	69
Cellulose Nitrate	62	65	57
Filter Paper (Whatman #1)	40	42	37
B. ^{131}I			
Toluene - 10% BBS-3 Solubilizer	100	81	63
Cellulose Nitrate	83	67	46
Filter Paper (Whatman #1)	85	68	48

Table 3. Efficiency of ^{125}I and ^{131}I in liquid scintillation counting.

Duplicate samples were dissolved in counting solution or dried onto discs as in Table 2. The discs were placed on the bottom of vials filled with toluene-PPO,7g/L. In the Beckman LS-150 system, ^{14}C and ^{3}H isosets were used to count ^{131}I and ^{125}I respectively. The efficiencies determined were quite similar to those with ^{14}C and ^{3}H programs on the Nuclear Chicago Mark II Liquid Scintillation System used to generate the spectra shown in figures 1 and 2. The efficiency figures for the Nuclear Chicago Isocap system were obtained from the appropriate fixed channel in the preset $^{131}I/^{125}I$ program. The higher efficiency of ^{125}I (than the LS-150) is attributable to greater efficiency of the phototubes in the Isocap; the lower efficiency of ^{131}I to a narrower channel.

In liquid scintillation counting the two problems still exist, and are if anything more significant than in "γ" (NaI) scintillation counting, as Table 3 shows. The solid supports cause very significant decreases in counting efficiency which as Figure 1 and Table 4 show is due to absorption with ^{125}I and to both problems with ^{131}I:

	Channel		
	"^{131}I"	"^{125}I"	Ratio
Toluene - PPO 7g/L - 10% BBS Solubilizer	376,000	134,000	0.36
Cellulose Nitrate	314,000	113,000	0.36
Filter paper (Whatman #1)	313,000	132,000	0.42

Table 4. Effect of Sample Geometry on ^{131}I channels ratio. Samples of ^{131}I in solution or on discs (see Table 3 - legend) were counted in the fixed $^{131}I/^{125}I$ channels of a Nuclear Chicago ISOCAP system.

The loss of ^{131}I efficiency with samples dried onto cellulose nitrate filters was at least partly due to loss of energetic βs or γs from the vial before they could interact with the scintillator: PPO. The lack of spectral distortion in Figure 1 and the unchanged $^{131}I/^{125}I$ channel ratio (Table 4) is strong evidence. Suspending the disc in the vial rather than letting it lie flat on the bottom might well improve the situation. The greater absorptive properties of the filter paper (6) through attenuating the β and Compton electron spectra of ^{131}I may have actually increased the counting efficiency. The resultant pulse bright shift, although hard to discern in Fig.1 because of the broad spectrum of ^{131}I is evident in an increased $^{131}I/^{125}I$ channels ratio (Table 4).

With the lower energy Compton Spectra of ^{125}I, absorption of β energy by the solid supports is the predominant cause of decreased efficiency. In Figure 2, the resulting spectral distortions are evident as the peak from the 27 kev (E_{max}) k-X-rays becomes less defined as efficiency drops. We have not yet had the opportunity to determine whether the changes in channels ratios resulting from this pulse height shift fall into the same domain on a channels ratio-quench correction curve as does impurity quenching. Our prejudice, based on experience with absorption of ^{14}C βs by solid supports (6) is that the two relationships will be dissimilar. Monitoring and correcting for ^{125}I absorption in samples on solid supports promises then, to be fairly complex.

CONCLUSIONS

In "γ" counting (by scintillation of NaI crystals in response to γ rays), the measurement of γ emission by liquid scintillation is subject to decreased efficiency from inadequate attenuation of the radiation by the detector or from absorption of radiation by the sample. An inadequate yield of Compton electrons and of X-rays (from isotopes decaying by electron capture) is difficult to detect and will be significantly effected by the geometry of the sample in the counting vial. Samples on solid supports therefore may be measured with considerably less efficiency than identical samples in solution. It seems at least at the present time, when the problem has not yet been adequately explored, that the effects of absorption of Compton electron energy by solid supports is a complex phenomenon which will be difficult to estimate in any specific group of samples.

These findings obviously result in a dictum to those who employ liquid scintillation counting in measuring iodine isotopes or other γ emitting nuclides: it is of paramount importance that standard curves be determined with samples prepared in exactly the same manner as unknown samples are. Otherwise there may be sometimes unpredictable systematic errors in measurement of the substance of interest.

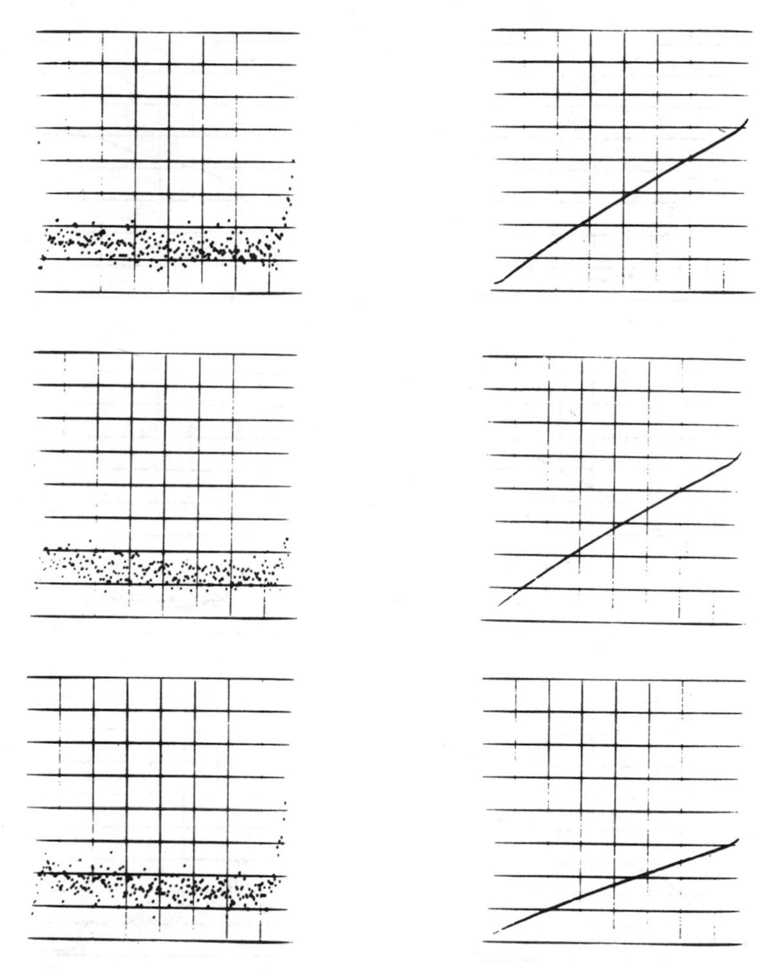

Figure 1. Spectra of ^{131}I: Effect of solid supports obtained with a Nuclear Chicago Mark II liquid scintillation counter interfaced to a Nuclear Chicago 4096 multichannel analyzer (See Ref. 3).

Raw data are displayed on the left; integrals of the raw data on the right. In descending order: 376,000 DPM ^{131}I-new/rare NaI in//-toluene-PPO,7g/L-N5 solubilizer 10%: 512 counts/channel/1.5 min. full scale - dried onto a 20 mm cellulose nitrate filter disc placed on the bottom of a vial filled with toluene - PPO, 7g/L 512 counts/channel/2 min. full scale - dried onto a 20 mm. filter paper disc (Whatman #1) 512 counts/channel/1.5 min. full scale.

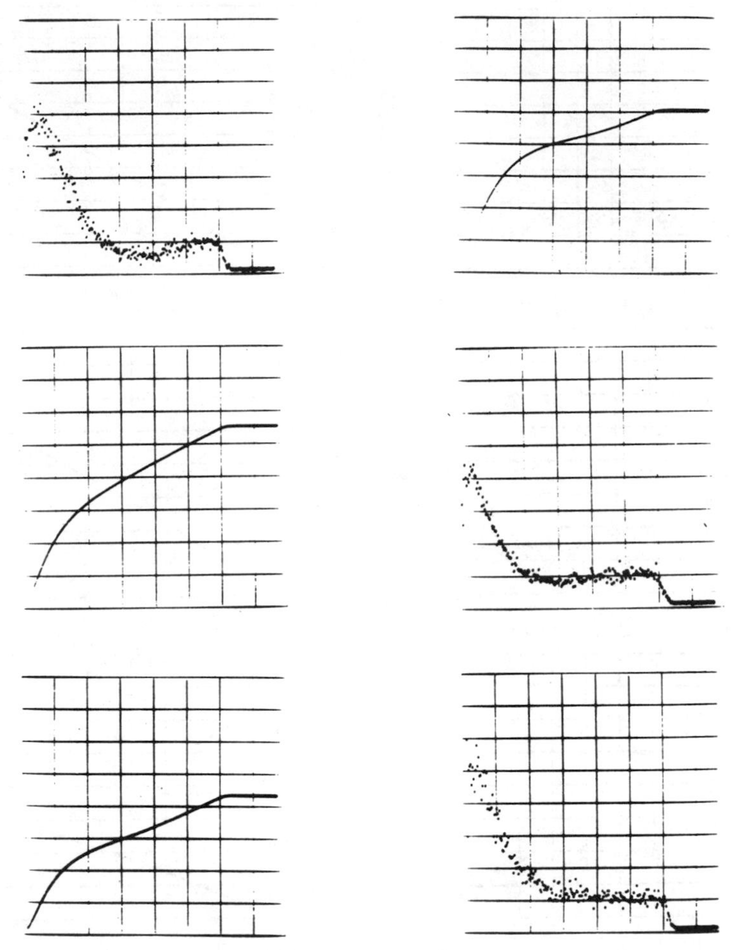

Figure 2. Spectra of ^{125}I: Effect of solid supports.

Raw data and integrals of the raw data. Samples of ^{125}I-insulin (362,000 DPM) were as described in the legend of Fig. 1: in descending order:

- toluene - PPO, 7g/L - 10% Biosolv-BBS-3
  512 counts/channel/2 min. full scale
- on a 20 mm Cellulose nitrate disc in toluene
- PPO, 7g./L 512 counts/channel/2 min. full scale
- on a 20 mm filter paper (Whatman #1) filter paper disc in toluene - PPO, 7g./L. 256 counts/channel/2min. full scale.

REFERENCES

1. E. D. Bransome, Jr. and S. E. Sharpe, III, Anal. Biochem. 49; 343 (1972).
2. J. B. Birks, Discussion in Liquid Scintillation Counting (Crook, M. A., P. Johnson, and B. Scales, eds) Heyden and Son, London, 1972, p. 20.
3. S. E. Sharpe, III and E. D. Bransome, Jr., in Liquid Scintillation Counting: Recent Developments (P. E. Stanley and B. A. Scoggins, eds) New York and London : Academic Press Inc. (this volume).
4. J. Ashcroft, Anal. Biochem. 37; 208 (1970).
5. P. Ting, Clinical Brief 6. Beckman, Inc. Scientific Instruments Div. Fullerton, Cal. (1973).
6. E. D. Bransome, Jr. and M. F. Grower, Anal. Biochem. 38; 401 (1970).
7. P. Holmbert and R. Rieppo, Int. J. Appl. Radiation Isotopes 24; 99 (1973).

REFERENCES

1. E. D. [illegible] [illegible] [illegible]
[illegible] 45 [illegible]
2. [illegible] D[illegible] [illegible]
Contin[illegible] [illegible] [illegible]
[illegible] 19[illegible]
3. [illegible] and [illegible]
Liquid Scintillation [illegible]
[illegible] and [illegible] New York
and London [illegible]
4. [illegible] Anal. [illegible] 19[illegible]
5. [illegible]
[illegible]
6. [illegible]
[illegible]
7. [illegible] and [illegible]
[illegible] (19[illegible]).

SURFACTANTS BEHAVE AS SCINTILLATORS IN LIQUID SCINTILLATION COUNTING

S. E. Sharpe, III and E. D. Bransome, Jr.
Division of Metabolic and Endocrine Disease
Medical College of Georgia, Augusta, Georgia USA

A number of solubilizers, proprietary mixtures of non-ionic and ionic surfactants have been used to make samples in aqueous solutions miscible with toluene, the organic solvent usually employed in liquid scintillation counting. We have discovered that such surfactants may increase the scintillation yield of quenched samples emitting sufficiently energetic photons, and may themselves act as efficient scintillators. Data on Biosolv-BBS-3 (distributed by Beckman Instruments, Inc.) are in press at the time of this writing (1). A detailed examination of our own series of surfactant combinations of known composition is in preparation (2). For this paper, we have selected our N-5 solubilizer, one with surfactant and scintillant properties very similar to those of the commercial BBS-3 preparation to illustrate the behavior of many surfactants, and the practical implications for liquid scintillation counting.

Materials and Methods

All samples were counted (except for reference standards) in aqueous solution and were air-equilibrated.

Compounds	Isotope	Principal Emission On Decay	E_{max} (kEV)	Activity (approx.) (DPM)
H_2O	^{3}H	β	18	220,000
Cytosine	^{14}C	β	156	55,000
Insulin	^{125}I	γ k-X-ray	35 27	362,000
NaI	^{131}I	γ β	360 610	420,000

Table 1. Samples counted.

The absolute radioactivity of each sample was determined by comparing external standard channels ratios obtained with a Beckman LS-150 system to known quench correction curves. N-5 Solubilizer was mixed in our laboratory using commercially available nonyl-phenoxyethanol (94%) and sodium dihexylsulfosuccinate (6%). Spectra were obtained using a Nuclear Chicago 4096 Multi-channel analyzer interfaced to a Nuclear Chicago Mark II Liquid Scintillation System. The interface was set at analyzer Channel A and the attenuator at 0. Standard ^{3}H settings were used to count ^{3}H samples and ^{125}I samples and ^{14}C settings were used to count ^{14}C and ^{131}I with the upper discriminator set to infinity. After raw data were obtained, they were stored in the analyzer memory and the integral obtained and displayed. The resulting plots of spectral shape also indicate relative detection efficiency of identical amounts of radioactivity counted under different conditions: The greater the efficiency, the higher the curve will intersect the ordinate (counts per channel).

RESULTS AND DISCUSSION

Figure 1 (a reproduction of Figure 2 in reference 1) shows that BBS-3 can serve as an efficient primary scintillator with βs above 100 kEV maximum energy (E_{max}). The emission spectrum of the fluorescence from BBS-3 is similar to those of the scintillators in general use (1). The higher excitation threshold is a handicap however, particularly with samples which are appreciably quenched; detection efficiency drops precipitously as the photoelectron spectrum shifts below 100 kEV.

The effects of N-5 solubilizer on the spectrum of ^{3}H, shown in figure 2 are therefore anticipated: N5 acts as a quencher, shifting the spectrum toward the low energy channels and decreasing efficiency with ^{3}H samples. Therefore, a standard quench correction curve will indicate the loss of counting efficiency. (1).

Figure 3 illustrates the prediction that with more energetic βs, a surfactant will act as a scintillator even in the presence of a more efficient primary scintillator such as PPO.

From the loss of efficiency indicated by the lower ordinate heights integral plots N5 did act as an impurity quencher. The raw spectral data show that there was a loss (as expected) of low energy photons when N-5 was added to a toluene PPO solution. It appears that there was a significant shift of the fluorescence spectrum toward higher energies.

The effect of this phenomenon on ^{14}C quench correction is shown in Figure 4. If samples are being counted in a solution containing surfactant, it is therefore necessary to set up a quench correction curve with standards containing a similar amount of surfactant. Otherwise the radioactivity of quenched samples will be overestimated. A solution of 10% N-5 in toluene may be a practical "cocktail" for ^{14}C counting, since the counting efficiency of the air-equilibrated sample shown in Figure 3 was 51% (28,000 DPM) vs. 80% for the N5-PPO, 7g/L sample (44,000 DPM).

The behavior of surfactants as scintillators for counting iodine isotopes has been of more than idle interest as our other paper in this volume (3) and elsewhere (4) testify. ^{125}I is counted at good efficiency in 10% N5 solution (66%), slightly less than the 73% with a comparable solution of 10% BBS-3 (4). The solubilizers act as quenchers as they do with ^{3}H. There is no evidence of a significant scintillator effect of solubilizers in the presence of PPO.

Although the comparison of the counting efficiencies of toluene/N5 (6%) and toluene alone ($< 1\%$) indicates that some of the Compton electrons from the ^{125}I γ or k-X-rays are energetic enough to excite the surfactants to fluorescence.

The spectra in figure 5 show that in the absence of PPO it is no longer possible to differentiate the photopeaks resulting from the 35 kEV γ and the 27 kEV k-X-ray. We have not yet examined the effect of surfactants on ^{125}I quench-correction curves.

^{131}I is counted with good efficiency in Toluene 10% N5 or toluene - 10% BBS-3: at virtually 100% efficiency with PPO, and at 87-97% efficiency without PPO. The high energy β of ^{32}P is counted with similar ease and resistance to quenching (See Figure 1). The broad spectra of ^{131}I in Figure 6 are attributable to the broad Compton Spectrum of the 360 kEV E_{max} γ. As with the ^{14}C spectrum (Figure 3) the presence of solubilizer has shifted the spectrum toward the higher energy ranges: the low energy peak is less conspicuous and the integral is less hyperbolic and more linear.

Without PPO present, there is an accumulation of counts at the lower end of the spectrum (Figure 7) much as there was with ^{14}C - but few counts have fallen below detection limits. Some of this shift is attributable to the Cerenkov effect (5) which allows ^{131}I to be counted at high efficiency. The counting efficiency of ^{131}I insulin in toluene alone which we obtained (45%) may be spuriously low, since we cannot be sure all of the sample was in solution.

CONCLUSIONS

The use of nonionic-anionic surfactant combinations in liquid scintillation counting mixtures will render quench correction curves prepared with dissimilar standards invalid if the βs emitted are more energetic than ^{3}H. High energy βs can be counted at high efficiency in a 10% solution of nonyl phenoxyethanol (94%) and sodium dihexylsulfosuccinate (6%) in toluene alone.

Further investigation of the structure - activity relationships of micelles of water and surfactants in scintillation solvents to fluorescence yield will we hope, result in a new and less expensive kind of scintillation "cocktail".

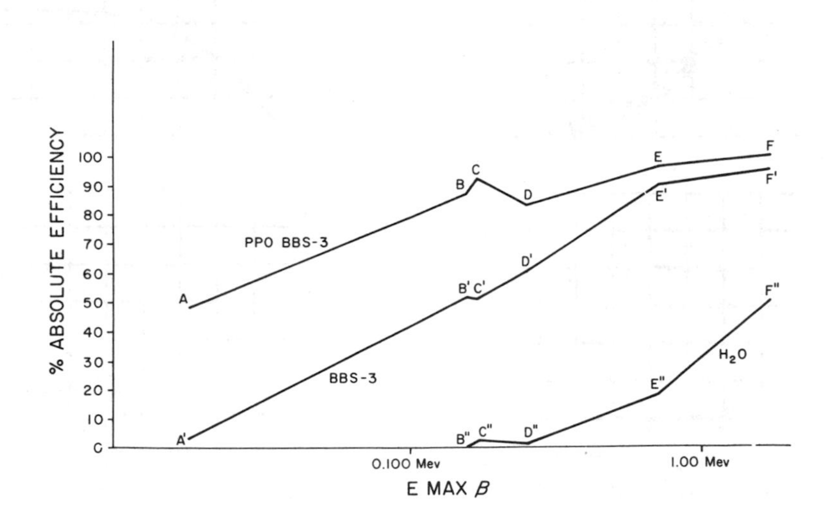

Figure 1. Relationship of detection efficiency to E max in various solutions of toluene-based β-emitting isotopes.

A. ^{3}H B. ^{14}C C. ^{35}S D. ^{45}Ca E. ^{36}Cl F. ^{32}P

PPO-BBS: 7g PPO/L Toluene 90%: BBS-3 10%
BBS: toluene 90%: BBS-3 10%
H_2O: This latter curve represents the Cerenkov efficiency of these isotopes in a system without anomalous refractive dispersion. (from Reference 1).

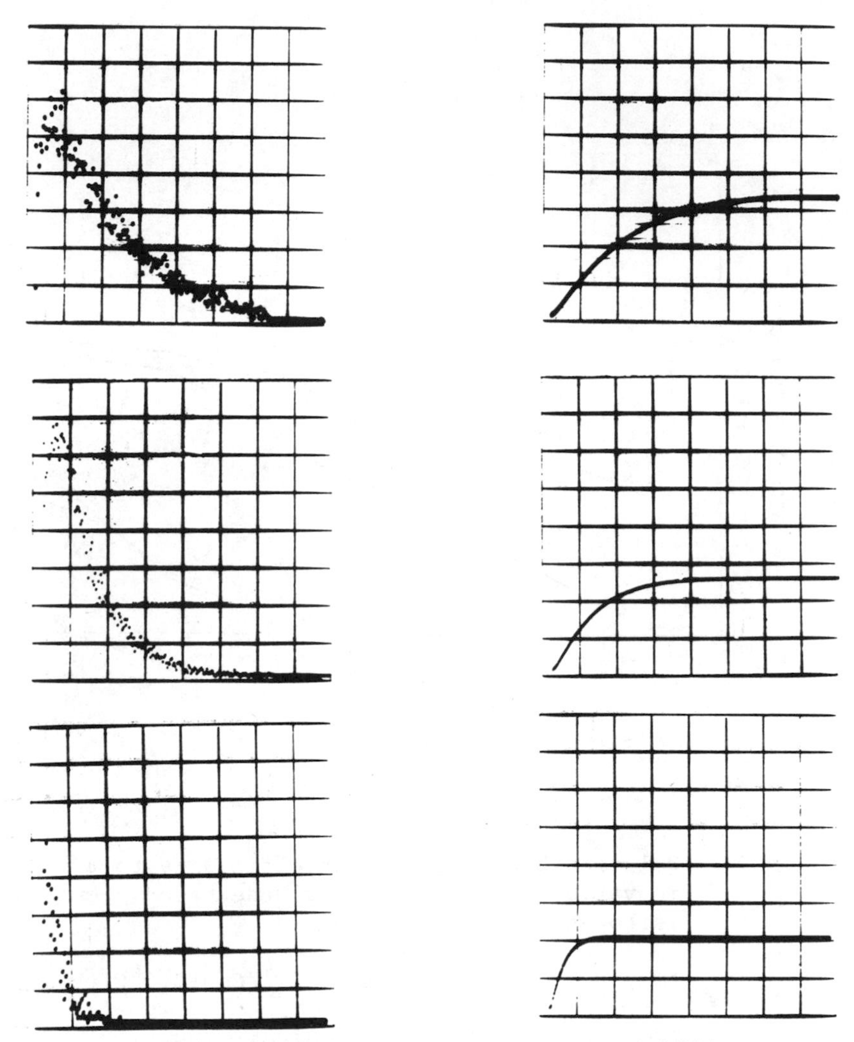

Figure 2. Spectra of ^3H: effect of N-5 solubilizer (10% by volume).

Raw data are shown on the left and integrals of the spectra are on the right. In descending order samples are

- toluene - PPO, 7g/L (^3H-Toluene) equilibrated with N_2.
- toluene - PPO-7g/L - 10% N5
- toluene - 10% N5

Full scale for the two samples containing PPO was 256 counts/channel/2 min. For N 5 - toluene full scale was 64 counts/channel/2 min.

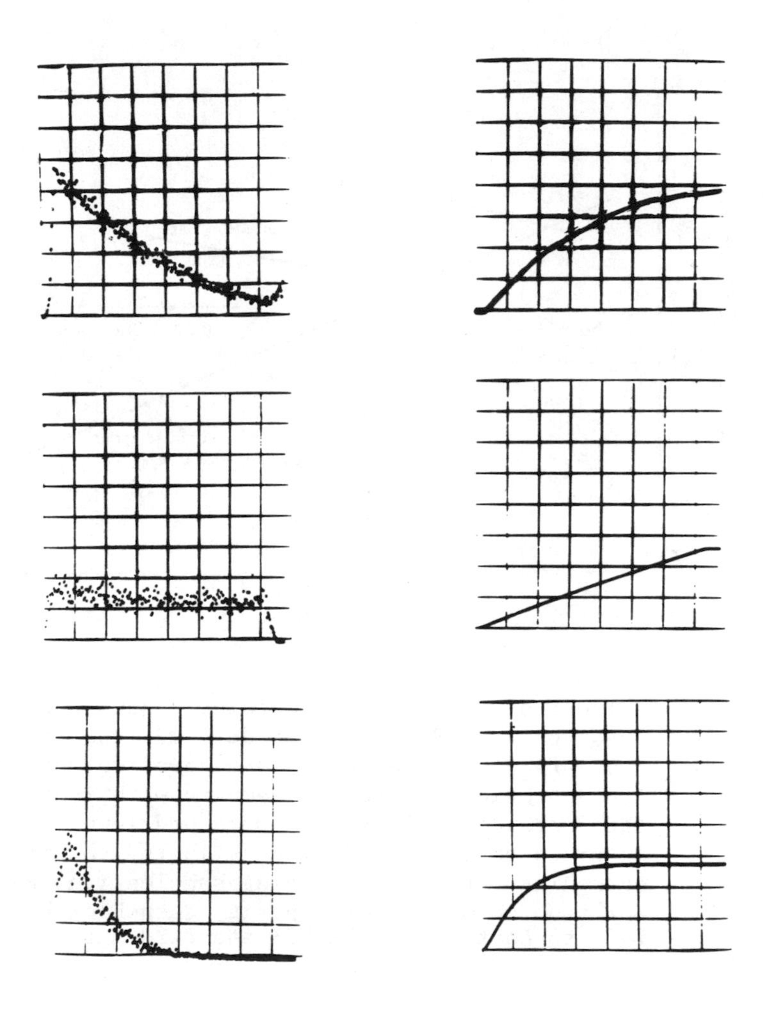

Figure 3. Spectra of ^{14}C: effect of N-5 solubilizer (10% by volume).

Raw data are on the left, integrals on the right

- toluene - PPO, 7g/L (^{14}C toluene) equilibrated with N_2
- toluene - PPO-7g/L - 10% N5
- toluene - 10% N5

Full scale for the samples was 512 counts/channel/2 min.

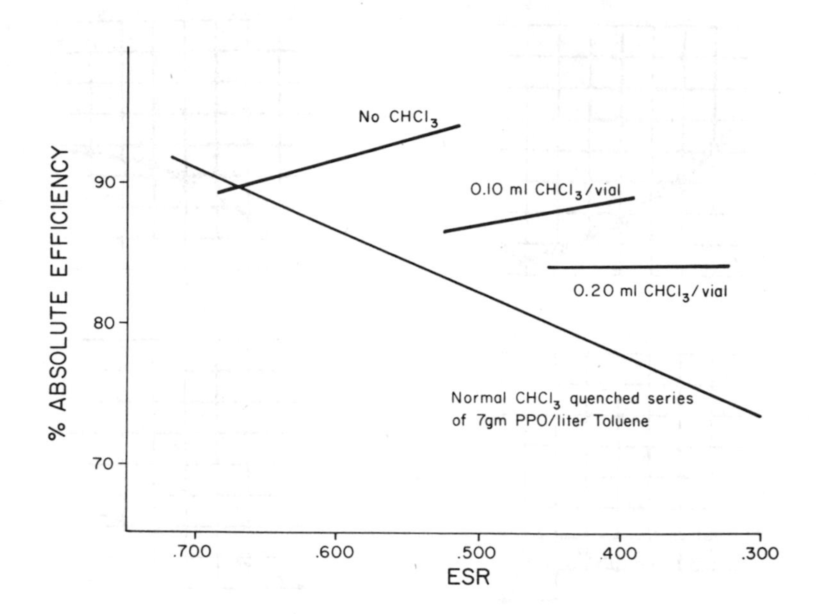

Figure 4. Effect of BBS-3 on a ^{14}C-toluene quench correction curve. The data are derived from Fig. 4, but are plotted without consideration for the concentration of BBS-3. Samples with increasing BBS-3 concentrations of course had lower ESRs inasmuch as BBS-3 itself contributed to impurity quenching. BBS-3 containing samples (indicated for each condition of impurity quenching by $CHCl_3$ fail to conform to the ESR standard curve (Ref.1)

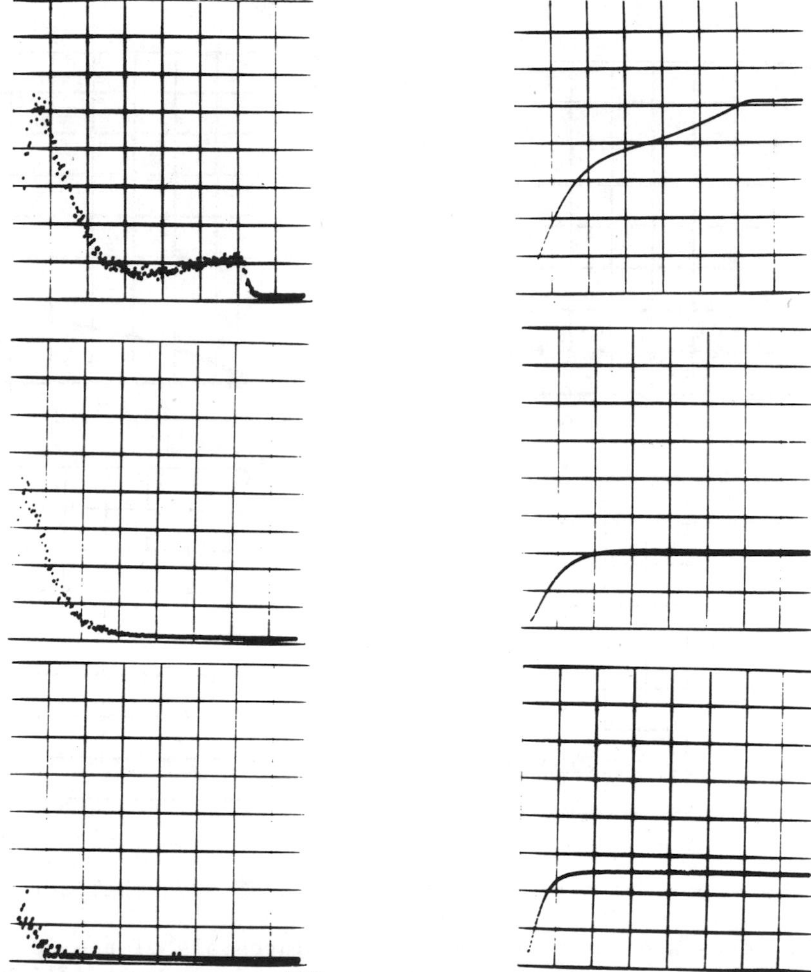

Figure 5. Spectra of ^{125}I: effect of N-5 solubilizer (10% by volume).

Raw data are shown on the left and integrals of the spectra on the right. In descending order the samples are:

- toluene - PPO, 7g/L - 10% N5 (256 counts/channel/2 min. full scale) counting efficiency 66%
- toluene - 10% N5 - (128 counts/channel/2 min. full scale counting efficiency: 6%
- toluene alone. (64 counts/channel/2 min. full scale) counting efficiency < 1% (See Figure 2).

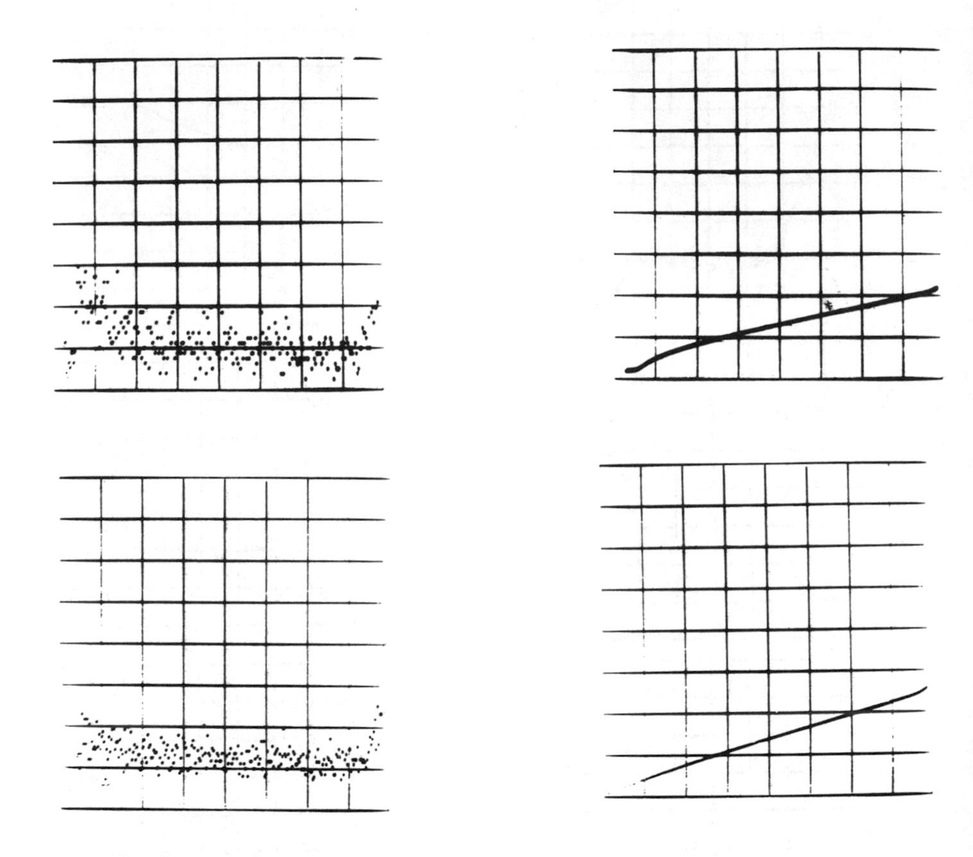

Figure 6. Spectra of ^{131}I: effect of N-5 solubilizer (10% by volume).
Raw data are shown on the left and integrals of the spectra are on the right. The upper sample represents ^{131}I-NaI in toluene - PPO 7g/L. and the lower toluene - PPO-7g/L-10% N5. Full scale in both cases is 512 counts/channel/2min. The lesser efficiency of the sample without solubilizer is attributable to incomplete solution of the sample.

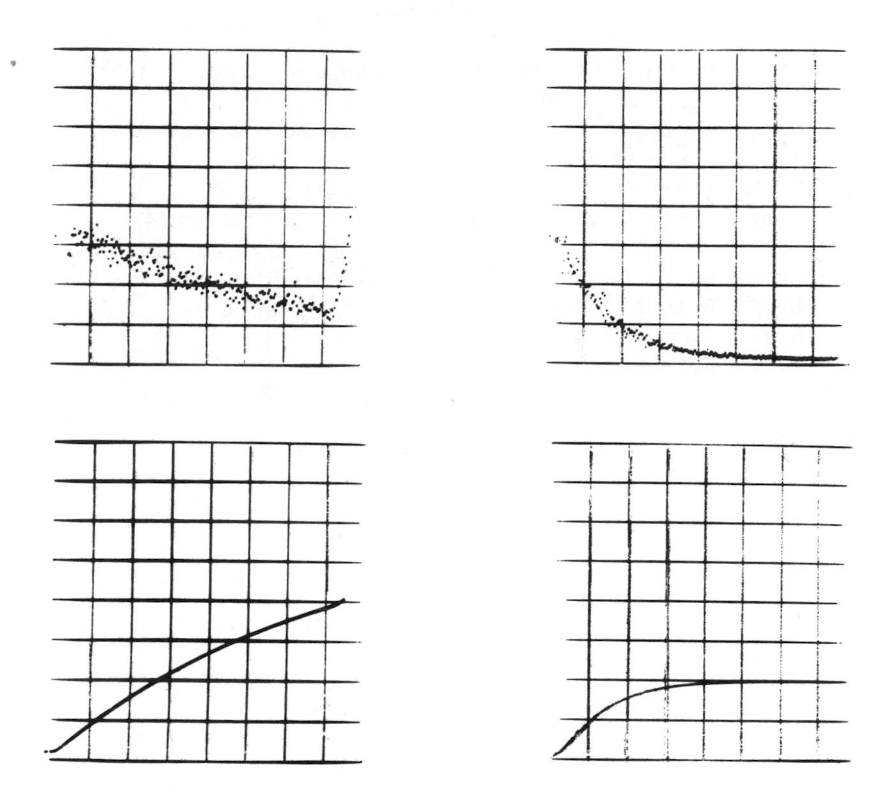

Figure 7. Spectra of ^{131}I in toluene: effect of N-5 solubilizer.
Raw data are at the top, the integral plots below. The left represents a sample in toluene - 10% N5 counted at 96% efficiency (256 counts/channel/2 min. full scale). On the right, an incompletely dissolved sample of ^{131}I in toluene alone.

REFERENCES

1. S. E. Sharpe, III and E. D. Bransome, Jr., Anal. Biochem. In press.
2. S. E. Sharpe, III and E. D. Bransome, Jr., In preparation.
3. E. D. Bransome, Jr. and S. E. Sharpe, III, in Liquid Scintillation Counting: Recent Developments (P. E. Stanley and B. A. Scoggins, eds) New York and London, Academic Press Inc. (this volume).
4. E. D. Bransome, Jr. and S. E. Sharpe, III. Anal. Biochem. 49; 343 (1972).
5. H. H. Ross and G. T. Rasmussen.

SPARK COMBUSTION OF ^3H AND ^{14}C LABELED SAMPLES SUITABLE FOR LIQUID SCINTILLATION COUNTING

John E. Noakes
Geochronology Laboratory
The University of Georgia, Athens, Georgia U.S.A.
Walter Kisieleski
Argonne National Laboratory
Argonne, Illinois U.S.A.

Abstract

A new method of combusting labelled ^3H and ^{14}C samples is accomplished through continuous spark ignition. Liquid, wet tissue and powdered samples are combusted in disposable steel planchets. Radioisotope collection is in liquid scintillation cocktails as ^3H_2O and $^{14}CO_2$. Quantitative collection and minimum memory is enhanced through controlled combustion rates. Comparison of the spark ignition method is made with presently available commercial combustion equipment.

Introduction

The liquid scintillation spectrometer is a versatile instrument for the measurement of low-energy beta emitters, especially tritium and carbon. The instrument, utilizing computers for data computation and presentation has proven especially valuable in biological and clinical application. On the other hand, biological materials are most difficult to prepare as true solutions for liquid scintillation counting and therefore present unique sample preparation problems.

Numerous methods and techniques have been developed in an attempt to solubilize diverse biological materials for scintillation counting. Generally these direct dissolution methods are limited by the small amount of material that can be dissolved in the more efficient cocktails. Furthermore, the severe quenching with highly colored biological materials results in decreased counting efficiency. Chemiluminescence is also a constant problem since many biological materials, when prepared for counting, generate light spontaneously even in the absence of radioactivity.

To overcome problems of sample solubility, color quenching and chemiluminescence, a more universal preparation technique for highly diverse biological materials can be achieved if the sample is completely oxidized, and the oxidation products are quantitatively dissolved in a scintillator to produce a minimum constant quenched sample. The most common example is the combustion and oxidation leading to collection of tritium as water and carbon-14 as carbon dioxide (1-3).

Recently there have been significant advances in the field of instrumental methods for sample combustion of biological materials as applied to liquid scintillation counting. Among the new developments are instruments that give greater precision and accuracy, increased ease of operation, and, most important, are completely automated, thereby removing operator error (4-9).

We presently would like to report on a new oxidation method that utilizes a continuous spark ignition for sample combustion. This technique offers certain advantages not existing in the presently available commercial units.

Experimental

A) INSTRUMENT DESIGN. The combustion chamber of the continuous spark ignition system is perhaps the most unique new aspect of this combustion method. The novel aspect is the employment of a continuous electrical spark to ignite the sample and sustain ignition throughout total sample combustion. Figure 1 shows the construction of the chambers in detail. The sample to be combusted is placed in a 3/32-inch low shouldered steel planchet of approximately 1¼-inch diameter. The planchet when placed in the combustion chamber rests on and makes electrical contact with a negative ground electrode. A stationary positive electrode is positioned about ¼ to ½ inches above the top most portion of the sample. This electrode is constructed of a hollow 1/8-inch steel tubing with a 1/32-inch wall thickness and extends to the outside of the chamber by a seal connector through the wall of the pyrex combustion chamber. The electrode is spiraled in a 1-inch diameter coil and insulated from the instrument by passing within a ¼-inch thick walled ceramic tube of 1½-inch I.D. The electrode is finally connected to a oxygen flow metering device and an electrical oscillating circuit. The ceramic tubing is attached to a heat gun in such a manner that hot air of >200°C can be blown through the ceramic tube, over the coiled steel electrode and exhausted on the surface of the pyrex combustion chamber. The pyrex combustion chamber

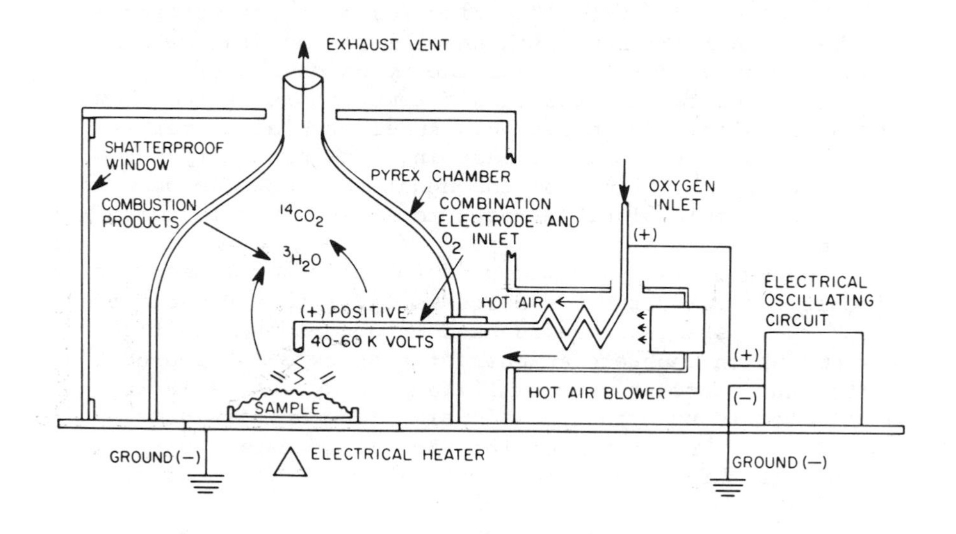

Figure 1. Spark Ignition Combustion Chamber

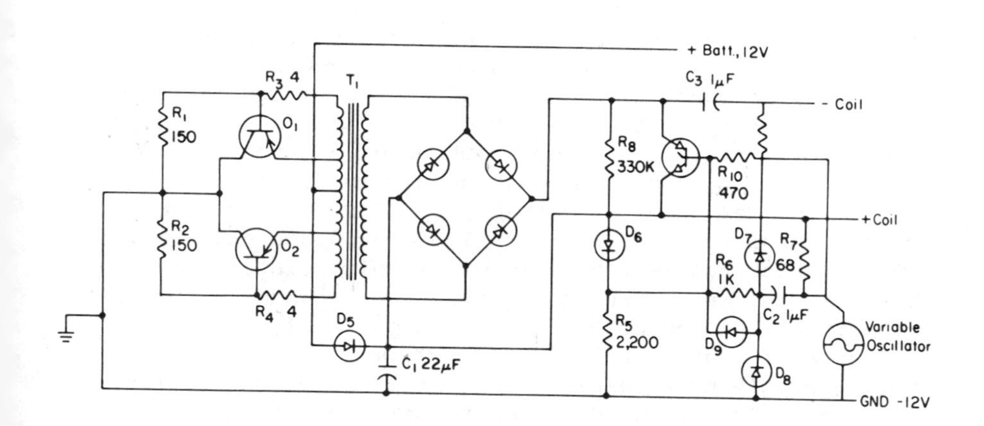

Figure 2. Spark Ignition Electrical Oscillator Circuit

is semi-hemispheric in shape with a gas exhaust portal at the top. It has a 3-inch O.D. at its maximum width with ¼-inch walls and a vyton "O" ring for airtight sealing at its base. A 1/16-inch thick sheet metal housing with a shatterproof window fits over the pyrex chamber for the purpose of containing the raised temperature of the chamber and as a safety consideration. An electrical heater is also placed in the chamber base in close proximity but electrically isolated from the negative electrode on which the steel sample planchet rests to enhance sample combustion rates.

Figure 2 is a schematic of the electrical oscillating circuit. A 12-volt D.C. source is fed into a capacitance discharge system such as is used in automotive ignition systems which generate an energetic spark of 40,000-60,000 volts and a current of low milliamp range. A variable oscillator is used to give a spark frequency discharge of 25,000 cps. The output of the electrical oscillator circuit is through a 12-volt coil which is directly connected to the combustion chamber.

Figure 3 shows the overall schematics of the equipment and procedures used to combust, transfer, separate and collect the ^{14}C carbon dioxide gas and ^{3}H water vapor originating from sample combustion. The combustion gases are transferred from the combustion chamber to the tritium water vapor trap through a heated metal transfer line to minimize the formation of water condensate. The tritium trap consists of spiraled tubing cooled to +5°C by a glycol refrigerant. The ^{14}C carbon dioxide gas leaving the tritium trap is further scrubbed of any residual H-3 water vapor by passing through an exchange column on its way to the carbon dioxide trap. The tritium exchange column consists of a 1/16-inch I.D. roughened steel walled tubing which has wetted walls of organic scintillator. To flush the water condensate from the exchange column and primary collecting column to the sample vial, two 8 ml aliquots of Searle-Analytic PCS (Phase Combining System) were used giving a sample volume of 16 ml of scintillator solution. The carbon dioxide gas next passes into a carbon dioxide collecting column consisting of a gas bubbler emersed in a solution consisting of a mixture of 5 ml of 2 methoxyethyl amine and 3 ml of toluene in which the carbon dioxide is absorbed as a soluble carbamate. A secondary exchange column of similar construction proper-

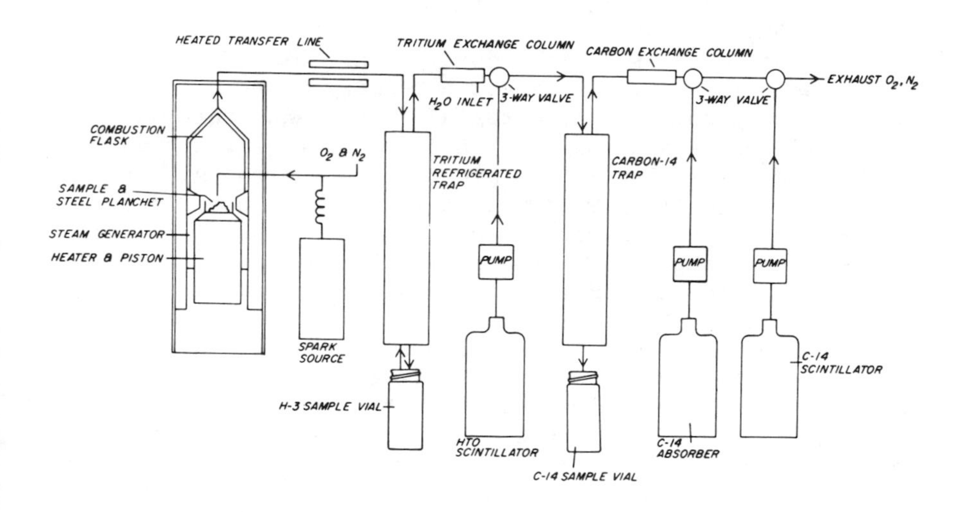

Figure 3. Spark Ignition Combustion System

ties as the tritium exchange column is used but its surfaces are wetted with the amine cocktail described above. The primary carbon collecting column is emptied after complete sample combustion by a valve at the bottom of the trap. A scintillator of toluene and PPO & POPOP concentrate is used in two 4 ml aliquots to clean the exchange column and primary collecting column of residual carbamate-amine solution. This cleaning solution along with the 8 ml of amine-toluene cocktail make up a total of 16 ml of cocktail used for the carbon sample vial.

B) INSTRUMENT OPERATION. Samples to be combusted in either a wet, dry or frozen condition are placed directly on a steel planchet. It is not necessary to wrap the sample in any combustible material to sustain ignition as the continuous spark ignition insures constant ignition conditions. The planchet is placed on a piston which is shown in Figure III. This piston raises and lowers from the bottom of the combustion chamber and serves as an entry port for the sample. The top of the piston is the negative electrode plate which makes electrical contact with the steel planchet. A heater within the piston is used for rapid elevation of sample temperature to enhance dehydration.

The initial sequence for sample combustion once the sample is within the combustion chamber and the combustion sequence initiated is a 10 second instrument activation time in which four functions are carried out. First, the heater within the piston elevator is activated. Second, the air-heat gun is turned on and hot air is directed over the coiled positive electrode and onto the surface of the pyrex combustion chamber. The third activity is that of hot nitrogen gas passing through the heated positive electrode and onto the surface of the sample for the purpose of surface drying the sample and elimination of any highly volatile combustible constituents which might be associated with the sample. The fourth step is that of loading the carbon dioxide collecting column with 8 ml of a mixture of 2-methoxyethyl amine and toluene. At the end of the ten second period, the hot nitrogen gas flow ceases, and the heated oxygen at a temperature $>200^{\circ}C$ and flow rate of 400 cc/minute is directed into the chamber through the positive electrode and over the sample to be combusted. The spark ignition system is activated and a continuous spark is visible passing down the same pathway as the heated oxygen onto the surface of the sample. Wherever the spark discharge strikes the sample a visible flame is present. The spark strikes and moves about the surface of the sample, igniting it at the closest point of the sample to the tip of the positive electrode. Combustion takes place at a uniform rate over the topmost surface of the sample. When wet or frozen samples are combusted, the rate of combustion can be controlled by the applied heat from the heater within the base of the sample holder. During the final stages of sample combustion, the spark will seek out and ignite the last vestiges of the uncombusted sample on the planchet. The spark ignition system can either be manually operated or can be automatically programmed. At the termination of the combustion step a nitrogen gas flow is reactivated to sweep out the last trace amounts of combusted gases to the appropriate collecting systems.

The ^{3}H-water vapor collecting column containing the condensed water vapor is emptied of its contents by two 8 ml aliquots of PCS (Phase Combining System) which empties into a sample vial. The ^{14}C carbon dioxide collecting column is emptied by first allowing the 8 ml of cocktail amine to drain into the carbon sample vial and then washing the collecting column with two 4 ml aliquots of concentrate scintillator into the sample vial. At this stage, the instrument's piston lowers automatically and the instrument is ready to receive another sample. A recent innovation to further reduce the instrument's memory and spillover has been the incorporation of a steam generator into the combustion chamber and a water inlet on the tritium exchange column. The two components have not been fully tested; therefore, the data presented in the section "Results and Discussion" do not include their use.

Analytical Results

The analytical capabilities of the spark ignition system were tested in the following manner. Radioactive standards of known activity of ^{14}C and ^{3}H labelled thymidine were pipetted onto #2 Whatman filter paper of 25 mm diameter having an average weight of approximately 70 mg. Filter paper was selected as the combustible sample as this material has been the chosen sample used by commercial combustion systems when reporting their recovery data. The amount of recovered radioactivity, after sample combustion, was determined by counting the sample vials containing the collecting cocktails used in the spark ignition system in a Mark II liquid scintillation counter. The liquid counter was calibrated for counting efficiency using the liquid scintillation cocktails of the spark system with known activity of ^{14}C and ^{3}H labelled toluene. Measurements were also made for instrument memory and spillover for each sample. Spillover measurements were carried out by collecting carryover radioactivity in the opposite collecting column. Memory measurements were made by burning a non-labelled filter paper after combustion of each radioactive standard and collecting the residual radioactivity in the designated collecting column. Data presented in Table I is an average value for three combusted ^{14}C and ^{3}H thymidine-filter paper standards.

Table I. Tritium and Radiocarbon Recovery from Combusted Radioactive Thymidine Filter Paper Samples.

Radioactive Standard	Cocktails	Average Recovery (%)	Spillover (%)	Memory (%)	Counting Efficiency (%)
^{14}C Thymidine	2-methoxy ethyl amine toluene PPO & POPOP	98.5±1.0	.07	.05	68
^{3}H Thymidine	PCS (Phase Combining System)	97.8±1.5	.8	.2	47

The ability of the spark ignition system to combust a wide spectrum of sample materials was tested in the following manner. Samples were selected representing typical materials that would be used in medical and biological studies and that would possess high quenching properties or would need to be combusted in order to be counted in a liquid scintillation counter. The samples were prepared in the dry, wet or frozen condition and in varying weights. Combustion times shown in Table II represent the actual time period for complete spark combustion and do not include the 10-second nitrogen flushing of the combustion chamber prior to ignition.

Table II. Combustion Times for Selected Samples in Frozen, Wet or Dry States.

Sample	Combustion Time	Sample	Combustion Time
paper (dry) 100 mg	8 sec.	crab meat (dry) 170 mg	19 sec.
paper (wet) 100 mg	12 sec.	crab meat (wet) 140 mg	13 sec.
wood (dry) 50 mg	10 sec.	crab meat (frozen) 110 mg	21 sec.
shell (crab) 80 mg	25 sec.	liver (dry) 200 mg	15 sec.
fat (dry) 210 mg	10 sec.	liver (wet) 60 mg	10 sec.
fat (wet) 160 mg	16 sec.	liver (frozen) 115 mg	31 sec.
fat (frozen) 185 mg	21 sec.	liver (wet) 730 mg	3 min.

A comparison of the capabilities of the spark ignition system to combust samples comparable to commercial combustion systems on the market today was carried out in the following manner. Rat fecal pellets were selected as the sample material to be used in the study as they fit the requirements of a typical biological sample that would require combustion prior to liquid scintillation counting. Fecal pellets were made up into two bulk samples, one labelled with ^{3}H and the other ^{14}C. The tritiated feces were fabricated by feeding one group of rats ^{3}H labelled steroids. The radiocarbon labelled fecal material was made by feeding a second group of rats ^{14}C labelled dipeptide. The fecal material from each group was separately collected, homogenized as an aqueous slurry and freeze

^{14}C RAT FECAL SAMPLES

PACKARD MODEL 305 DATA

SAMPLE #	TOTAL D.P.M.	MEMORY D.P.M.	MEMORY %	SAMPLE WEIGHT (mgm)	D.P.M./mgm
1.	209327	617	0.30	339.6	644
2.	224610	1177	0.52	361.1	653
3.	222460	745	0.33	385.6	605
4.	199746	892	0.47	338.2	591
5.	187905	808	0.43	377.3	520
	AVERAGE % MEMORY =		0.41	AVERAGE D.P.M./mgm =	602.6

INTERTECHNIQUE OXYMAT

SAMPLE #	TOTAL D.P.M.	MEMORY D.P.M.	MEMORY %	SAMPLE WEIGHT (mgm)	D.P.M./mgm
1.	157183	487	0.310	252.3	623
2.	173559	1180	0.678	297.7	583
3.	209167	962	0.460	325.3	643
4.	196880	885	0.450	326.5	603
	AVERAGE % MEMORY =		0.475	AVERAGE D.P.M./mgm =	613

CONTINUOUS SPARK IGNITION PROTOTYPE DATA

SAMPLE #	TOTAL D.P.M.	MEMORY D.P.M.	MEMORY %	SAMPLE WEIGHT (mgm)	D.P.M./mgm
1.	206939	99	0.047	349.2	593
2.	213440	102	0.048	392.7	544
3.	262217	168	0.064	422.2	621
4.	303243	118	0.039	390.0	778
5.	172604	133	0.077	368.9	468
	AVERAGE % MEMORY =		0.055	AVERAGE D.P.M./mgm	600.7

PACKARD MODEL 306 DATA

SAMPLE #	TOTAL D.P.M.	MEMORY D.P.M.	MEMORY %	SAMPLE WEIGHT (mgm)	D.P.M./mgm
1.	179413	101	0.056	308.8	581
2.	180172	82	0.046	294.4	612
3.	206592	97	0.047	322.8	640
4.	190032	76	0.040	321.0	592
	AVERAGE % MEMORY =		0.047	AVERAGE D.P.M./mgm =	606

Table III. Combustion Results for ^{14}C Rat Fecal Samples

dried. To assist in handling the samples, the material was pelletized.

Three commercial combustion instruments were selected for the comparison tests. These instruments were selected on the basis of present use in the liquid scintillation field today, performance and availability. The three instruments selected were the Packard 305, Packard 306, and the Intertechnique Oxymat. The procedure used to combust the ^{14}C and ^{3}H labelled pellets in the Packard 305 was to wrap the fecal pellets in filter paper and press the sample into a single, paper coated pellet. The sample preparation for burning pellets in the Packard 306 instrument was to wrap the pellets in tissue paper and place the wad in a Combusto-Cone paper basket specially designed to fit into a platinum heater coil within the instrument. Pellets combusted in the Oxymat instrument were placed into a polycarbonate plastic cup with a snap-on lid. Preparation for combustion of the pellets in the spark ignition system required only placement of the fecal pellets onto a steel planchet. Tables III and IV show the comparative combustion results of the ^{14}C and ^{3}H labelled fecal pellets as dpm/mg with the four instruments. Instrument memory, recorded as dpm, is with background subtracted.

A note should be made that the ^{3}H spillover rates as shown in Table IV for the Packard 306 are probably a factor higher than with normal operation. This was apparently due to a holdup in the water column collector that was not apparent at the time of operation.

Discussion and Summary

The prototype continuous spark ignition system reported in this paper offers a unique method for combusting either ^{14}C or ^{3}H labelled samples. Important features designed into the instrument which enhance its capabilities are as follows. The samples can be combusted in the dry, wet or frozen state, thereby reducing presample preparation time. Maximum sample specific activity is sustained by the lack of the need to add foreign combustible material to sustain ignition. Pre-nitrogen gas drying and heating of the sample, prior to ignition, is used to maximize combustion safety and to condition the sample for ignition. Cost per sample combustion is held to a minimum by using low-cost disposable steel planchets to contain the sample during combustion. Spark ignition over the upper dried surface of the sample enhances a uniform rate of combustion

^{3}H RAT FECAL SAMPLES

PACKARD MODEL 305 DATA

SAMPLE #	TOTAL D.P.M.	MEMORY D.P.M.	MEMORY %	SAMPLE WEIGHT (mgm)	D.P.M./mgm.
1.	18726	172	0.92	258.8	74.1
2.	24034	174	0.72	322.3	76.2
3.	19691	124	0.63	273.7	73.4
4.	20267	155	0.76	270.8	76.5
5.	22228	110	0.50	309.0	73.3
	AVERAGE % MEMORY =		0.71	AVERAGE D.P.M./mgm. =	74.69

INTERTECHNIQUE OXYMAT

SAMPLE #	TOTAL D.P.M.	MEMORY D.P.M.	MEMORY %	SAMPLE WEIGHT (mgm)	D.P.M./mgm.
1.	18399	147	.80	251.2	73.2
2.	18507	207	1.12	263.6	70.2
3.	18822	139	.74	299.9	62.8
4.	19200	184	.96	258.6	74.2
	AVERAGE % MEMORY =		.91	AVERAGE D.P.M./mgm. =	70.1

CONTINUOUS SPARK IGNITION PROTOTYPE DATA

SAMPLE #	TOTAL D.P.M.	MEMORY D.P.M.	MEMORY %	SAMPLE WEIGHT (mgm)	D.P.M./mgm.
1.	28707	266	0.93	334.3	86.7
2.	27898	180	0.65	305.7	91.8
3.	27282	22	0.08	295.9	92.3
4.	30580	60	0.20	334.3	91.7
5.	27820	3	0.01	314.7	88.4
	AVERAGE % MEMORY =		0.37	AVERAGE D.P.M./mgm =	90.18

PACKARD MODEL 306 DATA

SAMPLE #	TOTAL D.P.M.	MEMORY D.P.M.	MEMORY %	SAMPLE WEIGHT (mgm)	D.P.M./mgm.
1.	20028	86	.43	263.4	76.1
2.	27015	66	.24	337.2	80.1
3.	28675	98	.34	326.6	87.8
4.	21759	65	.30	276.7	78.6
	AVERAGE % MEMORY =		.33	AVERAGE D.P.M./mgm =	80.6

Table IV. Combustion Results for ^{3}H Rat Fecal Samples

for the total sample. Combustion rates can be further controlled by the amount of external heat applied to the sample during combustion. Nuclide collection is in separate vials ready for counting, thereby allowing optimum counting channels to be used in the liquid counter.

Present capabilities of the instrument allow for the combustion of samples up to 40 millimoles of carbon dioxide for ^{14}C samples and up to 85 millimoles of water for ^{3}H samples. Analytical recovery of the ^{14}C and ^{3}H labelled thymidine filter paper samples (Table I) show results in the 97-98% range with a precision of ±1.5%. Memory and spillover values recorded for these standards were .05% and .07% for ^{14}C and .2% and .8% for ^{3}H. Counting efficiencies of the collecting cocktails for the combusted labelled thymidine samples showed 68% for ^{14}C and 47% for ^{3}H.

Further modifications in the instrument are being conducted to increase the collection efficiency and to lower the amount of memory and spillover, especially for ^{3}H. Paramount in this work is the use of a steam generator with the combustion system and a H_2O exchange column. Continuous repetitive sample combustion time is not feasible to estimate at this time, as the instrument is still in a prototype design. However, from the combustion results shown in Table II, it can be seen that the combustion times are rapid for a wide spectrum of samples, and would appear to be comparable to commercial combustion instruments sold on the market today.

References

1. W. Schnoniger, Mikrochim Acta, 123 (1955).
2. G. G. Steel, Int. Journ. App. Rad. & Isot., 9,94,(1960).
3. W. D. Conway and H. J. Grace, Anal. Biochem., 9, 487, (1964).
4. L. G. Hubner and W. E. Kisieleski, Atompraxis, 16, 1-5, (1970).
5. J. I. Peterson, Anal. Biochem., 31, 204, (1969).
6. J. I. Peterson, F. Wagner, S. Siegel, and W. Nixon, Anal. Biochem., 31, 189, (1969).
7. N. Kaartinen, Packard Tech. Bul. 18, (1969).
8. L. Hunt and B. Bastomsky, Clinical Chem., 17, #10, (1971).
9. E. Rapkin and A. Reich, Amer. Lab., 35, (1972).

AN ADVANCED AUTOMATIC SAMPLE OXIDIZER - NEW HORIZONS IN LIQUID SCINTILLATION SAMPLE PREPARATION

L. J. Everett, N. Kaartinen and P. Kreveld
Research Manager, Consultant and Australian Sales Manager, Packard Instrument Co., Inc.

Abstract: Analysis of several animal tissues using the Packard Model 306 and two commercial tissue solubilizers (Soluenes) produced statistically identical results. Counting of tissue in an Insta-Gel H_2O suspension produced less counts per gram. The excellent performance of the Sample Oxidizer permitted: sequential analysis of labeled samples; dual analysis of ^3H plus ^{14}C singly labeled samples, with statistically accurate results. Sample preparation to final count may be achieved in 20-30 minutes. Tissue solubilizer digestions, to stable final count required 12-24 hours.

Introduction: It has long been the goal of the researcher using liquid scintillation counting techniques to reduce biological samples to a form which could be counted on a liquid scintillation spectrometer with zero quench. For tritium and carbon-14 labeled samples this is possible by converting water or carbon dioxide into labeled benzene or toluene, but it is not practical in a quantitative sense. Thus, based upon reality, the researcher is willing to compromise and settle for liquid scintillation counting cocktails giving a constant quench. One of the easiest ways to achieve this goal is to combust tritium and carbon-14 labeled samples producing tritiated water and carbon-14 dioxide. These materials are colorless and hence may be introduced into appropriate cocktails with only slight quenching appearing in the tritium cocktail and a constant quench or a decrease in quench of the carbon-14 cocktail.

Historically, mass combustion of biological samples was initiated by Kelly et al.[1] using the Schöniger combustion technique[2]. This combustion technique was further explored by Kalberer and Rutschmann[3] and extended further to a microtechnique by Gupta[4]. Sample preparation with the macrotechnique was cumbersome and neither the macro nor microtechnique provided a physical means of separating tritium from carbon-14. It is well known that the liquid scintillation spectrometer was developed for separating instrumentally the carbon-14 beta energy appearing above the tritium spectrum from the tritium spectrum containing a portion of the carbon-14 spectrum. If adequate activity is present within a dual labeled tritium and carbon-14 sample, instrumental methods work quite well. If sample activity is low, adequate counting statistics cannot be collected within reasonable periods of time.

On the practical side, the researcher has long known the advantages of dual labeled techniques to eliminate the ever-present biological variation.

Kaartinen[5] developed an experimental technique based upon a flowing oxygen flask combustion and gas scrubbing principle to provide physical separation of tritium as tritiated water from carbon-14 dioxide. This development was followed rapidly by the Peterson device[6] which used catalytic combustion followed by a physical technique of separating the combustion by-products. These ideas led to the Packard 305 Sample Oxidizer and the Intertechnique Oxymat. Performance of these instruments was discussed at the 1970 International Conference on Organic Scintillators and Liquid Scintillation Counting, San Francisco, USA, by Sher et al.[7] and Tyler et al.[8] respectively.

The second generation Packard 306 Sample Oxidizer, offering simple pushbutton operation, has been examined critically and compared with tissue solubilizer digestion and suspension counting techniques. This study was undertaken to determine data quality; preparation effort; speed, from preparation to data availability.

Materials and Methods: Groups of four 160-180 gram female rats were treated by oral untubation with either 0.5 ml of [U-^{3}H]-amino acid solution or 0.5 ml of [U-^{14}C]-amino acid solution. The animals were allowed to remain undisturbed without food or water for four hours following treatment. The animals were anesthetized and sacrificed by decapitation. Tissues were removed directly from each carcass and placed immediately into a Dewar containing liquid nitrogen.

A third group of rats served as controls.

Samples were processed in the following manner. Dry ice was added to a Waring Blender jar while the motor was in operation until a powder was formed cooling the jar and powdered dry ice swirled around inside the blending jar. The liquid nitrogen was quickly decanted from the Dewar. The motor of the blender was turned off. The frozen tissue was added, the jar capped, and the blender turned to high speed. This technique reduced the frozen tissue to a fine powder. These tissue samples served as the source of all biological samples for subsequent data appearing in this report.

Biological samples were processed for liquid scintillation counting by three sample preparation techniques. These were combustion on the Packard Model 306 Sample Oxidizer, digestion in Soluene-100® or Soluene-350 and counting of the powdered tissue in Insta-Gel® water suspension.

Powdered animal tissue samples were weighed into tared Combusto-cones and burned in the Sample Oxidizer. The combustion timer was set for 1.2 minutes to permit both large and small samples to be processed at the same combustion time. The volumes of solutions used were: 15 ml of Monophase-40™ added to the tritium collection vial and approximately 8 ml of Carbo-Sorb™ (dial setting 9) plus 12 ml of Permafluor®-V (dial setting 15) added to the carbon collection vial. A 0.35 ml aliquot of improved Combustaid™ was added to each of the frozen tissue samples other than fat to initiate a rapid combustion. Fat samples were combusted in the presence of Whatman* cellulose powder to eliminate the possibility of any incomplete combustion (soot formation).

A 1 ml aliquot of Soluene-100 or Soluene-350 was added to each of a large group of liquid scintillation vials. The vials were capped with a polyethylene lined cap and tared. Aliquots of frozen tissue were added directly to the Soluene. The vial was reweighed to obtain a weight of tissue. Samples were digested both at room temperature with agitation, and at 50°C. Following complete digestion, 10 ml of toluene scintillator solution containing 4 grams PPO and 0.25 gram dimethyl POPOP per liter were added to each vial to complete the cocktail.

Other portions of tissue samples were weighed into a tared 10 ml aliquot of Insta-Gel. The vial was capped and the vial contents shaken vigorously. This disperses and wets the pulverized tissue with Insta-Gel. Four milliliters of water were added, the vial capped and shaken vigorously producing a suspension of particulate material in a rigid gel.

All samples were counted on a Packard Model 3390 Tri-Carb Liquid Scintillation Spectrometer. The tritium was counted in a wide window (^{3}H pushbutton). The carbon labeled samples were counted in a wide window (^{14}C pushbutton).

Results and Discussion: The collection of quality liquid scintillation data from biological samples depends upon having an adequate knowledge of various parameters of the liquid scintillation cocktail and sample. Various animal tissues have

*Whatman is the registered trademark of W&R Balston Ltd., Maidstone, Kent, England.

been subjected to combustion using the Sample Oxidizer. An analysis of both the tritium and carbon-14 backgrounds demonstrates the background of the resulting cocktails is independent of tissue weight over the examined range of 0 to 0.6 grams. Tissues examined were beef muscle, pork liver, and beef fat (0 to 0.2 grams).

The rat tissue samples of control animals were examined in the same way. Liver control samples indicated a tritium content of 219±18 DPM/gram of tissue with weights ranging from 0.1 to 0.58 gram. Rat muscle control exhibited a background of 176±28 DPM/gram. The rats were maintained in a laboratory where we store trace quantities of tritiated water and tritiated toluene. It now became apparent that our control animals and presumably our treated animals contained trace amounts of tritium within the tissue. This sensitivity of tritium detection points out one of the advantages of the Sample Oxidizer.

A second factor which must be evaluated in rapid processing of biological samples is the length of time required to produce the sample and then the length of time required to achieve constant counting conditions. The analytical weighing time to prepare the above samples is nearly identical for each of the three procedures of sample preparation since two weighings are required for the production of each sample. Both combustion and suspension of tissue samples in Insta-Gel water are very rapid. Digestion in Soluene-100 or Soluene-350 requires approximately two hours for 0.1 to 0.12 grams of tissue at 50° C and overnight digestion with periodic agitation is required for these same reagents at room temperature.

The Model 306 Sample Oxidizer tritium cocktail described above reached a background equilibration in approximately six minutes. Following that point in time, only one observation out of 24 fell outside the two standard deviations. This represents a rapid stability of background by placing a room temperature cocktail into a subambient temperature-controlled Tri-Carb

Scintillation Spectrometer. A tritium sample combusted utilizing the Sample Oxidizer falls within expected counting statistics at the same period in time. Thus, tritium samples may be processed on a Sample Oxidizer with adequate counting data some six to 10 minutes following actual combustion of the sample.

A background sample derived from the carbon side of the Sample Oxidizer was constant for one minute counts from the point of time where it was introduced into the Tri-Carb. This cocktail contains Carbo-Sorb (an organic amine) as a trapping agent. Since amines are strong quenchers to the liquid scintillation process, the quenching of a carbon sample should decrease until the sample reaches temperature stabilization. Previous experiments in our laboratory had demonstrated the minimum time to achieve this temperature stabilization was approximately 45 minutes. To investigate this 45 minute wait, a carbon sample was burned on the Sample Oxidizer and counted continuously using one minute counts until stabilization was achieved. The stabilization occurred at 15 minutes.

These count rate stabilization times for the Sample Oxidizer cocktails are very significant since samples may be counted 15 minutes following sample preparation. This permits the investigator to obtain meaningful data and make necessary corrections in critical experimental projects. In the past this has not been possible.

A Soluene-100 or Soluene-350 plasma sample exhibits a reasonable degree of chemiluminescence. Initial counts of 3,000 CPM are quite common. These drop off quite rapidly in an exponential manner. In our experiment the background dropped below 100 CPM at approximately 30 minutes after the sample was introduced into the liquid scintillation spectrometer. For high specific activity samples a 100 CPM background may require one or two days for total stabilization of the background counts. This background is normally attributed to chemiluminescence but exists in the carbon-14 channel as well as the tritium channel. Duplicate samples prepared from the same reagents may produce chemilumenescence in one vial but not the second. Thus, additional time is required to be sure that labeled samples have definitely stabilized.

Insta-Gel water suspension of pulverized tissues achieves a stable background within five minutes. Since quantitative data was not achieved with this technique, temperature stabilization times for emulsion of radioactive tissue samples were not measured.

The various rat tissues from single labeled samples, either tritium or carbon-14, were burned on a Sample Oxidizer and the activity obtained converted to a specific activity expressed as DPM/gram. This data is presented in Table I. Most of the labeling is about as one would expect following oral administration of a labeled material. It may be interesting to note that the fat samples derived from pooled omental and renal fat were quite high in terms of apparent labeled activity.

The standard deviations of the tissues range from very low to moderately low. As one might predict, the gastrointestinal tract variation is higher than others. This is certainly a very difficult area to sample and more than one animal is represented in the sample. The standard deviation of the data appears to have some correlation with the ability to thoroughly pulverize the sample. In samples which contained chunks of the original tissue, the standard deviation generally exceeded 2%. In samples which had no chunks remaining within the tissue, the standard deviation was generally 2% or less. In one case, for the tritium brain tissue, the standard deviation was 0.25%. This is indeed excellent reproducibility for tissues. Tissue weights range from approximately 40 mg to 400 plus mg.

Most investigators will probably concede that a Sample Oxidizer is capable of producing quality data. A knowledge of how the results of experimental data of this technique compared with tissue solubilizers and a third technique, suspension in a gel emulsion, allows the investigator to evaluate the possible techniques. A comparison of this data is presented in Table II. Careful examination of the data in Table II demonstrates very significantly that the two Soluene tissue solubilizers give data equal to the 306 Oxidizer.

In only one case the Soluene appeared to give a higher specific activity than the Sample Oxidizer. This is for carbon-14 labeled rat muscle. As stated previously, it was demonstrated that the control rats sacrificed some days before the treated animals contained traces of tritium contamination. The data obtained from the Sample Oxidizer would contain no tritium in the carbon-14 vial. A statistical analysis of a wide group of carbon-14 tissues combusted demonstrated one count per minute less for the possible tritium spill into the carbon-14 vial than was observed in a composite of blanks containing no tissue material. Thus, the 306 Oxidizer sample would contain no tritium whereas the Soluene-100 and Soluene-350 samples would definitely contain the tritium as well as the carbon-14 present within the sample. This higher value does not occur in the carbon-14 labeled rat gastrointestinal tract. On further examination it appears that the Sample Oxidizer exhibits a lower percent standard deviation than do the tissue solubilizers.

Insta-Gel water suspension gives less than the theoretical amount of tritium and carbon activity. The rat muscle suspensions, which create only slight quenching, are very reproducible using this technique. However, this strongly demonstrates that reproducibility does not mean correct data. The tissue samples were simply pulverized in the presence of powdered dry ice. The technique might still offer considerable promise if the particle size could be further reduced.

The specifications for the Model 306 Sample Oxidizer (both tritium and carbon-14 recoveries are 99±1%; tritium memories and carbon-14 memories less than 0.05%; carbon-14 residual remaining in the tritium vial of less than 0.02%; and tritium spillover into the carbon-14 vial of less than 0.001%) suggest many unique things can be done with the Sample Oxidizer. In the past, due to the presence of reasonable memories, spillover, and residual activities, it has been desirable to burn blank samples between each radioactive sample which is combusted. Therefore, based upon the above specifications, an experiment was devised to burn labeled samples, containing single label and widely varying weights, one after another with no blank combustion between the samples. All data would be calculated to DPM/gram. Thus, a large sample followed by a small sample should produce a higher than expected DPM/gram for the small sample. The data from this experiment is included in Table III. Examination of the tritium data column demonstrates a 0.414 gram sample on line 2 followed by a 0.060 gram sample. The lesser sample does not have a higher specific activity. In the same column there is a 0.485 followed by a 0.111 sample. Likewise, a 0.239 followed by a 0.095. Neither of these latter samples has a higher specific activity than the preceding sample. A similar discussion may be presented for the carbon-14 data column. In the case of the 0.463 followed by the 0.079 there does indeed appear to be an increase. However, this value falls very close to the two sigma limit of the average. More significantly, the pile-up of memories should yield a larger average DPM/g for samples burned one after another than should a series of samples burned with blanks between each sample. Therefore, a group of samples was burned along with blanks between each sample. The tritium specific activity from the rat muscle was 186,500±2,930 (1.57%) DPM/gram. The carbon-14 specific activity was 15,140±320 (2.11%) DPM/gram. These values are experimentally identical to the values obtained by burning samples one after another.

Knowing that it is possible to burn samples one after another, with insufficient cross-talk or memory to produce a statistical difference in the results, it was theorized that singly labeled tritium and carbon-14 samples could be burned in combination to produce DPM data for each of the single labeled isotopes. In principle, this is the only way the Sample Oxidizer may be checked to determine adequate performance from labeled biological tissues. Thus, singly labeled tritium samples were weighed into Combusto-cones. Carbon-14 labeled samples were weighed into a second Combusto-cone. The two cones were stacked one within the other, Combustaid was added, and the samples were burned. The weight of tritium sample to carbon-14 sample was varied widely.

In various weight ratios once again all data was calculated to a specific activity of DPM/gram. This data is presented in Table IV. An analysis of the data indicates the combustion technique using the Model 306 Oxidizer is independent of sample size and produces constant results for tritium activity and carbon-14 activity for widely varying combinations of samplè weights. This technique thus provides a means for analyzing single labeled samples in tritium-carbon combinations to yield DPM per unit weight or volume with a high degree of accuracy. Sample preparation time is rapid and results consistent.

A comparison of the data obtained by burning samples alone with no blanks between samples; by burning samples alone with a blank between each sample; and burning samples in combination (tritium and carbon-14 being burned simultaneously) gives statistically identical results.

The Model 305 manufactured by Packard, the Oxymat (Reich modification of Peterson system), and the Harvey oxidizer have significant memories and spillover. These spillovers and memory would likely preclude any dual combustion of singly labeled tritium and carbon-14 samples or sequential combustion of labeled samples with no blanks between. Thus, the Packard Model 306 Sample Oxidizer does indeed open up new horizons in liquid scintillation sample preparation with wide flexibility, easy sample preparation, and quick accumulation of accurate counting data.

References

1. R. G. Kelly, E. A. Peets, S. Gordon and D. A. Buyske. Anal. Biochem. 2, 267 (1961).
2. W. Schöniger, Mikrochim. Acta 1, 123 (1955).
3. F. Kalberer and J. Rutschmann, Helv. Chim. Acta 44, 1956 (1961).
4. G. N. Gupta, Anal. Chem. 38, 1856 (1966).
5. N. Kaartinen, Personal communication.
6. J. I. Peterson, Anal. Biochem. 31, 204 (1969)
7. D. W. Sher, N. Kaartinen, L. J. Everett and V. Justes, Jr. in Organic Scintillators and Liquid Scintillation Counting, p. 849 (D. L. Horrocks and C. T. Peng, Ed). New York and London: Academic Press (1971).
8. T. R. Tyler, A. R. Reich and C. Rosenblum in Organic Scintillators and Liquid Scintillation Counting, p. 869 (D. L. Horrocks and C. T. Peng, Ed). New York and London: Academic Press (1971).

TABLE I

Specific Activity (DPM/gram) of Rat Tissues Separately Labeled with Either [^{3}H] or [^{14}C]-Amino Acids (Sacrifice Four Hours After Oral Intubation)

Rat Tissue	^{3}H (X10^3)	^{14}C (X10^2)
Brain	193.5± 0.5 0.25%	166.0± 3.6 2.19%
Fat	219.7± 4.4 2.01%	463.0±31.0 6.72%
GI Tract	559.5±43.1 7.71%	1190.0±26.0 2.18%
Heart	231.5±6.2 2.68%	337.0± 3.4 1.00%
Kidney	351.0±19.5 5.55%	749.0±12.7 1.69%
Liver	491.7± 7.5 1.53%	___________*
Muscle	195.5± 2.4 1.24%	150.0± 1.1 0.74%
Spleen	321.7± 6.4 2.00%	777.0±29.3 3.77%

*Sample was not completely ground; large chunks of liver made sampling impossible.

TABLE II

A Comparison of Calculated Specific Activity (DPM/gram) of Rat Tissues Using Different Sample Preparation Techniques

Sample Preparation	Rat Muscle		Rat GI Tract	
	^{3}H ($X10^{3}$)	^{14}C ($X10^{2}$)	^{3}H ($X10^{3}$)	^{14}C ($X10^{2}$)
306 Oxidizer	195.5±2.4 1.24%*	150.0±1.1 0.74%	559.5±43.1 7.71%	1190± 26 2.18%
Soluene-100	195.7±6.2 3.15%	165.0±13.6 8.27%	560.2±57.3 10.2%	1169± 74 6.35%
Soluene-350	197.0±5.7 2.88%	160.0± 3.7 2.28%	571.5±37.6 6.58%	1217± 54 4.45%
Insta-Gel--H_2O	150.0±3.0 2.01%	103.6±3.6 3.76%	279.2±33.6 12.0%	772±116 16.2%

*Percent Standard Deviation was calculated from data before rounding off; direct data manipulation may yield a slightly different value.

TABLE III
Constant Specific Activity from Single Labeled Animal Tissue with No Blank Combustion Between Samples

Tritium Rat Muscle Sample Wt. (grams)	Tritium Rat Muscle Activity DPM/gram	Carbon-14 Rat Muscle Sample Wt. (grams)	Carbon-14 Rat Muscle Activity DPM/gram
0.1112	186,000	0.0847	15,130
0.4145	191,800	0.2798	14,880
0.0600	180,900	0.1086	15,270
0.3019	183,000	0.4633	14,820
0.0839	183,800	0.0798	15,950
0.2890	187,100	0.2749	15,030
0.4855	189,700	0.1829	15,320
0.1118	182,100	0.0815	14,880
0.2399	187,400	0.1417	14,950
0.0959	184,100	0.2264	15,350
Average	185,600		15,160
Std. Dev.	3,500		340
% Std. Dev.	1.89		2.24

TABLE IV
Analysis of Singly Labeled Tritium and Carbon-14 Samples by Combustion of Both Samples Simultaneously

Tritium Rat Muscle Sample Wt. (grams)	Tritium Rat Muscle Activity DPM/gram	Carbon-14 Rat Muscle Sample Wt. (grams)	Carbon-14 Rat Muscle Activity DPM/gram
0.0494	187,200	0.1978	15,100
0.3295	186,900	0.0698	15,180
0.2893	189,800	0.2436	14,900
0.2698	189,700	0.1804	14,850
0.2153	187,400	0.0655	15,590
0.0777	188,300	0.3368	15,290
0.0400	180,500	0.0819	15,170
0.4103	188,200	0.2244	14,940
0.0755	184,400	0.0902	15,240
0.0523	189,100	0.4266	14,750
Average	187,200		15,100
Std. Dev.	2,820		250
% Std. Dev.	1.51		1.66

APPLICATION OF LIQUID SCINTILLATION SPECTROMETERS TO RADIOCARBON DATING

Henry A. Polach

The Australian National University
Radiocarbon Dating Laboratory
Canberra

Abstract

Low-level countrate characteristics of liquid scintillation spectrometers are evaluated, specifically in relation to radiocarbon dating. The ^{14}C age determination precision (and limits of determinable ages) are evaluated in terms of a *Relative Factor of Merit* and sample size. A statistical analysis of reproducibility of age determinations and background stability is given. It will be shown that *unmodified commercially available* liquid scintillation spectrometers can produce ^{14}C age determinations of high precision, and that L.S. equipment performance is not the limiting factor in the evaluation of validity of ^{14}C ages.

Introduction

The early application of the liquid scintillation counting technique to radiocarbon dating was primarily motivated by the desire to develop this technique for use as a research tool in archaeology. The first account of its application is given by Arnold in 1954 (1) and reminiscences of the early and now historical efforts were presented at the Liquid Scintillation Counting conference held in 1957 at the North Western University (2). In his summary of the mildly successful results, Arnold (3) concluded that the only thing that could be said with certainty about the future of liquid scintillation counting, as applied to radiocarbon dating, was that it awaited an elegant solution of the chemical synthesis problem for the conversion of specimen carbon to suitable organic liquids which preferably should be scintillation solvents.

The application of benzene for this purpose had its conception in the early work of many researchers and accounts of evolution of ideas and progress made can be found in the work of Tamers (4), Noakes (5) and Polach (6). The reader interested in chemistry of the benzene synthesis and purity of the final product is referred to these papers. This paper will deal particularly with the evaluation of L.S. spectrometer equipment performance, bearing in mind the specific requirements of the ^{14}C dating method. Counting vial design considerations are not discussed here: although this is important in low-level counting, the matter is discussed elsewhere (4), (12), (14).

Low-level countrate characteristics of L.S. spectrometers

Modern commercially available spectrometers using two bialkali quartz faced photocathodes, selected for high quantum efficiency and low thermal noise and working on the coincidence pulse summation principle, have a guaranteed ^{14}C detection efficiency of >96% and in this respect alone are directly comparable to gas proportional detectors. The performance of L.S. Spectrometers is commonly expressed as the *Figure of Merit* (E^2/B) where E is % efficiency and B is background in cpm. Manufacturers of spectrometers quote an E^2/B of >450 for ^{14}C above ^3H and >250 for whole of the ^{14}C spectrum for a 20 ml low ^{40}K glass counting vial, filled with 15 ml of toluene. However, the use of E^2/B is inappropriate as an expression of *Merit* for low-countrate determinations, as the background (B) is volume dependent and the % efficiency (E) is not related to the sample volume in the formula used. Loevinger and Berman (7) have introduced the volume dependence of the *Figure of Merit* (M) into their formula, but an even more appropriate approach, particularly suitable for the evaluation of equipment performance related to age calculations, was proposed by Felber (8). One of the components of the Felber expression, $N_0/(N_B)^{\frac{1}{2}}$, (N_0 = Modern dating Reference Standard cpm; N_B = background cpm), is used commonly by all dating laboratories when evaluating equipment performance based on Poisson distribution of countrate alone.

L.S. - spectrometer evaluation for ^{14}C dating

A radiocarbon age (A) is calculated using the decay equation $A = \tau \ln (N_O/N_S)$ (1)

where τ represents the mean life, taken by agreement to be 8033 years (9), and N_O represents the counting rates of standard and N_S of sample. The radiocarbon dating modern reference standard is accepted as 95% of the observed activity of the National Bureau of Standards Oxalic acid (.95 NBS Ox). The absolute activity of .95 NBS Ox is given as 13.53 ± .07 DPM g^{-1} Carbon* (10)) which corresponds to 10.978 ± .06 DPM ml^{-1} benzene.

Of particular interest in radiocarbon dating is the confidence with which we are able to discriminate between samples of similar age. Two natural limits exist: (i) the age of the youngest sample which can be distinguished from the modern reference standard (Amin) and (ii) the age of the oldest sample which can be distinguished from the background (Amax). Both are of significance, but to evaluate equipment performance the theoretical maximum detectable age is the most appropriate. From the theory of extreme age calculations (8), Calf and Polach (12, this vol.) give the following equation

$$Amax = 8033 \ln (t/k)^{\frac{1}{2}} + 8033 \ln [N_O/(N_B)^{\frac{1}{2}}] \quad (2)$$

where t = equal time spent counting the sample, standard or background; k = 18 for 3σ criterion; k = 8 for 2σ criterion**. For a given 'k' and 't', the first term of equation (2) is a constant, the values of which are listed in Table I.

---

*Karlen et al. (11) report 13.56 ± .07 DPM g^{-1} Carbon. Both determinations are in excellent agreement.

**The 2σ criterion gives a probability of .975 of identifying a countrate greater than background (13).

Table I: Values for 8033 $\ln(t/k)^{\frac{1}{2}}$ for t = 1 to 4 x 10^3 minutes and k = 8 and 18

Time in minutes	2σ criterion k = 8	3σ criterion k = 18
1000	19,390	16,100
2000	22,180	18,900
3000	23,800	20,550
4000	24,960	21,700

After substituting the appropriate value for the first term of equation (2), the theoretical maximum age is dependent on the value of $N_O/(N_B)^{\frac{1}{2}}$, which Calf and Polach (12, this vol.) called the *Relative Factor of Merit*.

Relative Factor of Merit and Optimum Sample Size

To estimate the Relative Factor of Merit (F) for any L.S. spectrometer used for dating, N_O and N_B must be measured at chosen equipment parameters. The important equipment variables are (i) Photomultiplies high voltage, (ii) amplifier gain, (iii) lower and upper discriminator settings (which define the 'limited coincidence window'). The window efficiency E, must be evaluated directly from the observed countrate of N_O by relating it to the absolute determinations of the standard used for dating. The background N_B must be measured for the same vial design, same solvent, solute mix, and volume as N_O.

Prior to selection of optimum parameters a series of tests must be carried out and F must be evaluated for a range of background and modern sample pairs of different sizes. An experiment of this kind conducted at ANU using a Beckman LS-200 spectrometer is described.

The modern reference standard and background were counted at fixed high voltage*, and amplifier gain set at 360. The settings of the lower and upper

* The Beckman Spectrometer EHT is fixed by the manufacturer and cannot be readily varied without modifying the equipment.

discriminators were adjusted to correspond to balance-point operations (14) for a given ^{14}C efficiency window. The observed countrates therefore characterise uniquely and reproducibly the spectrometer's performance at the chosen settings.

To reduce the background of the 20 ml low ^{40}K glass counting vial, used for this series of experiments, to best obtainable values for a given sample size, the glass vial was masked above the level reached by the top of the scintillation liquid*.

The experiment yielded the results list in Table II.

The background of the progressively masked vial was found to be directly related to the sample volume (V), within the range of efficiencies studied, according to the linear equation:

$$cpm_B = mV + b \qquad (3)$$

Where m and b are the regression parameters.

In addition to determinations at .95E, .85E, and .75E, listed in Table II, the ln(F) for .65E and .55E were also evaluated. They were so similar to those given for the .95E window that they are not tabulated. The results show that the reduction of *limited coincidence* window width, corresponding to lower efficiencies does not give a higher *Relative Factor of Merit*. Indeed, the highest was obtained at .85E.

The net countrates for N_0 were also linearly related to sample size. From the data given in Table II it can be readily shown that ln(F) is dependent on the regression parameters of the linear equations for N_B and N_0 in the following manner:

$$\ln(F) = \ln(m_2V) - \tfrac{1}{2}\ln(m_1V + b) \qquad (4)$$

where m_1 and b are slope and intercept for N_B versus V, and m_2 is slope for N_0. From this it is clear that rate of increase of F diminishes as V increases, as Table III illustrates for 0.85E data.

Values of Amax in Table III indicate the relative gain from the ^{14}C point of view when sample size is increased. An increase from 1 to 10 ml raises Amax

---

*Black paper, aluminium foil or black matt paint are suitable.

Table II: Evaluation of Beckman L.S. Spectrometer performance for radiocarbon dating.

V	.95E			.85E			.75E		
	N_B*	N_0	ln(F)	N_B	N_0	ln(F)	N_B	N_0	ln(F)
1	5.4	10.4	1.50	2.9	9.33	1.7	2.6+	8.2+	1.63
2	7.0	20.9	2.07	3.8	18.7	2.26	3.4+	16.5+	2.19
5	11.5+	52.1+	2.73	6.4+	46.7+	2.91	5.6+	41.2+	2.86
10	19.0	104.3	3.17	10.8	93.3	3.35	9.4+	82.3+	3.29
15	26.5+	156.4	3.41	15.2+	140.0	3.58	13.2+	123.5+	3.53
20	34.0	208.6	3.58	19.6	186.6	3.74	17.0	164.7	3.63

V = ml sample benzene with dry scintillant weighed in

Efficiency calculated using .95 of the *observed* countrate modern oxalic standard whose .95 of absolute activity is 10.978 ± .06 DPM ml^{-1} C_6H_6

F = Relative Factor of Merit = $N_0/(N_B)^{\frac{1}{2}}$

* Linear equations relating sample volume to background in shielded vials were found to be:

cpm (.95E) = 1.5 V + 4.0

cpm (.85E) = 0.88 V + 2.0 (USED IN TABLE 3)

cpm (.75E) = 0.755 V + 1.85

cpm (.65E) = 0.58 V + 1.75

cpm (.55E) = 0.48 V + 1.6

\+ values determined by measurement throughout the range of .75E on N_B and N_0. Correlation coefficient r = >0.99 for the regression equation for this data. Other data (i.e. .95, .85, .65, and .55 E) could therefore be soundly inter- and extra- polated from measured values at 5 and 15 ml benzene.

Table III: Evaluation of *Merit* of counting samples of increasing sizes for radiocarbon dating. 0.85E, Beckman L.S.-200, calculated from values given in Table II.

V	ln(F) (Calc')	Amax (years)
1	1.70	37,490
2	2.26	41,990
5	2.91	47,210
10	3.35	50.680
15	3.58	52,560
20	3.74	53,850
50	4.23	57,790
100	4.59	60,660

[In Table III the following values were used:

$m_1 = 0.88$; $b = 2.0$; $m_2 = 9.33$; ($k = 8$; $t = 3000$; i.e. 1st term of eq. (2) = 23,800).]

Table IV: Sample sizes submitted for dating.

% of FULL REQUIREMENT*	2	2-5	6-10	11-50	51-99	100	200	500	Total
No. of samples	8	9	33	206	167	164	164	25	774
% of samples (Rounded off)	1	1	4	27	22	21	21	3	100

* % FULL REQUIREMENT BASED ON 3.5g Carbon = 100% yielding ca. 4 ml C_6H_6 counting sample.

from 37.5 Ky to 50.7 Ky, while a further increase to 100 ml only raises Amax further to 60.7 Ky.

We can conclude that counting sample volumes from 5 to 10 ml at ca. 85% efficiency would give the best return for our labours. This opinion is reinforced by the range of actual sample sizes submitted for dating to the ANU laboratory over the past 3 years (Table IV).

The results indicate that ca. 45% of samples submitted for dating are more than adequate for 4 ml benzene counting, 49% are just adequate and 6% were inadequate. Only 25% of samples submitted could have been applied to 10 ml benzene counting. This strongly supports the choice of the 5 ml counting vial (4,14).

Long term performance of L.S. spectrometers

Few radiocarbon dating laboratories have published data on long term stability of their counting equipment. Whilst background and reproducibility tests are available for L.S. spectrometers [e.g. (4,5)] and the results given are very satisfactory, only one detailed case study is available to me, and hence I propose to analyse the performance of the L.S. counting equipment of the ANU ^{14}C dating laboratory.

The equipment, a Beckman LS-200, was set up early in 1968. Lacking experience, we chose on an intuitive basis to work at ca. 70% efficiency (relative to .95 NBS Ox), with a sample size of 4 ml benzene, to which we added 1 ml of toluene in which PPO and POPOP were dissolved at concentrations corresponding to 5g/1 and 0.05g 1 of final solvent (14). Operating at balance-point, with ¼" masking placed on the periphery of the tubes (12, this vol.) the following results were achieved: N_O = 30.77 cpm; N_B = 5.5 cpm; F = 13.1 with Amax = 42,400 years (k = 18; t = 4000 Min). Tests on stability then carried out made us state that: 'the degree of stability is 200 times greater than observed or anticipated chemical quench variations, and 10 times greater than observed equipment-induced countrate variations' (14). Now, after 4 years of continuous operation, we can examine how well the equipment really behaved, but before we do so, some points as to counting procedures must be made clear.

Eleven vials are placed within the automatic charger in the following order: S_1 S_2 S_3 S_4 B_1 S_5 S_6 S_7 S_8 H B_2, where S_{1-8} are sample counting vials, B_{1-2} are flame sealed backgrounds and H is flame sealed ^{14}C spiked 'hot' standard of 800 cpm. The vials are cycled at 20 minute intervals, with the exception of hot standard, which gives 2000 counts in ca. 2.5 minutes and is then cycled out. When the relationship of individual background of each counting vial to the mean background value of the two sealed background vials is established, the ^{14}C modern reference standard (NBS Ox) is counted in our standard sample vials, and its relationship to the 'hot' standard is calculated. The cycling method thus provides quasi-simultaneous background and modern countrate checks, as well as automatic short term checks on stability. Long term reproducibility checks of the spectrometer are also made easier as the data output is stored on punch tape. Cycling operations commenced in March 1968 and continued until January 1970; the equipment was then moved to its present position and cycling operations recommenced June 1970 and continue without change of operating parameters today. To evaluate performance of the equipment for low-level counting, the reproducibility of age determinations and the stability of background has to be analysed.

^{14}C Counting:reproducibility of age determinations

We had a choice to analyse either the reproducibility of the Oxalic reference standard measurements, the 'hot' sealed ^{14}C standard, or the reproducibility of duplicate age determinations.

The oxalic standard, for reasons of isotopic fractionation (15, 16), is not satisfactorily reproducible and should not be used for purposes of equipment performance evaluation. To analyse the 'hot' would be interesting but it is arguable that is is irrelevant to R.C. dating as we are dealing with activities 20x higher than are actually attributed to the modern reference standard. We therefore chose to analyse the reproducibility of duplicate age determinations.

From the start of L.S. operation in 1968 until 1972, when all duplicate measurements were analysed, 67 sample pair comparisons were available for analysis. These consisted of 39 inter-lab. cross checks and 28 other cross checks which were duplicates of our own measurements. All repeat age determinations were carried out on portions of the same sample. All duplicate measurements carried out at the ANU lab.are listed, none were knowingly omitted. The analysis was carried out on results as reported to sample submitters, (for full details, code numbers and ages ± errors, see Polach (15)) Results are presented in Table V.

To better evaluate equipment performance, we subdivided the duplicate determinations into two groups: those carried out between other laboratories and ANU (Table V, group A), and ANU-duplicates of our own determinations (Table V group B). It is apparent that the broadness of distribution (ca. 11% more values lie between 1 and 2 S.D. and ca. 6% more beyond the 2 S.D. limit) is entirely confined to group A, the inter-laboratory cross check age determinations, whilst ANU-duplicates of our own determinations, group B, are almost ideally distributed around the mean.

We can therefore conclude that other factors than equipment performance influenced the distribution of international cross checks and that based on ANU-duplicates alone, the equipment performance was as good as could be expected (14).

^{14}C Counting: reproducibility of background

Whilst the long term stability for ^{14}C countrates for samples submitted for dating is shown to be excellent, it would be wrong to assume that background counting is just as stable. Indeed, the spectra for ^{14}C sample activity and corresponding background are vastly different (the background counts increase continuously towards the lower energy region) and operating at the ^{14}C balance-point, with demonstrated stability of ^{14}C counts, does not imply that background

Table V: Analysis of pair comparisons of duplicate ^{14}C age determinations.

WITHIN $\pm$ S.D.	INTER LAB X-CHECKS B				ANU-DUPLICATES B			
	ACTUAL		IDEAL		ACTUAL		IDEAL	
	FREQ.	%	FREQ.	%	FREQ.	%	FREQ.	%
0-1	20	51.3	26.8	68.3	19	68.0	19.2	68.3
1-2	15	38.5	10.6	27.2	8	28.5	7.6	27.2
2-3	4	10.2	1.6	4.2	1	3.5	1.2	4.2
SUM OF PAIRS	39	100	39	99.7	28	100	28	99.7

would exhibit the same degree of stability.* In selecting operating parameters the background stability has therefore to be carefully evaluated, as in low-level counting age determinations a high statistically reproducible background is acceptable, whilst a low changing background would spell disaster.

An analysis of the continuous 20 minute output for our flame sealed background vial No. 5 (B_1) forming part of our background record B-001 to B-258 (i.e. June 1970 to August 1972) has now been performed, and is presented in Table VIa, b, c. All data as punched on tape were analysed and none for any reason whatsoever were rejected. We could not have wished for better results.

Measurement errors and ^{14}C dating errors

We have been concerned with the error of measurement based on L.S. spectrometer performance (also of concern is the laboratory error due to variation in purity of benzene discussed in this volume (17)). It is appropriate however to recognise that these errors today form the minor and diminishing component in establishing the true ^{14}C age of the sample. More important in ^{14}C dating are errors resulting from the age of the material at time of deposition, the chemistry of the sample and the chemistry of the environment in which the sample lay buried or exposed, the variations in concentration of radiocarbon due to past variation in production rate and/or variations due to equilibration and distribution of radiocarbon within its storage reservoirs.

All of these, in the ^{14}C dating experience, are the most significant ones and are subject to continuous studies. An excellent review is given by Neustupny (18).

---

* We were able to test that if ^{14}C operations were carried out with very much reduced high voltage when ^{3}H efficiencies equalled zero, that backgrounds were very low, but extremely unstable.

Table VI: Analysis of Background Readings

VIa) Main Characteristics

TOTAL TIME = 123,000 MINUTES	POISSON DEVIATION = 10.59
TOTAL COUNT = 689,372	GAUSSIAN DEVIATION = 10.70
MEAN CPM = 5.6047	CHI SQ = 6286 with 6150 DF

VIb) Percentile Distribution of 20 min Counts Round Mean

±S.D.	20 MIN. COUNT LIMITS	No REC.	%
+ 4	154.45	12	0.19
+ 3	143.85	130	2.11
+ 2	133.27	856	13.92
+ 1	122.68	1952	31.74
MEAN = 112.09 ± 10.59 COUNTS; No REC = 6150			
- 1	101.50	2234	36.32
- 2	90.92	849	13.80
- 3	80.33	108	1.76
- 4	69.74	9	0.15

VIc) Distribution Summary

SIGMA LIMITS	ACTUAL %	IDEAL %
WITHIN 1 SIGMA	68.1	68.3
1 to 2 SIGMA	27.7	27.2
2 to 3 SIGMA	3.9	4.2
3 - 4 SIGMA	.34	.3
BEYOND 4 SIGMA	NIL	

Nevertheless, and quite correctly, much detailed attention continues to be given to evaluation of accuracy of measurement based on equipment performance. Studies as presented by Geyh (19), introducing the concept of *limiting accuracy* which is primarily determined by the instability of electronic measuring equipment and secondarily by the purity of synthesised gas or liquid as prepared for counting, lead to the introduction of *longest justifiable counting time.* Studies such as carried out by Polach and others [e.g. (20,21,22)] introduced the reporting of *pertinent statistical data* allowing the evaluation of results that hitherto could be reported only as 'greater than' ages; further introduced the concept of running *background blanks* in studies of age determinations close to the limit of countrate detection in the presence of suspected contamination as well as counting a number of *chemical fractions* of the same sample rather than prolonging the counting time of individual age determinations to achieve greater validity and precision. Equally significant is the concept of the *reduced activity* (sample activity/background equivalent activity) plot offered by Currie (23) as a means of planning dating activities and rapidly assessing the capabilities of any specific age determination equipment and procedure.

Notwithstanding the limitations imposed by the above statement, I am nevertheless presenting here a plot of ages and errors showing their dependence on sample size and counting time alone (Fig.1), as the limitation imposed by these remain determinable, whilst the previously mentioned limitations are subject to interpretation and are the basis of the contributions which a radiocarbon dater can make to chronological investigations.

The family of curves presented in Fig. 1 are based on the performance of the Beckman LS-200 spectrometer on which all ANU routine dating operations have been carried out since 1968. The errors are based on 3000 minutes countrate determinations, chosen because for critical measurements, 3000 minutes counting time is indicative of the maximum time we generally spend counting a single sample. The fundamental basis for error assessment, in this diagram, is the theoretical

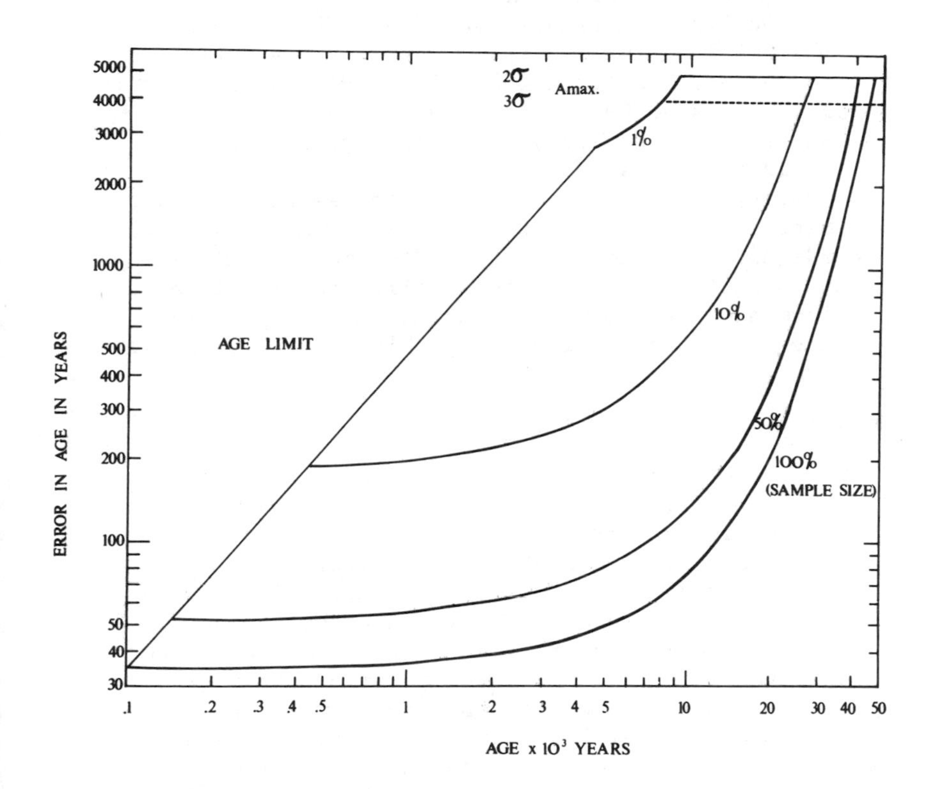

Figure 1: Precision of age determination is dependent on sample size, age and counting time. (Counting time = 3000 Min; 3.5 g carbon = 100% Age limits are imposed by the 2σ and 3σ detection criteria).

Poisson variation of countrates. Fig. 1 becomes more significant if we remember the distribution of sample sizes as submitted to the ANU ^{14}C dating laboratory over the past 3 years (Table IV).

An analysis of sample ages (Table VII) reported by the ANU lab. in the last 3 years of its dating activity indicates that 45% fall within the age limits of 1-10 thousand years, and that beyond this the demand on our laboratory is equally great for younger and older samples.

Referring back to Fig. 1 we can see that the performance of the equipment meets the demand made upon it adequately: the ability to discriminate minimum ages is extremely good; the ability to discriminate maximum ages is adequate; the minimum sample size that can be counted with acceptable errors is ca. 250mg carbon.

Conclusions

We have established that commercially available liquid scintillation spectrometers will in the hands of ^{14}C dating experts give a highly satisfactory long term performance. The ability to determine ^{14}C ages within the range of 100 to 30,000 years with maximum precision conforms well with the average demand made upon the method. Extension of precise dating beyond 30,000 years is limited by the relationship of the signal to noise ration of L.S. spectrometers. In order to achieve higher equipment resolution near the maximum age limit, an increase in sample size alone is not appropriate. Aware of the limitations imposed by the average sample size submitted for ^{14}C dating, we recommend a 5 to 10 ml counting vial as most appropriate. This coupled with the use of vial material with a high *Figure of Merit* and selection of best operating parameters to reduce background (12,24, this vol.) makes the usage of liquid scintillation spectrometers for dating purposes very attractive. The quotation taken from Arnold and used in our introduction therefore no longer applies. We no longer 'await the

Table VII: Sample ages evaluated over the last 3 years of activity in the ANU ^{14}C dating laboratory.

Age limits x 10^3 y.	<.5	.5-1	1-10	10-20	20-30	>30	Total
No. of determin.	70	49	260	61	74	66	580
%	12.0	8.4	44.9	10.5	12.8	11.4	100

elegant solution to our problems' as is so vividly demonstrated by the fact that the majority of newly established ^{14}C dating laboratories have adopted the liquid scintillation counting technique.

Acknowledgements

I wish to thank Professors John C. Jaeger, Jack Golson and Donald Walker, The Australian National University, who throughout the past years have encouraged and supported the activities of the ANU ^{14}C dating laboratory. Particular thanks go to my colleague, Dr. John Chappell, ANU, as he contributed so much through discussion and critical reading of this manuscript, and to the staff of this laboratory, John Gower, John Head and Maureen Powell whose continuous effort made this work possible.

References

1. J.R. Arnold, Science *119*, 155 (1954).

2. C.A. Bell, Jr. and F.N. Hayes, Ed., Liquid Scintillation Counting, Proc. of the Northwestern University, August 1957. London: Pergamon Press, (1958).

3. J.R. Arnold *in* Liquid Scintillation Counting, p.129 (Carlos G. Bell, Jr. and F. Newton Hayes, Ed.) London: Pergamon Press (1958).

4. M.A. Tamers *in* Radiocarbon and Tritium Dating, Proc. 6th Internatl. Conf., Pullman, Washington, p. 53 (1965).

5. J.E. Noakes, S.M. Kim and J.J. Stipp *in* Radiocarbon and Tritium Dating, Proc. 6th Internatl. Conf., Pullman, Washington, p. 68 (1965).

6. H.A. Polach and J.J. Stipp, J. Appl. Radiation and Isotopes, *18*, 359 (1967).

7. R. Loevinger and M. Berman, Nucleonics, *9*, 26 (1960).

8. H. Felber, Report of the Austrian Ac. of Sc., *170:2*, 85 (1962).

9. H. Godwin, Nature, *195* : 4845, 984 (1962)

10. Int. Commiss. Rad. Units and Measurements, *in* Measurement of Low-Level Activity, ICRU Report *22*, p. 47 (1972).

11. I. Karlen, I.U. Olsson, P. Kallberg and S. Killicci, Arkiv Geofysik, *4*, 465 (1964).

12. G.E. Calf and H.A. Polach *in* Liquid Scintillation Counting: Recent Developments. (Philip E. Stanley and Bruce A. Scoggins, Ed.) New York: Academic Press (this volume).

13. W.J. Kaufman, A.Nir, G. Parks and R.M. Hours *in* Tritium in the Physical and Biological Sciences, IAEC, Vienna, p. 250 (1962).

14. H.A. Polach, Atomic Energy in Australia, *12:*3. 21 (1969).

15. H.A. Polach *in* 8th Internatl. Conf. on Radiocarbon Dating Proceedings, Lower Hutt, New Zealand, 688 (1972).

16. H.A. Polach and H.A. Krueger, 8th Internatl. Conf. on Radiocarbon Dating Proceedings, Lower Hutt, New Zealand, 718 (1972).

17. I. Fraser, H.A. Polach, R.B. Temple and R. Gillespie *in* Liquid Scintillation Counting: Recent Developments (Philip E. Stanley and Bruce A. Scoggins, Ed.) New York : Academic Press. (this volume).

18. E. Neustupny *in* Radiocarbon Variations and Absolute Chronology, (Ingrid U. Olsson, Ed.) Stockholm: Almqvist and Wiksell, p. 23 (1970).

19. M.A. Geyh *in* Radiocarbon and Tritium Dating, 6th Internatl. Conf., Pullman, Washington, 29-35 (1965).

20. H.A. Polach, J. Chappell and J.F. Lovering, Radiocarbon, *11*:2, 245 (1969).

21. I. McDougall, H.A. Polach and J.J. Stipp, Geochem. Cosmochim. Acta. *33*, 1485 (1969).

22. J. Chappell and H.A. Polach, Quaternary Research, *2*:4, 244 (1972).

23. L.A. Currie *in* 8th Internatl. Conf. on Radiocarbon Dating, Lower Hutt, New Zealand, 598-611 (1972).

24. P.E. Hartley and V.E. Church *in* Liquid Scintillation Counting: Recent Developments. (Philip E. Stanley and Bruce A. Scoggins, Ed.) New York : Academic Press (this volume).

PURITY OF BENZENE SYNTHESISED FOR LIQUID SCINTILLATION C-14 DATING

I. Fraser[+], H.A. Polach[*], R.B. Temple[+], and R. Gillespie[+]

[+] The University of Sydney
Department of Physical Chemistry
Sydney, Australia

[*] The Australian National University
Radiocarbon Dating Laboratory
Canberra, Australia

Abstract

Radiocarbon dating laboratories using the liquid scintillation counting technique synthesise benzene by trimerization of acetylene over commercially available vanadium or chromium activated cracking catalysts. The purity (i.e. absence of quenching agents) of the synthesised benzene is one of the significant factors influencing the reproducibility of C-14 age determinations. However, a recent study has revealed that two laboratories, using the same catalyst and the same synthesis procedures, can obtain a significantly and apparently systematically different level of benzene purity. To evaluate the significance of this we have now analysed the purity of benzene prepared for routine dating by seven established radiocarbon dating laboratories, whose synthesis techniques vary considerably in procedural detail and four of whom use a different type of catalyst.

Generally, the results indicate the presence of the same impurities at approximately the same concentrations in all samples analysed. Acetaldehyde, acetone, butane, ethanol, ethylbenzene, hexane and toluene were identified by gas liquid chromatography as the main impurities present; it could be established that these impurities (varying in concentration from <20 ppm to <400 ppm) arise, in our own samples, entirely from the acetylene synthesis step preceding benzene catalysts. Of the impurities identified, only acetone appeared in one of the samples analysed in quantities sufficient to be considered a quenching agent. Quantitative experiments were hence carried out to determine its quenching effect at typical equipment settings used for radiocarbon age determinations.

Introduction

The efficient application of the liquid scintillation counting technique to radiocarbon dating requires the quantitative chemical conversion of specimen carbon to liquid scintillant carbon. Whilst a number of procedures were investigated and reviewed by Arnold (1), the most elegant and today the predominantly used technique is based on the conversion of specimen carbon to carbon dioxide (De Vries,(2), Rafter,(3)),with subsequent production of acetylene (Barker,(4)) and polymerization of acetylene to benzene (Noakes et al (5)).

Developments and improvements to the acetylene and benzene synthesis steps (Noakes et al (6), Polach and Stipp(7)) lead to a conversion efficiency of 97 $\pm$ 2% from specimen to benzene carbon. This, coupled with increasing availability of improved commercial liquid scintillation spectrometers, made the technique as used today very attractive. Its evolution and the establishment of low-level liquid scintillation counting for ^{14}C and ^{3}H age determinations are however based on the efforts of many other researchers, e.g. Tamers et al (9), Starik et al(9), Pietig and Scharpenseel (10), McDowell and Ryan (11), Tamers (12), Polach (13), and others.

Purity of Benzene

It was recognised by all of the workers cited above that it is essential for the synthesised benzene to be highly pure, and procedures such as external standardization (Higashimura et al (14)) and the more common channel-ratio analysis were used, often together in order to detect possible quenching. However, Noakes et al (6) made an attempt to analyse the purity of synthesised benzene, and identified acetone as primary contaminant of their experimental samples, with water and ethyl benzene present in trace amounts. They considered that the acetone originated in the cylinder acetylene used for their experiments (acetylene is stored, dissolved in acetone and adsorbed on charcoal), and that it was not a product of the synthesis. They further established that the ethyl benzene was associated with low benzene yields and was present in undetectable amounts when benzene yield approached 90%.

In a recent study, however, Polach et al (15) drew attention to the possible presence of impurities in the synthesised benzene which could cause severe quenching of fluorescence if used for liquid scintillation counting. It seemed particularly significant that following the same synthesis procedures one researcher could obtain reproducibility ultra-pure benzene samples, whilst another researcher consistently obtained samples which contained ethane, butane, acetaldehyde, acetone, ethanol, cyclohexane, toluene, etc., often at significant levels (100 to 400 ppm). As mass spectrometric determinations of the acetylene used for the benzene synthesis indicated that it was always of equally high purity, these results were taken to indicate that minor differences in operating procedures may affect the purity of the sample benzene significantly.

Experimental Methods

In order to study the effect of synthesis procedures in the benzene purity better we have analysed samples prepared for us according to their own normal procedures by the following established radiocarbon dating laboratories: Australian Atomic Energy Commission (Calf); Australian National University (Polach); The British Museum (Burleigh); Illinois Geological Survey (Coleman); University of Bonn (Scharpenseel); University of Sydney (Gillespie); Washington Geological Survey (Pearson).

The samples were examined by gas-liquid chromatography using Pye Series 104 equipment specially fitted with a 50 m capillary column containing poly-propylene glycol. This was operated at temperatures between 60 and 125°C.

A few samples were also examined with a Perkin-Elmer Series 800 ambient temperature instrument using a 3 m column packed with pure Al_2O_3.

The latter gave much better resolution for lower hydrocarbons, i.e. ethane butane etc. but at the expense of greatly increased scanning time. All impurity levels were determined by comparison with standard samples. Several other types of column were tried unsuccessfully.

Results and Discussion

The results of the current analysis are presented in Table I.

The major impurities identified are ethane, butane, acetaldehyde, acetone, ethanol, hexane, cyclohexane, toluene, and ethylbenzene. These were present essentially at the same concentrations in the Gillespie and Polach samples, and in slightly reduced amounts in the Burleigh samples. The others show a notable absence of ethane, butane, acetaldehyde and hexane; the Calf sample contained no ethyl benzene; the Pearson sample contained no acetone, and the Scharpenseel sample contained significantly more of unidentified product 'X', which appears from mass spectra to be either propene or acetonitrile.

None of the samples, in contrast to the previous study referred to (Polach et al (15)), were ultra pure*. It is obviously significant that the new preparations have essentially all those impurities that were present in the original determinations of the Gillespie samples. As among those who submitted samples for analysis, Calf is the only other researcher who uses exactly the same synthesis technique as Gillespie and Polach, we expected his samples to contain either none of the impurities

---

* 18 purity analyses on benzene residues, as used for dating by Polach, were carried out; only two had toluene at 20-40 ppm concentration; the others were ultra pure, only the benzene peak being present.

(as obtained previously by Polach, see footnote), or the same level of impurities, as obtained now by both Gillespie and Polach. The fact that the Calf samples showed an immediate level of impurities closely related to the Burleigh, Coleman, Pearson and Scharpenseel samples (all of whom use different catalysts and procedures)** is interpreted by us as indicating that neither the synthesis procedures nor the catalyst used affect the level of benzene sample purity very much.

One common factor in the three sets of samples containing high levels of impurity is that they were freshly prepared for this analysis: whilst the other four sets may also have been prepared freshly, their common characteristic is absence of volatiles. This was thought to be particularly significant and an impure sample was allowed to stand at room temperature sealed in a screw top vial for 2 days and then reanalysed; all impurities except toluene and ethylbenzene were lost. If stored in a refrigerator at ca 3°C the impurities were lost more slowly (up to 14 days) and when frozen in a deep freeze the sample appeared to retain its impurities. It therefore seems likely that the apparent levels of purity reflect conditions of aeration during encapsulation rather than differences in carrying out the benzene synthesis.

To check origins of impurities two further experiments were carried out.

Cylinder acetylene whose purity was established by gas-chromatography was used, and the conditions of benzene synthesis and regeneration of the Noakes' catalyst (i.e. pressure, time, cooling, flushing with dry air, moist air, no flushing) were varied. The resulting benzene purity analysis indicated only the presence of traces of toluene, ethanol and ethylbenzene, confirming Noakes' conclusions

** Calf, Gillespie and Polach used vanadium-activated silica/alumina catalyst supplied by Noakes. Burleigh used ICI cracking catalyst. Coleman used Vanadium activated cracking catalyst No.V-0701 T $\frac{1}{8}$" supplied by Harshaw Chemicals. Scharpenseel used chromium activated "catalysator-Neu" supplied by Nienburgh-Weser, Hanover.

that if pure acetylene is used, the synthesis products do not contain acetone. However ethanol may also act as a quenching agent if present in sufficient concentration.

The acetylene generated by hydrogenation of lithium carbide was then re-analysed. This confirmed the difficulty in establishing mass spectrometrically the presence of impurities in the acetylene gas at the levels they naturally occur in our samples. However, as it is clearly established in the foregoing that the impurities are carried in the acetylene, the careful purification of acetylene must remain an essential part of the synthesis procedures and the doubts expressed as to its needs and merit by Polach et al (15) are no longer valid.

In the quantities found to be present, acetone was thought to be the only impurity which might be capable of producing detectable quenching.

To measure the quenching effect for concentrations of acetone observed (i.e. 0-180 ppm), aliquots of acetone were added to a counting sample spiked with ^{14}C to ca. 3000 c.p.m. Table II lists the results.

For a level of 100 ppm of acetone the effect was nil, and only ca 0.2% for the 150 ppm level. This is detectable at count rates of 3000 c.p.m. but at count rates of 30 c.p.m. (i.e. modern C^{14} standard value) and less, it would be fully masked by statistical variations. The merit of operating at the balance-point as suggested by Arnold(16) and applied to low-level counting by Leger and Tamers(17) and Polach(13), is clearly seen, as little change in count rate is observed due to the spectral shift due to quenching.

A value as high as 250 ppm of acetone was never observed in any of the samples tested and whilst its effects might be significant they need not be considered. Should such a level of contamination however occur, the associated significant change in channel-ratio or external standardization count rate would warn of its presence.

Conclusions

The conditions used for benzene synthesis and the type of cracking catalyst used do not affect the purity of benzene synthesised from acetylene. With the exception of toluene and ethylbenzene and traces of ethanol, no impurities are formed during the polymerization step if pure acetylene is used for the benzene synthesis. It is therefore concluded that these are present in the acetylene generated from the lithium carbide by hydrolysis and the acetylene purification step must not only be retained but possibly should be improved.

The impurities are volatile; aeration of produced benzene reduces the levels of impurities to negligible proportions. Should however by accident the impurities attain a >150 ppm level, then their effect might become significant if not detected by channel ratio or external standardization analysis.

Acknowledgements

We wish to thank all those colleagues who made synthesised benzene available for analysis.

References

1. J.R. Arnold, in Liquid Scintillation Counting, p.129 (C.G. Bell and F.N. Hayes Eds.), Pergamon Press (1958).

2. H.L. De Vries, Appl. Sci. Res. Section B, 5, 387 (1955).

3. T.A. Rafter, New Zealand J. of Science and Technology, Section B, 36, 363 (1954).

4. H. Barker, Nature, 172, 631 (1953).

5. J.E. Noakes, A.F. Isbell and J.J. Stipp, Geochim. Cosmochim Acta, 27/7, 797 (1963)

6. J.E. Noakes, S.M. Kim and J.J. Stipp, Radiocarbon and Tritium Dating Proceedings VI Internatl.

6. (Continued)
Conf. Pullman, Washington (1965).

7. H.A. Polach and J.J. Stipp, Internatl. J. Appl. Rad.Isotopes, 18, 359 (1967).

8. M.A. Tamers, R. Bibron and G. Delibrias, in Tritium in the Physical and Biological Sciences, I.A.F.A., Vienna, 1, 303 (1962).

9. I.E. Starik, K.A. Arslonov and I.R. Klener, Radiochimiya, 5, (1963).

10. F. Pietig and H.W. Scharpenseel, Atompraxis, 10, 71, (1964).

11. L.L. McDowell and M.E. Ryan, Radiocarbon 7, 174 (1965)

12. M.A. Tamers, Radiocarbon and Tritium Dating Proceedings VI, Internatl. Conf. Pullman Washington, 53 (1965).

13. H.A. Polach, Atomic Energy in Australia, 12, 21,(1969)

14. T. Hagishimura, O. Yamada, N. Nohara and T. Shider, J. Appl. Rad. Isotopes, 13, 308 (1962).

15. H.A. Polach, J. Gower and I. Fraser, 8th Internatl. Conf. on Radiocarbon Dating, Lower Hutt, New Zealand, 144, (1972).

16. J.R. Arnold, Science, 119, 155 (1954).

17. C. Leger and M.A. Tamers, J. Appl. Rad. Isotopes, 15, 697 (1967)

TABLE I

Major Impurities in benzene samples in ppm

	A	B	C	D	E	F	G
Ethane	120-400	150-350	0-100				
Butane	20-70	25-60	0-30				
Acetal-dehyde	0-30	∿10					
Acetone	20-180	105-130	0-20	0-40	0-30		45
Ethanol	0-40	0-10	0-40		∿10	∿10	
Hexane	0-80	60-70	0-20	0-20			
"X"	0-3	2-5	0-6				80
Cyclohexane	0-100	80-90	0-370	0-50		∿20	
Toluene	40-950	60-100	400-1100	20-800	20-50	∿95	600
Ethyl-benzene	0-250	30-40	120-180	60-120		∿80	70
No. of Samples	30	2	3	6	3	1	1

A. Gillespie
B. Polach
C. Burleigh
D. Coleman
E. Calf
F. Pearson
G. Scharpenseel

TABLE II

Effect of Acetone quenching on sample countrate

Acetone ppm	Decrease in countrate
0	0[b]
50	0[b]
100	0[b]
150	0.2%[a]
250	0.5%[a]

a. Mean of 2 determinations individually prepared.

b. Mean of 4 determinations individually prepared.

Sample counting times ∿ 3000 cpm and individual sample counting times were c.a. 60 minutes.

THE PURITY OF COMMERCIAL LIQUID SCINTILLATION FLUORS AND THE EFFECTS OF IMPURITIES ON PERFORMANCE

Kent Painter and Michael Gezing
Amersham/Searle Corporation
Arlington Heights, Illinois, USA

ABSTRACT

As a large user of liquid scintillation chemicals, we have noted considerable variation in the quality of liquid scintillation fluors supplied by commercial manufacturers. We have examined in detail the purity of two common fluors, PPO and Bis-MSB. The limitations of conventional methods of analysis are discussed along with suggestions for analyzing fluors to determine scintillation efficiency and to detect color quenching impurities. Examples of results of several quenched and unquenched cocktails demonstrate the effects of impurities on optimum performance.

INTRODUCTION

In a critical experiment the research chemist should never presume that ordinary laboratory reagents are pure. Likewise in the liquid scintillation laboratory the analyst should ensure that all chemicals meet the purity criteria established for the experiment.

In 1957 Hayes noted that "In any liquid scintillator system, purity of the components is a very important consideration, because of the possible quenching or light absorption due to contaminants" (1). Eisenberg (2) has noted the variability in PPO quality, and Radin (3) repurified a batch of PPO to remove impurities which darkened in solution with Hyamine*. Schram (4) has stressed the importance of purity of components in scintillator systems. Horrocks (5) has described impurities in a commercial supply of Butyl-PBD.

*Trademark Rohm and Haas

Our interest in the purity of commercial fluors stems from our massive use in the laboratory and for manufacture of highly unquenched liquid scintillation standards which are used to determine optimum performance of liquid scintillation spectrometers. In our purchases over the last several months we have received fluors containing bits of wood, a ground-up rubber glove and nearly 5% by weight glass wool. Several batches of fluor when dissolved in toluene gave turbid solutions which could not be clarified.

On one occasion when preparing a concentrated solution of 250 g/l PPO which was supplied by a reputable manufacturer, the resultant solution was so heavily color-quenched it could have been mistaken for urine. After several weeks' delay, the supplier reluctantly checked a sample in his quality control laboratory and discovered to his amazement that the batch contained a very pronounced yellow-colored impurity.

On another occasion we analyzed a batch of Bis-MSB which was found to be grossly impure and heavily color-quenched. As a result of our complaint, the manufacturer recalled the batch and is no longer preparing this fluor. We know of another manufacturer who relies solely on melting point (with a 3°C. range) and visual inspection of dissolved samples for "foreign matter" in quality control of fluors.

It should be emphasized at the outset that the average liquid scintillation analyst would incur, at most, perhaps a 5-10% variation in efficiency due to fluor impurity. However, those of us who study liquid scintillation, who prepare calibration standards, must carefully test not only the fluors, but also the solvents and vials. Based upon our experience with impure commercial fluors, we find it quite astonishing that a recent survey of new liquid scintillators to determine optimum relative pulse heights could be undertaken using chemicals which were neither analyzed nor repurified (6).

MATERIALS AND METHOD

Bis-MSB: A five-liter batch of Triton X-100* scintillant was prepared using freshly distilled toluene: Triton X-100 (2:1 v/v) which contained 6 g/l PPO. To one-liter samples of scintillant was added 1.5 g Bis-MSB

*Trademark Rohm and Haas

from each supplier. The samples were stirred simultaneously under identical conditions until all solutes had dissolved. From each batch ten samples were prepared by adding 10.0 ml of each scintillant and 100 µl Toluene-^{3}H (350,000 dpm) internal standard to a counting vial which contained 5.0 ml distilled water. The samples were shaken, allowed to equilibrate four hours at room temperature and placed in the counter for four hours prior to counting. Average results of ten samples from each batch are shown in Table I.

PPO: A six-liter batch of Triton X-100* scintillant was prepared using toluene: Triton X-100 (2:1 v/v). To one-liter samples of the batch was added 2 g PPO. The samples were further prepared as described above.

DISCUSSION

Farmer and Bernstein (6) have previously demonstrated that infrared spectral analysis is an unreliable measure of scintillator performance. Our results indicate that neither fluorescence nor ultraviolet absorption alone are good measures of liquid scintillation fluor performance. In the former case the presence of quenching impurities could give a lower apparent fluorescence yield, even though the ultraviolet absorption spectrum indicates a pure sample. Conversely, the presence of ultraviolet-absorbing non-fluorescent impurities could result in falsely high absorption values. A ratio of absorbance/fluorescence appears to correlate well with counting efficiency. However, the ultimate test is the application for which the materials are to be used as compared to a standard known to be pure.

We observed tremendous variation in appearance and dissolution time of fluor crystals, particularly with Bis-MSB. It required only 30 minutes to dissolve one Bis-MSB sample, but 3 hours for another which, incidentally, was found to be the worst performer. Bis-MSB samples showed a significant 7% difference in performance.

The seven commercial PPO samples varied only 1.8% in tritium counting efficiency, even at 2.0 g/l where small variations in purity affect efficiency greatest (8); this was not a significant difference in the experiments we performed. Both fluorescence emission and ultraviolet absorption spectra appeared normal, but one sample ex-

*Trademark Rohm and Haas

TABLE I: Evaluation of Commercial Bis-MSB

Supplier	Relative Tritium Counting Efficiency*	Relative Fluorescence**	Relative Absorbance***	Relative Absorbance/Fluorescence
B	100	95.7	100	100
A	99.6	95.1	92.8	93.8
C	97.2	100	94.1	90.5
D	96.8	96.8	89.8	89.1
E	93.2	95.1	78.2	79.0

*Nuclear-Chicago Mark II at tritium balance point.
**Exitation 374 mμ; Emission 424 mμ (Perkin Elmer MFP-2A).
***Absorbance 353 mμ (Cary 14).

hibited a rather pronounced yellow-colored impurity which absorbed in the region of 670-770 mμ.

SUMMARY

The term "Scintillation Grade" so often used by suppliers has never been defined for specific chemicals. One supplier defines "Scintillation Grade" nebulously as meeting "rigorous quality standards established after careful consideration of the actual requirements of the individual reagents, and their role in the total system".

We have shown that there can be significant variation in the purity of commercial liquid scintillation fluors and that the impurities affect performance.

Standards of "Scintillation Grade" should be established for industry to follow. I propose that at our next conference we consider standards for the purity of toluene, xylene, p-xylene, dioxane, PPO, Butyl-PBD, POPOP and Bis-MSB.

REFERENCES

1. F. N. Hayes *in* Liquid Scintillation Counting, p. 87 (C. G. Bell and F. N. Hayes, Eds.). New York: Pergamon (1958)
2. F. Eisenberg *in* Liquid Scintillation Counting, p. 123 (C. G. Bell and F. N. Hayes, Eds.). New York: Pergamon (1958)
3. N. S. Radin *in* Liquid Scintillation Counting, p. 108 (C. G. Bell and F. N. Hayes, Eds.). New York: Pergamon (1958)
4. E. Schram, Organic Scintillation Detectors, p. 41. London: Elsevier (1963)
5. Donald L. Horrocks, Beckman Instruments, personal communication
6. E. C. Farmer and I. A. Bernstein, Nucleonics *10*, p. 54. (1952)
7. J. B. Birks and G. C. Poullis *in* Liquid Scintillation Counting, p. 25 (M. A. Crook, P. Johnson and B. Scales, Eds.). New York: Grune and Stratton (1970)
8. D. L. Horrocks *in* The Current Status of Liquid Scintillation Counting, p. 25 (E. D. Bransome, Ed.) New York: Grune and Stratton (1970)

RECENT ADVANCES IN SAMPLE PREPARATION

Yutaka Kobayashi and David V. Maudsley
The Worcester Foundation for Experimental Biology
Shrewsbury, Massachusetts 01545 U.S.A.

Although new concepts in the preparation of samples for liquid scintillation counting have not emerged, some improvements have been made. We now have, for example, a range of solubilizing agents which are more effective and less quenching than Hyamine hydroxide and we can count several milliliters of aqueous material where, previously, we were restricted to a few microliters. In the area of combustion techniques we have traveled a route which started from tedium with a touch of hazard and is now firmly headed for automated convenience. What has also changed is the application of liquid scintillation counting to samples generated by current analytical techniques such as gel electrophoresis and radioimmunoassay. Certainly a new dimension to radioisotopic techniques has been added by the recognition that energetic beta-emitters such as ^{32}P can be counted directly in a liquid scintillation counter by Cerenkov radiation.

A sample for liquid scintillation counting consists of 4 components: 1) the solvent 2) scintillator 3) sample 4) the sample container or counting vial.

Solvents

Toluene remains the best general purpose solvent with regard to energy transfer efficiency, solubility of the scintillator, availability, uniformity and cost. The para and meta xylenes are more efficient solvents than toluene giving a 10% increase in pulse height. Since the various xylenes have different energy transfer efficiencies ensuring the uniformity of solvent composition adds to the cost of this system although xylene is used as the major solvent in some commercial preparations of ready-to-use cocktails. The solvent is one of the least efficient components of the counting system since 90 to 95 percent of the energy theoretically available for photon production is lost during the energy transfer between the beta particle and the solvent. It is unlikely, however, that a new solvent will be discovered which exhibits markedly better energy transfer characteristics than those already available.

Scintillator

Although PPO is still widely used, a better choice is butyl PBD. The major difference is that butyl PBD is more resistant than PPO to concentration or self-quenching. This is illustrated in Fig. 1 which shows the effect on counting efficiency of increasing concentrations of scintillator on a tritiated toluene based counting solution quenched with 0.5 percent carbon tetrachloride. The concentration quenching observed with PPO can be partially alleviated by the addition of a secondary scintillator as discussed below. Butyl PBD shows essentially a flat response with an optimum concentration of around 25g/liter. Butyl PBD is, therefore, used in much higher concentrations than PPO and is more expensive to use.

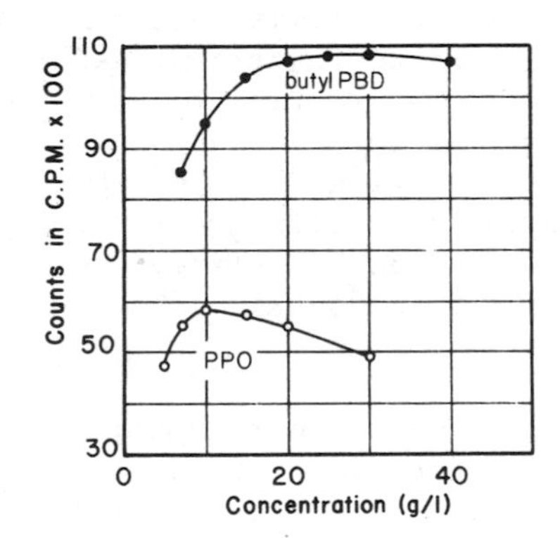

Fig. 1 Effect of increasing concentrations of PPO and butyl PBD on counting efficiency of tritiated toluene containing 0.5% CCl_4.

The most frequently used primary scintillators emit light with a wavelength in the region of 365 nanometers. For many years scintillation counters were equipped with photomultiplier tubes with a cesium-antimony photocathode exhibiting an S-11 spectral response curve. This is illustrated in Fig. 2 which shows that these tubes are maximally sensitive at 440 nanometers. It was common practice, therefore, to include a secondary scintillator in the sample to function as a wavelength shifter emitting light at approximately 410 nanometers. Alterations in design, however, have produced photomultiplier tubes which are most sensitive around 390 nanometers. The spectral response of two of these bialkali tubes is shown in Fig.2.

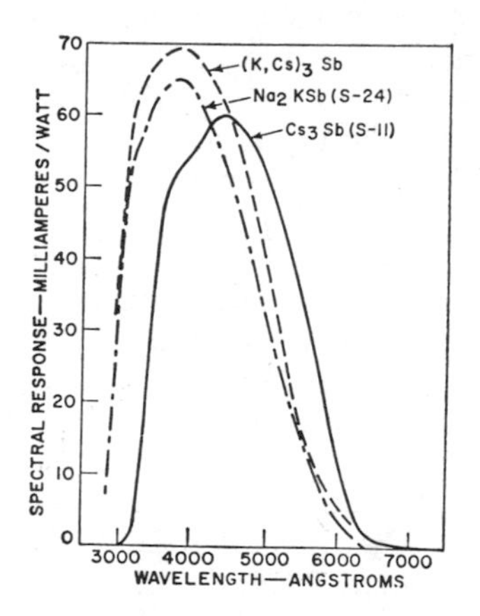

Fig. 2 Spectral characteristics of 3 photomultiplier tubes used in liquid scintillation counters (Courtesy of RCA)

It can also be seen that inclusion of a secondary scintillator, if it emitted light above 410 nanometers would actually reduce efficiency. PBBO, 2-(4'-biphenylyl)6-phenylbenzoxazole, emits light at 397 nanometers and is potentially one of the more useful of the secondary scintillators currently available for optical matching with the bialkali tubes. Whether or not a secondary scintillator should be included in a counting solution depends, therefore, on several factors and Bush and Hansen (1) have suggested that a secondary scintillator should be used only when:

1. the sample contains a compound which directly quenches the primary scintillator
2. the concentration of primary scintillator produces concentration quenching
3. the counter demonstrates better response at longer wavelengths
4. the counting sample has significant absorption in the near ultraviolet.

Sample Vials

The standard sample container has been the 22 ml low potassium glass vial. Recently low potassium vials have become available as well as a series of plastic vials. The usefulness of the smaller vial is illustrated in Fig.3 which shows the effect of increasing the volume of the counting solution on the counting efficiency of tritium in a standard 22 ml vial and a 12 ml low potassium glass vial. The optimum volume for this counter is between 8 and 13 ml for the 22 ml vial and between 4 and 7 ml for the minivial. The small reduction in tritium efficiency of about 5 percent is due to altered geometry i.e. the increased distance between the smaller sample vial and the photomultiplier tubes. In our laboratory we have switched almost completely to the use of minivials. The saving in cost is considerable since the vials themselves are cheaper and the amount of counting solution required is much less than for the 22 ml vial.

The background characteristics of some of these vials were investigated. Tritium background was determined in a wide tritium window where the ^{3}H efficiency was 60%. The carbon-14 background was determined in a narrow window set above the ^{3}H-spectrum i.e. at the conventional double-isotope setting. In this particular counter the carbon-14 efficiency above ^{3}H was 63%. Ten samples of each vial were counted twice for ten minutes each time and the results are summarized in Table I.

An interesting observation is that the major contributor to the carbon-14 background is the counter itself. The primary sources of these empty chamber counts are

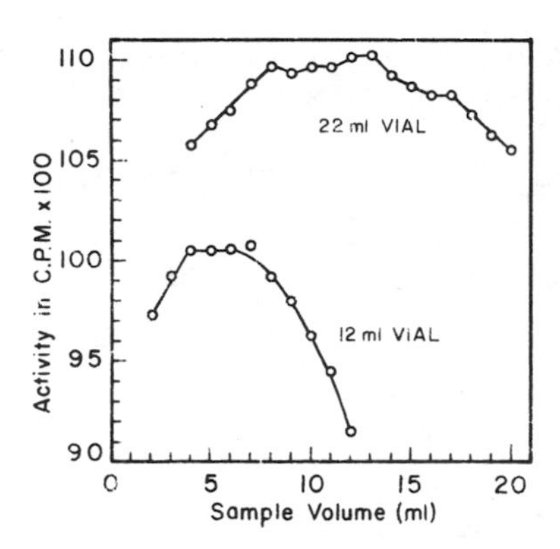

Fig. 3 Effect of counting solution volume on counting efficiency of tritium in toluene containing 8g butyl PBD/l in 12 and 22 ml low K glass vials.

contributions from the glass faces of the photocathode and from what is known as electroluminescence and arcing on the envelope of the photomultiplier tubes. Glass has a tendency to electroluminesce when subjected to an electrical gradient. This condition occurs between the pins of the photomultiplier tube which carry the high voltage and is increased by increasing humidity. In this respect, the refrigerated systems have an advantage over ambient units because the air within the cooled units is dehumidified by the cooling. Another possible source of empty chamber background is the lucite light pipe which is sometimes used to improve optical coupling between the sample and the photomultiplier tube. The lucite could be a source of Cerenkov radiation, but this is only conjecture.

For tritium the counting chamber is a minor contributor and most of the counts are due to the vial itself. Accordingly, there is a substantial reduction in the tritium background when plastic vials are used in place of glass vials. The E^2/b value for tritium in this particular counter is increased from around 130 to 330 by

Table I

SOURCES OF BACKGROUND COUNTS IN AIR-QUENCHED SAMPLES OF TOLUENE CONTAINING 0.8% BUTYL-PBD

Counting Conditions	^{3}H Channel	C-Channel
Tritium counting efficiency	60%	0.05%
Carbon-14 counting efficiency	20%	63%
System	**C.P.M.±S.D.**	**C.P.M.±S.D.**
22 ml low-K glass vial + 10 ml counting solution	27.19±1.91	18.57±1.72
12 ml low-K glass vial + 5 ml counting solution in polyethylene holder	17.70±1.25	16.33±1.82
20 ml polyethylene vial + 10 ml counting solution	11.76±0.91	18.40±1.53
7 ml polyethylene vial + 5 ml counting solution in plastic holder	8.94±0.95	15.34±0.77
Empty counting chamber	4.42±0.46	15.41±0.91

switching to miniplastic vials. For carbon-14 there is little or no change in background when plastic vials are used. The solvent contributes very little to the background.

Sample Preparation

Samples may be prepared by methods which result in homogeneous solutions or by procedures which give rise to heterogeneous systems (Table II). Cerenkov counting is also included here even though it does not involve any sample preparation.

Homogeneous Systems

Whenever possible samples should be counted as homogeneous systems. The isotope is then in intimate contact with the solvent and the scintillator, there is no self-absorption and various methods of quench correction such as channels ratio and the external standard ratio methods can be applied. The five subgroups listed under homogeneous counting are self-explanatory. In this area of

Table II

SAMPLE PREPARATION METHODS FOR LIQUID SCINTILLATION COUNTING

Homogeneous Samples

1. Toluene soluble samples
2. Polar solvent systems for aqueous samples
3. Systems for cationic and basic samples
4. Systems for anions, proteins and acidic samples
5. Combustion methods

Heterogeneous Samples

1. Samples on solid supports
2. Emulsions
3. Gels
4. Solid scintillators

Cerenkov Counting

1. Direct counting of beta emitters with energies greater than 263 Kev in water

---

sample preparation there has been little that is new. There are a variety of solubilizing agents now available which are more effective than Hyamine hydroxide. The most commonly used are NCS, Soluene-100, Protosol and the Biosolves. One of the problems that has arisen with the use of the solubilizing agents is the difficulty in determining whether the resultant sample is a true solution. In some cases the sample may become adsorbed onto the walls of the counting vial or small particles may be present which are not detected by a visual inspection. These phenomena are evidenced by a steadily decreasing count rate or a change in the sample channels ratio. Bush (2) has devised an ingeneous method for detecting the homogeneity of solution. For homogeneous solutions there is a curve linear relationship between the sample channels ratio and the external standard channels ratio. The external standard channels ratio measures the counting efficiency of the solution whereas the sample channels ratio measures the efficiency of the sample irrespective of its solubility. For a partially soluble sample the sample channels ratio gives a different efficiency than the external channels ratio as illustrated in Fig. 4. Points which deviate from the curve were shown to be heterogeneous samples by virtue of adsorption or precipitation.

Another problem encountered with basic organic solubilizing agents is that the use of butyl PBD as the scintillator results in the development of a yellow color in the sample. The scintillator of choice, under these circumstances, is PPO. With most of these solubilizers the usual precautions against chemiluminescence should be observed i.e. neutralizing the digest with acid before the addition of the counting solution.

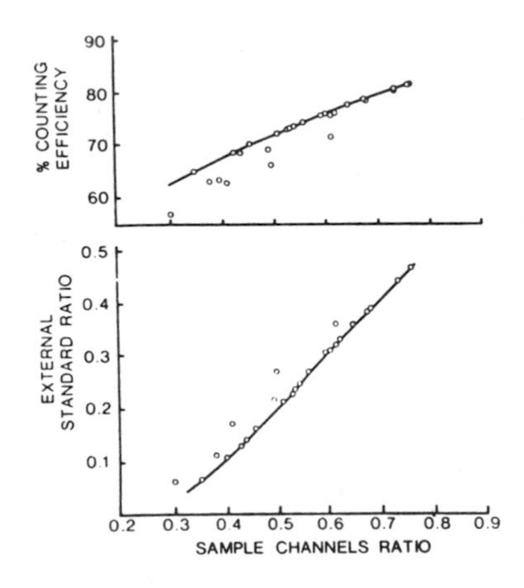

Fig. 4 Double-ratio plot of determining homogeneity of a counting solution (2).

Combustion Techniques

One solution to most problems associated with the sample preparation is to oxidize all samples to a common material such as carbon dioxide and water. Combustion techniques, therefore, have wide application. The procedure should also be rapid, simple and quantitative and automated combustion may prove to be such a method. Our experience with the early Packard apparatus, model 305, indicated that recoveries were near quantitative for ^{3}H and about 95% for ^{14}C. Almost any type of biological material can be combusted but the early instrument had mechanical problems and was difficult to maintain in working order. An improved and more automated version of the oxidizer, model 306, has recently become available. Recoveries are stated to be 99 ± 1% for both isotopes and memory is less than 0.05 percent. For dual isotope combustion the contamination of ^{3}H with ^{14}C is less than 0.02% and contamination of ^{14}C with ^{3}H is less than 0.001%.

The automated combustion apparatus originally designed by Peterson in 1969 forms the basis for both the new Packard and the Intertechnique combustion units, the latter called the Oxymat, (3). This is illustrated in Fig. 5 and represents, in essence, an automated version of the classical catalytic combustion method used in organic chemistry.

Heterogeneous Systems

Samples on Solid Supports

The important factor to remember when counting samples on support media such as filter paper or scrapings from thin layer plates is that the sample should be soluble or insoluble. Partial solubility will make any measurement meaningless because the soluble fraction is counted with 4π geometry whereas the insoluble portion is counted with 2π geometry. For experimental samples the contribution from each fraction will be unknown.

Emulsion Counting

One of the limitations of toluene-based counting solutions has been the inability to accommodate milliliter quantities of aqueous samples. The best known cocktail for this purpose has been Bray's solution which is a dioxane-based counting solution which could hold over

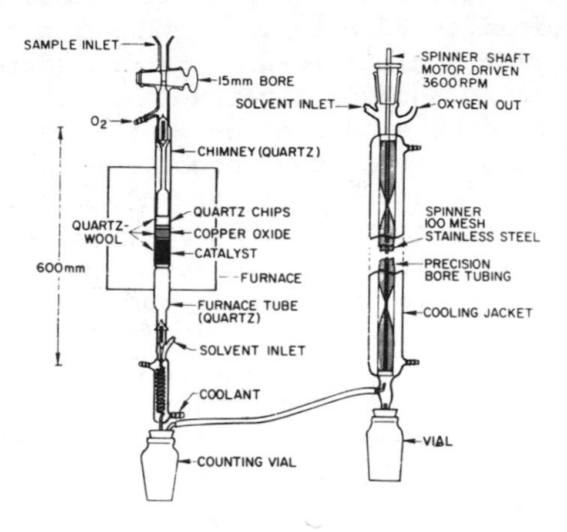

Fig. 5 The Peterson Combustion apparatus (3).

20 percent water at -5°C as a homogeneous sample (4). A variety of similar dioxane-based cocktails have been formulated over the years for counting aqueous solutions. The main drawback with these cocktails is that micro precipitation or adsorption on the vial walls may occur because salt solutions are not easily accommodated. The problem of counting of relatively large volumes of aqueous radioactive solutions has been met by the use of non-ionic emulsifying agents, the most popular being the surfactant Triton X-100. The chemical formula of Triton X-100 is shown in Fig. 6. A surfactant such as Triton X-100 is composed of a hydrophilic end which is attracted to water and to polar compounds and a hydrophobic end, which is attracted to non-polar solvents and compounds such as toluene. In the Triton series of surfactants, the hydrophilic end is, of course, the ethoxy end. The longer the polymer, the greater the solubility in water. Its solubility in water is inversely proportional to temperature - that is, the higher the temperature, the less water it can hold.

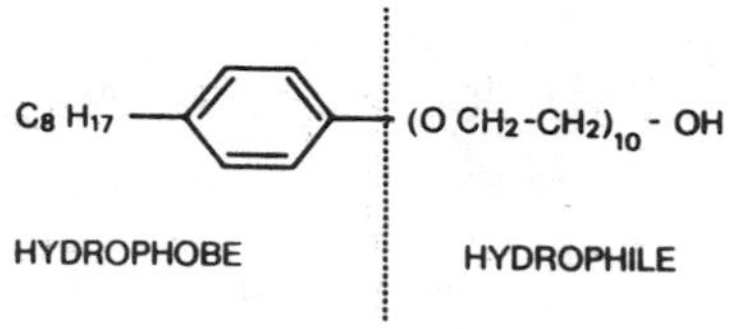

C_8H_{17} — (benzene ring) — $(OCH_2\text{-}CH_2)_{10}\text{-}OH$

HYDROPHOBE | HYDROPHILE

Fig. 6 Chemical formula of Triton X-100.

Increasing aqueous salt concentrations also reduces its miscibility. The Triton X-series are those which have 8 carbons on the hydrophobic end and the N-series are those which have 9 carbons.

The Triton surfactants have been formulated with toluene or xylene to hold as much as 50 percent water as a gel for counting. The Triton surfactants are probably used in the commercial products such as Insta-gel, Aquasol and PCS which are widely used for aqueous solutions. These chemicals are able to hold water as micelles as illustrated in Fig. 7. Water is held by the hydrophilic end of the Triton molecule while the hydrophobic end, represented by the tails, holds the micelle in the toluene. The water droplet represents the dispersed phase and the toluene, the external phase. The character of the micelle is a function of the amount of water being held and the micelle can be very small - less than five-hundredth of a micron - or can range over 1 micron in diameter. As the contribution of the dispersed phase increases, the optical

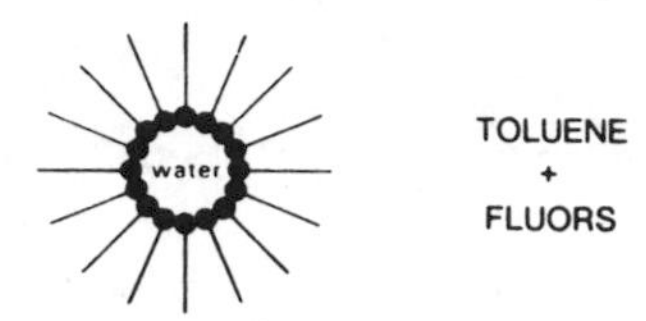

Fig. 7 A hypothetical model of a Triton X-100/water micelle in toluene.

characteristics change. In Table III the size of the micelle is related to the physical appearance of the emulsion. The maximum range of a tritium beta particle is 6 microns and the average range is 1.2 microns. This means that the tritium should be counted with solution efficiency in all but the milky emulsion. Since carbon-14 has about ten times the range of tritium, carbon-14 should be counted at solution efficiency with even the heaviest emulsions. Emulsion counting, then, appears to be a very reasonable way to prepare and count aqueous radioactive samples. However, the evidence is not at all clear that this is a simple technique. Emulsion counting requires strict adherence to a defined regimen if reliable data are to be generated.

Fig. 8 is a type of phase diagram prepared by Turner some years ago showing the effect of increasing water content on the counting efficiency of ^{14}C glucose and tritiated water in a toluene-Triton X-100 system. The percent value of the Y-axis refers to the efficiency of emulsion counting compared to that in a homogeneous solution.

Table III

PARTICLE SIZE AND EMULSION APPEARANCE

Macroglobules	Two Phases May Be Distinguished
Greater than 1 micron	Milky white emulsion
1 to 0.1 micron	Blue-white emulsion
0.1 to 0.05 micron	Gray semi-transparent
0.05 micron & smaller	Transparent

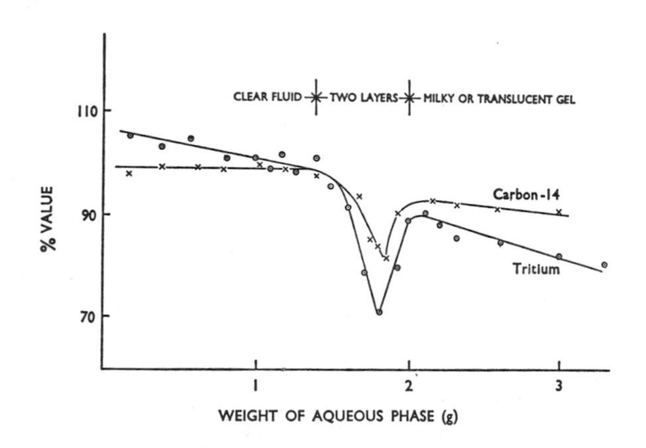

Fig. 8 Emulsion counting of D-glucose-^{14}C (U) and tritiated water using toluene/Triton X-100, 2:1 (v/v). Reprinted from (5) by permission of The Radiochemical Centre Ltd.

The pattern shown here is typical of emulsion counting systems. With a low water content, there is a clear phase followed by a region where phase separation occurs. As more water is added, the phase then changes into a milky or translucent gel. The clear phase has been described by Turner (5) and Benson (6) as a true solution while Van der Laarse (7) classifies all phases as suspensions of water in toluene. As the concentration of water is increased, the counting efficiency of tritium decreases because of the quenching effect of water. When the first phase change occurs, phase separation into an aqueous and organic layer is observed and there is a sharp fall in counting efficiency because the tritium stays in the aqueous phase whereas the scintillator remains dissolved in the organic phase. Only the tritium beta particles at the interface between the organic and aqueous layers have sufficient energy to penetrate the organic layer and interact with the scintillator. As more water is added, an emulsion is formed which consists of sub-micron water particles dispersed in the organic phase, as already described. Since the size of the water particles or micelles is less than the mean free path of the average tritium beta particle, the beta particles can interact efficiently with the scintillator with the consequent proportionate reduction by self-absorption in the number of beta particles escaping into the organic phase resulting in a drop in counting efficiency. For carbon-14 the sample can be counted at constant efficiency in the clear phase and near-constant efficiency in the emulsion phase due to the more energetic nature of carbon-14 beta particles.

A critical problem in emulsion counting that arises as a consequence of the partitioning of solutes between the organic and aqueous phases is the determination of counting efficiency. As a general rule, for emulsions, the proper internal standard can give the most accurate results. The internal standard should have the same partition coefficient for the emulsion system as that of the material being counted and, for tritium, this would normally be tritiated water. For ^{14}C, it should be an aqueous standard solution of the material being counted. In any case, the internal standard must not be labeled toluene or hexadecane which are insoluble in the aqueous phase. For a large number of samples, the sample channels ratio method for efficiency determination is convenient and useful if carefully done but the external standard cannot be used to monitor counting efficiency because the Compton electrons produced by the gamma source interacting with the sample vial and its contents are not attenuated by the dispersed water particles.

There are several properties of these emulsions that need serious consideration. First, the various phases are temperature dependent, as is the counting efficiency. One report documents the fact that the counting efficiency of a Triton X-100/toluene/water emulsion increased linearly 10 percent with decreasing temperature from about 17°C to a maximum at about 4°C (8). Lower temperatures did not increase the counting efficiency further. Also, emulsions were found to give reproducible results only if they were first heated to 40°C and then allowed to cool without shaking to 4°C and then stand for 2 to 4 hours. Emulsions prepared simply by shaking vigorously at room temperature, though clear in appearance were found to be unstable as judged by lower and more inconsistent counting data.

Some of the requirements for emulsion counting are:

1. Purity of the surfactant. The different batches of commercial surfactants are not always uniform and may contain luminescent impurities as well as quenching agents.
2. All samples should have the same volume and composition, i.e., the same aqueous sample volume in all cases.
3. All samples should be prepared in the same way, i.e., heat to 40°C and allowed to cool without further agitation.
4. All samples should be equilibrated a minimum of 2 hours in the counter to stabilize the gel phase and temperature before counting since the counting properties of emulsions are temperature-sensitive. Dark-adaptation will also reduce any phosphorescence or chemiluminescence.
5. Counting efficiency should be determined either by sample channels ratio or the internal standard method. External standardization should not be used.

Cerenkov Counting

Cerenkov radiation is generated when a charged particle travels through a medium faster than the speed of light through that medium. Theoretically, a beta emitter must have an energy greater than 263 to be detected by Cerenkov counting in water and this eliminates tritium, carbon-14 and sulfur-35. The minimum energy threshold, however, is a function of the refractive index of the medium as discussed elsewhere in this volume. The major application of Cerenkov counting for biological scientists is in the counting of ^{32}P in aqueous solution. The sample size is limited only by the capacity and geometry of the counting chamber. Disadvantages of the method are that the counting efficiencies are lower than would be obtained if the sample were counted by more conventional means and sensitivity to color quenching. Counting efficiencies can be improved by the addition of a wavelength shifter and color quenching can, in many instances, be removed by chemical treatment.

References

1. E.T. Bush and D.L. Hansen in Radioisotope Sample Measurement Techniques in Medicine and Biology, p.395 Int. Atomic Energy Agency, Vienna (1965)

2. E.T. Bush, Int. J. Appl. Radiation Isotopes 19, 447 (1968)

3. J.I. Peterson, F. Wagner, S. Siegel and W. Nixon, Anal. Biochem. 31, 189 (1969)

4. G.A. Bray, Anal. Biochem. 1, 279 (1960)

5. J.C. Turner, Int. J. Appl. Radiation Isotopes 19, 557 (1968)

6. R.H. Benson, Anal. Chem. 38, 1353 (1966)

7. J.D. Van der Laarse, Int. J. Appl. Radiation Isotopes 18, 485 (1967)

8. P.H. Williams and T. Florkowski in Radioactive Dating and Methods of Low Level Counting, p.703 Int. Atomic Energy Agency, Vienna (1967)

The authors gratefully acknowledge the generous travel grants from Searle Analytic, New England Nuclear Corporation and Mr. Calvin Fisher which made this trip to Sydney, Australia possible, and thank Mrs. Virginia Barber for her editorial and secretarial assistance.

SUSPENSION COUNTING OF ^{14}C IN SOIL, SOIL EXTRACTS AND PLANT MATERIALS BY LIQUID SCINTILLATION

J.K. Adu and J.M. Oades

Department of Agricultural Biochemistry and Soil Science,
Waite Agricultural Research Institute,
The University of Adelaide,
Glen Osmond, South Australia 5064

ABSTRACT

Standard ^{14}C-labelled soils or minerals were prepared by addition of standard ^{14}C-labelled benzoic acid. It was shown that addition of labelled benzoic acid as an internal standard to suspensions of these samples, stabilized by CAB-O-SIL in toluene-PPO-POPOP scintillant, yielded efficiencies which allowed 100% recoveries of activity in the solid samples, provided a) the solid samples were ground to <53 μm diameter and b) the weights of sample used were such that the optical density of the gel remained below about 0.9 at 450 nm (1 cm cell). The method was applied successfully to soil samples containing labelled microbial tissue and to freeze-dried coloured extracts of soil.

INTRODUCTION

The determination of ^{14}C in soil has generally involved oxidation of the organic materials to carbon dioxide using either wet (1,2) or dry (3,4) combustion. The $^{14}CO_2$ evolved has been trapped in alkaline solutions and counted as $Ba^{14}CO_2$ on planchets, or in suspension. Alternatively, aliquots of the alkaline solutions have been rendered miscible with scintillation fluors and the ^{14}C counted by liquid scintillation.

These methods require quantitative conversion of organic materials to $^{14}CO_2$, which is difficult to achieve. The trapping of CO_2 and subsequent counting as $Ba^{14}CO_2$ or as $^{14}CO_2$ in alkaline solution render the procedure tedious and subject to error at several stages.

The requirement for a reliable routine method for the determination of ^{14}C in soils and soil extracts led to a study of liquid scintillation counting of suspensions of soil particles. Suspensions of samples for counting by liquid scintillation have been stabilized by gelling agents such as aluminium stearate (5), Thixcin - a castor oil derivate (6) and CAB-O-SIL - a finely divided silica (7). While this system of counting has been applied to $Ba^{14}CO_3$ in particular, it has also been tried on ^{14}C-tagged, coloured biological materials such as liver and bacteria with success (8). Cheshire *et al.* (9) recently described application of the method to counting of ^{14}C in soils. However, the data presented in this present paper indicate that counting of ^{14}C in suspensions of soil particles is more complicated than suggested by these workers.

Direct counting of soil particles tagged with ^{14}C has several advantages. The necessity for quantitative combustion to CO_2 is eliminated. The method is simple, rapid and can be applied to small samples with low activity. The main problems anticipated were colour quenching, self-absorption and the effects of large quantities of heavy metals, e.g. iron.

MATERIALS

^{14}C-Labelled materials and reagents. Benzoic acid-^{14}C (specific activity 5.32×10^6 dpm/g) was obtained from Packard Instrument Company Inc. and used as a primary standard. Benzoic acid-^{14}C (specific activity 452 μCi/mg) from The Radiochemical Centre, Amersham, England, was dissolved in water diluted with unlabelled benzoic acid and the mixture recrystallized twice to ensure a uniformly labelled product, which was dried *in vacuo* over P_2O_5 to a constant weight. The specific activity of this secondary standard was established by counting equal weights of the two standards in 10 ml of toluene-PPO-POPOP scintillant (0.5% PPO plus 0.03% POPOP w/v in toluene). The vials were shaken vigorously and counted in a Packard Tri-Carb scintillation counter Model 3375 operating at 4°C, after allowing one hour for the equilibration of the vials with the temperature of the counter. Vials were counted for 5 min to give a standard deviation of 0.1% for the diluted secondary standard and 0.5% for the primary standard. The settings used were

20-1000 with 6% gain in the Red Channel and 20-100 with 6% gain in the Green Channel. Counts in the Red Channel were used to calculate the specific activity of the secondary standard. The efficiency of counting based on the primary standard was 96%. The secondary standard contained 20,000 dpm/mg.

PPO, POPOP, CAB-O-SIL and Instagel were obtained from Packard Instrument Company Inc. Triton X-100 (Rohm and Haas technical grade) was from Robert Bryce and Co., Australia; this product yielded results similar to those obtained by scintillation grade Triton X-100 provided it was centrifuged or filtered before use.

Mineral and soil samples. All samples were ground in a Siebteknik mill to pass through a 300-mesh sieve (<53 μm) and the colour was compared with a Munsell Colour Chart.

(a) Acid-washed sand (greyish white N 8/0)
(b) Red-brown earth (dull brown 7.5 YR 5/4)
(c) Ground-water rendzina (grey N 5/0)
(d) Goethite, α-FeOOH - yellow iron oxide (yellow 2.5 Y 7/8)
(e) Haematite, α-Fe_2O_3 - red iron oxide (red 7.5 R 4/8)

Samples (4.9 g <53 μm) were mixed with 100 mg of the benzoic acid-^{14}C secondary standard in the mill for 2 min, to ensure uniform distribution of ^{14}C in the sample. The samples then contained 400 dpm/mg and are referred to as "standard" samples.

Soils containing a range of ^{14}C-labelled compounds. To ensure that ^{14}C present in compounds other than benzoic acid could be counted in soil suspensions, a sample of the red-brown earth was incubated with [G-^{14}C]glucose for 10 days. The $^{14}CO_2$ released was trapped in 1N NaOH; 0.1 ml aliquots plus 0.9 ml H_2O were mixed with 10 ml of Triton X-100 scintillant [4 g PPO, 0.1 g POPOP in 1 litre 2:1 v/v toluene/Triton X-100; (10)] and the samples counted at 4°C after 24 hr equilibration at this temperature to eliminate chemiluminescence. After incubation, the soil was dried at 70°C overnight and passed through sieves with a range of mesh sizes. It was assumed that the ^{14}C in the incubated soil sample was present in microbial tissue, i.e. a wide variety of biological compounds.

A further sample of the red-brown earth was incubated

with [G-^{14}C]glucose for 14 days and then freeze dried. The incubated soil was subjected to a series of extraction and fractionation procedures (11). The various products were freeze dried, ground to <53 μm and counted for ^{14}C in the same manner as for standard samples.

^{14}C-Labelled plant material. ^{14}C-Labelled wheat straw (specific activity 171 μCi/g C) was obtained from Landwirtschaftlich-Chemische, Bundesversuchsanstalt, Vienna. This was diluted with unlabelled wheat straw (3 parts labelled to 7 parts unlabelled) and ground in the Siebteknik mill. A sample of the diluted straw (<53 μm) yielded 43,000 cpm/mg. 100 mg of this material was mixed with 4.9 g of <53 μm acid-washed sand to give 860 cpm/mg, and was counted as for standard samples.

METHODS

Counting of ^{14}C-tagged solid samples. Duplicate samples (5 to 100 mg) were weighed into scintillation vials. 500 ± 10 mg of CAB-O-SIL were added, along with 10 ml of toluene-PPO-POPOP scintillant (0.5% w/v PPO and 0.03% w/v POPOP in toluene). Caps with Polyseal cones were screwed tightly onto the vials, which were shaken for 2 min on a vortex mixer using maximum speed. This treatment yielded a transparent or an opaque gel (depending on the colour of sample) containing a good dispersion of the samples. Vials prepared in this way were either stored at 4°C or allowed to equilibrate with the temperature of the counter (4°C) for one hour before counting. Window settings were 20-1000 with 20% gain in the wider channel (Red) and 20-100 with 20% gain in the narrow (Green) channel. The background varied from 40-52 cpm during the period of study, which was more than 18 months.

RESULTS AND DISCUSSION

Stability of suspensions. The results obtained from counting suspensions formed using a range of weights and particle sizes of the red-brown earth showed no significant difference between the counts obtained 1 hr after placing in the spectrometer and those obtained after 3 weeks' storage at 4°C. The 500 mg of CAB-O-SIL formed a gel which was rigid enough to hold even the largest particles, as found by Ott *et al.* (7) and Lloyd-Jones (12). However, evaporation of the scintillant did occur in occasional vials (6,13).

Effect of sample size on recovery of activity. The recoveries of activity have been expressed as a percentage of the expected dpm based on the weight of "standard" sample added. The linear part of the curves (Fig. 1) was reproducible provided the particle size of the samples was the same. Below a critical weight of sample per vial the recoveries of activity could be used as per cent efficiencies. This was established by addition of standard benzoic acid to vials already containing the "standard" soil samples, i.e. using the ^{14}C-benzoic acid as an internal standard. When sample weights exceeded the critical value, aggregation of particles occurred resulting in self-absorption of ^{14}C β-particles (Table II), and the gels became obviously coloured and optically dense.

The best recoveries of activities, or alternatively the highest efficiencies, were obtained for the standard acid-washed sand. For both this sample and the plant material mixed with acid-washed sand there was a linear relationship between recovery of activity and weight of sample from 5 to 100 mg. The linear relationship existed up to 45 and 25 mg sample per vial for the rendzina and red-brown earth respectively. The most difficult sample was the haematite, where even with 10 mg of sample per vial it was not possible to obtain meaningful counts. With more than 25 mg of this sample the activity was lost, such that counts obtained were below background.

It will be shown later how the maximum weight of a particular sample which can be counted with close to 100% recovery of activity using an internal standard is dependent on the optical density in the 400-500 nm region.

Incubated soil samples. After 10 days' incubation of soil to which [G-^{14}C]glucose was added, release of $^{14}CO_2$ accounted for 70% of the activity of the original labelled glucose incorporated into soil. The activity obtained using up to about 25 mg of <53 μm sample accounted for 30% of the ^{14}C added before incubation (Fig. 3). The recovery of 100% of the activity initially added to the soil indicates that the counting of the $^{14}CO_2$ and ^{14}C in the soil was probably reliable. The efficiency *versus* weight curve

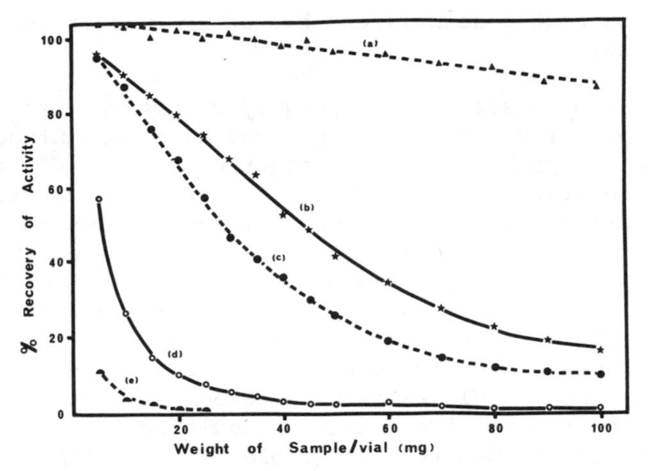

Fig. 1. Effect of weight of sample on recovery of activity. "Standard" samples - (a) Acid-washed sand; (b) Rendzina; (c) Red-brown earth; (d) Goethite, α-FeOOH; (e) Haematite, α-Fe_2O_3.

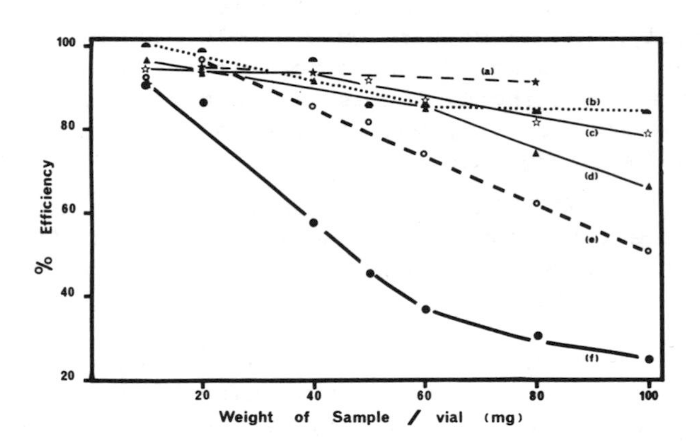

Fig. 2. Effect of particle size on efficiency of counting of ^{14}C in microbial tissue (red-brown earth). (a) 422-500 μm; (b) 353-422 μm; (c) 250-353 μm; (d) 124-250 μm; (e) 53-124 μm; (f) <53 μm.

(Fig. 2) is similar to the corresponding curve in Fig. 1. It is offset due to a difference in the particle-size distribution within the <53 µm range caused by the second grinding of the "standard" samples after addition of the labelled benzoic acid. Close to full recoveries of activity were obtained only for <53 µm samples with less than about 25 mg per vial (Fig. 3). With 100 mg per vial the recovery was 70%, the rest of the activity having been lost through either self-absorption, colour quenching or other factors. Thus, the chemistry of the ^{14}C-labelled compounds in the soil is not important, as the method works equally well with ^{14}C-labelled benzoic acid, microbial tissue or plant materials.

Attempts were made to determine ^{14}C in the samples to which ^{14}C-benzoic acid was added and also in soils after incubation with [G-^{14}C]glucose. The samples were combusted in oxygen in a conventional tube furnace, in a Fisher Carbon Induction furnace and in a Beckman Oxidiser. Recoveries varied from 70 to 90% and were never quantitative. It was originally considered that the combustion procedure could be used as a basis against which the suspension counting could be compared. The results indicate that the suspension counting is more rapid and quantitative than the combustion methods examined.

Effect of particle size. The average density of soil particles is 2.6 g/cm^3. It was therefore expected that variations in particle size would be reflected by changes in counting efficiency due to self-absorption of the ^{14}C β-particles (6). Using the weight of material which yields a layer of infinite thickness (~30 mg/cm), it can be calculated that the critical diameter of soil particles is in the range 60-120 µm to cover the density range 2.5-5.0 g/cm^3. The 300 µm diameter particles used by Cheshire *et al.* (9) were beyond this critical size range.

Even though there was a decrease in counting efficiency with reduction in particle size (Fig. 2), there was an increase in recovery of activity (Fig. 3). The lower efficiencies with increasing weight per vial obtained for the <53 µm particles were probably due to changes in the optical properties as a result of increase in colour intensity, in addition to self-absorption due to the formation of microaggregates within the gel.

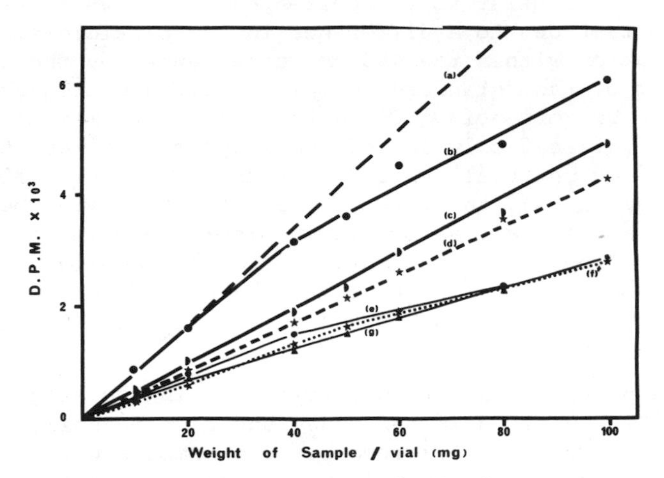

Fig. 3. Effect of particle size on recovery of activity in microbial tissue (red-brown earth). (a) 100% recovery of activity remaining in soil; (b) <53 μm; (c) 53-124 μm; (d) 124-250 μm; (e) 422-500 μm; (f) 353-422 μm; (g) 250-353 μm.

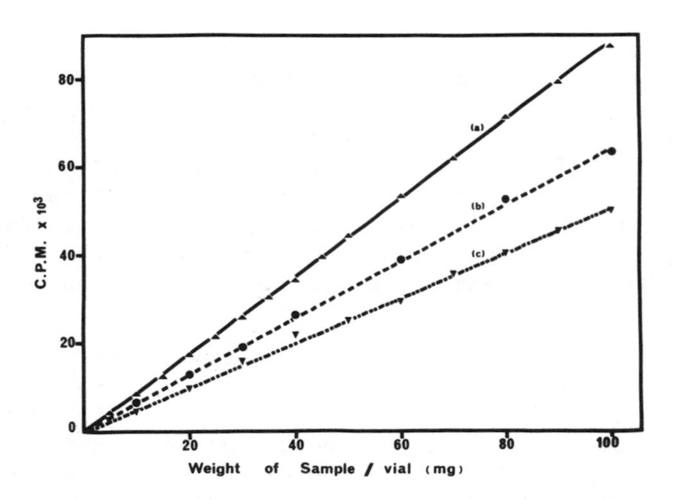

Fig. 4. Effect of particle size on counts (^{14}C/wheat straw/acid-washed sand mixture). (a) <53 μm; (b) 53-124 μm; (c) 124-250 μm.

High efficiencies were obtained for the larger particles because they were scattered sparingly in the suspension, so that when particles >53 μm were present the gel was not significantly coloured even with increasing weight of the sample. However, the efficiencies given using ^{14}C-benzoic acid as an internal standard were too high and resulted in poor recoveries of activity (Fig. 3). Only particles <53 μm yielded quantitative recoveries of activity and then only when the critical weight of about 25 mg sample per vial was not exceeded.

It was thought that the result of the effect of size might have been affected by segregation during sieving. The ^{14}C might have been concentrated in the clay fraction (<2 μm) resulting in high counts for the <53 μm samples. That this was not the case was shown when part of the 53-124 μm and 124-250 μm size fractions were ground to <53 μm and recounted. The recovery of activity increased and matched that of the original <53 μm fraction, i.e. the 60% recovery given by the larger particle-size fractions is real and not due to segregation of particles differing in activity during the sieving process.

The importance of particle size was also demonstrated with the plant material (Fig. 4). The activities for the same weight of sample per vial for 53-124 μm and 124-250 μm size fractions were 71 and 56% respectively of those given by the <53 μm fractions.

Self-absorption. The ratio of suspension counting efficiency to homogeneous internal standard efficiency has been denoted by a factor f, and this deviates from unity by an amount that depends only on self-absorption of the ^{14}C β-particles (8). The results in Table I explain why there were low recoveries from the larger particle sizes of the incubated soil (Fig. 3) and to some extent why quantitative recoveries of activities for the smallest particles could be obtained for <25 mg of the red-brown earth and <45 mg of the rendzina (Fig. 1 or 2). The table also shows that self-absorption increased with increasing weight of sample, while it decreased with particle size.

The low values of f obtained for particles >53 μm are evidence of the expected self-absorption of the weak β-spectrum of ^{14}C by the particles of the suspension because of their relatively large size compared with the range of

TABLE I. f^* VALUES FOR RED-BROWN EARTH AND RENDZINA

Weight in mg	Particle diameter (μm)						
	Red-brown earth						Rendzina
	422-500	353-422	250-353	124-250	53-124	<53	<53
10	0.64	0.46	0.42	0.59	0.65	0.98	1.01
20	0.47	0.36	0.43	0.55	0.60	0.98	1.02
40	0.44	0.38	0.37	0.51	0.57	0.92	1.20
50		0.38	0.35	0.52	0.55	0.84	0.84
60		0.36	0.35	0.51	0.57	0.89	0.93
80	0.31	0.33	0.33	0.53	0.51	0.71	0.93
100		0.33	0.33	0.50	0.57	0.71	0.83

$$^*f \text{ value} = \frac{\text{\% suspension counting efficiency}}{\text{\% homogeneous counting efficiency}}$$

^{14}C β-particles (8). Self-absorption in the rendzina was not as strong as in the red-brown earth. This could be part of the reason why efficiencies for this sample were higher than for the red-brown earth. The rendzina contains more clay than the red-brown earth, which would tend to make the density of particles lower.

It is quite clear that self-absorption is a significant factor unless particles of <53 μm diameter are used. This is close to the predicted critical size of particles as calculated from a layer of infinite thickness.

Effect of colour. Colour quenching is one of the serious problems in liquid scintillation counting. With soils and soil extracts it is a major problem and is mainly responsible for the low recoveries of activity and low efficiencies shown in Figs. 1 and 2.

This effect can be overcome in suspension counting up to a point by the use of internal standards to determine efficiencies. The cpm measured and corrected for these efficiencies will give 100% recoveries of activity until the critical amount of sample is exceeded when recoveries are not quantitative. The critical weights of sample for the rendzina, red-brown earth, goethite and haematite samples were 45, 25, 10 and <5 mg respectively.

Fig. 5 shows that these weights were really the weights of sample responsible for increasing the optical density of

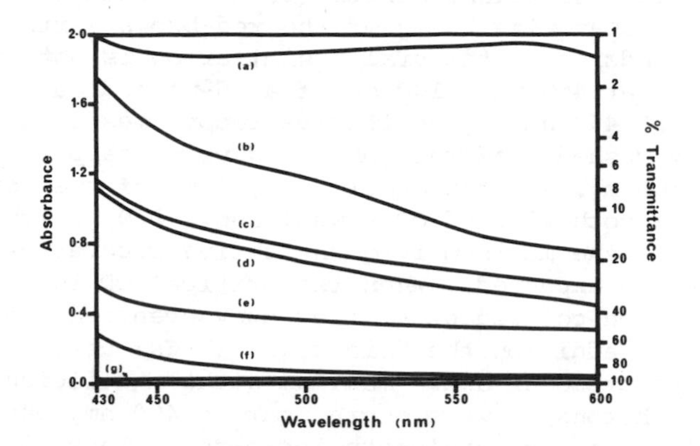

Fig. 5. Visible spectra for suspensions of ^{14}C-labelled samples in CAB-O-SIL-toluene-PPO-POPOP scintillant determined on a Unicam SP800 spectrophotometer in 1 cm cells. (a) 5 mg Haematite; (b) 5 mg Goethite; (c) 45 mg Rendzina; (d) 25 mg Red-brown earth; (e) 100 mg Acid-washed sand; (f) CAB-O-SIL alone; (g) Toluene-PPO-POPOP.

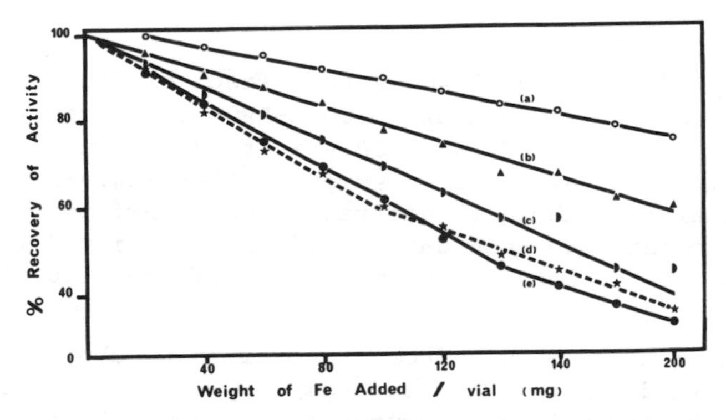

Fig. 6. Effect of iron in $Fe_2(SO_4)_3 \cdot 9H_2O$ on the recovery of activity. (a) Unground; (b) <250 µm; (c) <45 µm; (d) <250 µm; (e) <45 µm. a,b, and c with toluene-PPO-POPOP scintillant only; d and e with Triton X-100/scintillant mixture.

the CAB-O-SIL scintillant mixture to a critical level. The spectra show that with 25 mg of the red-brown earth and 45 mg of the rendzina in the vial a similar OD is obtained, e.g. ~OD 0.9 at 450 nm. 100 mg of acid-washed sand gave an OD of 0.42 at 450 nm. Quantitative recoveries of activity in the acid-washed sand were obtained using sample weights from 5 to 100 mg. On the other hand, 5 mg of the goethite or haematite both yielded OD's well above 0.9, which appears to be the maximum if quantitative recoveries of activities are expected. When the critical OD is exceeded, counting is non-coincident so that many events are not recorded. In addition the "bialkali" (K-Cs) used in the Tri-Carb spectrometer shows maximum quantum efficiency (25%) with photons of wavelength near to 400 nm, but this falls rapidly as the wavelength increases. Thus, increased absorbance in the 400 to 500 nm region eliminates the most efficient portion of the quantum efficiency curve. This means that problems of non-coincidence are compounded with quantum efficiency percentages of less than 10%.

To eliminate self-absorption it is necessary to reduce particles to <53 μm. This leads to lower counting efficiencies due to the increased colour or OD at 450 nm given to the gel by the dispersion of the finer particles. However, quantitative recoveries can be obtained using an internal standard until the optical density exceeds 0.9 at 450 nm.

Effect of iron. Fig. 6 shows the effect of iron on counting efficiency. The use of toluene-PPO-POPOP scintillant (0.5% w/v PPO and 0.03% w/v POPOP in toluene) gave a yellow colour with the basic ferric sulphate [$Fe_2(SO_4)_3 \cdot 9H_2O$)]. It was therefore thought that reduction in efficiency of counting might have been due to colour quenching. That this was not the case is shown in the different curves in Fig. 6.

Grinding to <250 μm reduced recoveries by about 15% of that obtained for the unground sample, while a second grinding to <45 μm again reduced the recovery by a further 13%. Thus, while the colour was kept constant with respect to the weight of sample used, the effect of iron was increased due to decrease in particle size. On the other hand, reduction in particle size has the effect of light scattering due to increase in total surface area (14).

This may account for the more drastic effect of Triton X-100 scintillant on counting efficiency than the toluene-PPO-POPOP scintillant. Though it maintained the greyish-white colour of the ferric sulphate, Triton X-100 dispersed the ferric sulphate better in the scintillant. The better dispersion increased the opacity of the gel and as the amount of the ferric sulphate was increased, scattering of the light emitted increased (14,15), resulting in lower counts.

With soil samples, however, this factor will not be of great importance unless samples very rich in iron oxides are studied, since the amount of iron required to reduce counts by 10% was about 20 mg of elemental iron per vial. Heavy metals such as iron are more likely to influence ^{14}C counting through colour quenching.

Examination of soil extracts and fractions. Various freeze-dried extracts and fractions of the red-brown earth incubated with [G-^{14}C]glucose were counted in suspension. The results (Table II) establish confidence in the procedure and its applicability to recalcitrant coloured materials. The 97.6% recovery of activity does not include some ^{14}C known to be present in the heavy liquid (1,2-dibromo-3-

TABLE II. DISTRIBUTION OF ^{14}C IN SOIL FRACTIONS AND EXTRACTS DETERMINED BY SUSPENSION COUNTING AND BY HOMOGENEOUS COUNTING USING INSTAGEL OR TRITON X-100

Fraction		% ^{14}C in sample	
		Suspension	Instagel/Triton X-100
Soil after incubation		100	
Light fraction		17.5	
Acid extract	>G-10	33.0	12.3
	<G-10	8.1	
Fulvic acid	>G-10	20.7	6.3
	<G-10	2.0	
Humic acid		3.4	
Acetylation extract			
Chloroform layer		2.8	
Aqueous layer		2.5	
Final residue		7.6	
Total recovery		97.6	

chloro-propane) used to float off the light fraction. Various attempts to count ^{14}C in this organic liquid indicated that at least a further 1% of the ^{14}C was present in compounds dissolved in the heavy liquid.

Attempts were also made to count ^{14}C in materials behaving as high molecular weight compounds on Sephadex G-10 from both the acid extract and fulvic acid extract. Aqueous aliquots of these samples were mixed with scintillant using either Instagel or Triton X-100. The percentage of ^{14}C in these fractions, determined using either of these emulsifiers (Table II), was about a third of the figure obtained by suspension counting.

The low recovery using the Instagel and the Triton X-100 could be explained by the fact that even though the system appeared to be homogeneous, it was virtually heterogeneous, due to the formation of microphases. The internal standards added to the system presumably existed in a different phase from the sample and were counted at considerably higher efficiencies than the samples. Suspension counting of the freeze-dried fractions must have yielded reliable quantitative results; otherwise the total recovery of ^{14}C in the various fractions would not have been close to 100%.

CONCLUSIONS

Liquid scintillation counting of ^{14}C in insoluble solid samples, or freeze-dried coloured extracts, in suspension is possible providing the samples are ground finely enough to minimize effects of β-particle absorption. The weight of finely ground sample which can be counted is limited by the colour it confers on the gel. The optical density (1 cm) of the gel at 450 nm must be less than 0.9; otherwise quantitative recoveries of activity will not be obtained.

ACKNOWLEDGEMENTS

The authors wish to thank Dr. P.E. Stanley of this department for his expert advice and encouragement. One of us (J.K.A.) is supported by an Australian Government Scholarship under the Special Commonwealth African Assistance Plan. The project was in part financed by a grant from the Australian Research Grants Committee.

REFERENCES

1. D.S. Jenkinson, J. Soil Sci. 16, 104 (1965)
2. H.E. Oberländer and K. Roth *in* Isotopes and Radiation in Soil Organic Matter Studies, pp. 251-261. Vienna : IAEA (1968)
3. E.C.S. Little *in* The Use of Isotopes in Soil Organic Matter Studies, pp. 371-374. (Rep. FAO/IAEA Technical Meeting, Brunswick-Völkenrode, 1963). Oxford : Pergamon Press (1966)
4. V.K. Mutatkar and G.H. Wagner, Soil Sci. Soc. Am. Proc. 31, 66 (1967)
5. B.L. Funt, Nucleonics 14(8), 83 (1956)
6. C.G. White and S. Helf, Nucleonics 14(10), 83 (1956)
7. D.G. Ott, C.R. Richmond, T.T. Trujillo and H. Foreman, Nucleonics 17(9), 106 (1959)
8. F.N. Hayes, B.S. Rogers and W.H. Langham, Nucleonics 14(3), 48 (1956)
9. M.V. Cheshire, H. Shepherd, A.H. Knight and C.M. Mundie, J. Soil Sci. 23, 420 (1972)
10. J.C. Turner, Int. J. Applied Rad. and Isotopes 19, 557 (1968)
11. J.M. Oades, 10th Int. Conf. in Soil Sci., Moscow, USSR (1974) - In press
12. C.P. Lloyd-Jones, Analyst 95, 366 (1970)
13. A.C. Angel, Anal. Chem. 40, 463 (1968)
14. H.J. Cluley, Analyst 87, 170 (1962)
15. S. Helf and C.G. White, Anal. Chem. 29, 13 (1957)

TEFLON VIALS FOR LIQUID SCINTILLATION COUNTING OF CARBON-14 SAMPLES

by

G.E. CALF* and H.A. POLACH+

* Australian Atomic Energy Commission, Lucas Heights, N.S.W., 2232.
+ Australian National University, Canberra, A.C.T. 2600.

Abstract

The design of Teflon counting vials for use in liquid scintillation counting of ^{14}C samples is discussed. When compared with the more commonly used glass vials, Teflon vials give a slight increase in the ^{14}C detection efficiency (E) and a significant reduction in the background count rate (B) resulting in an increase of about 30% in the Figure of Merit (E^2/B).

This increase in performance, when applied to radiocarbon dating, is equivalent to extending the maximum determinable age of a sample by about 2,000 years.

Introduction

Teflon vials of 20 ml capacity give a much higher Figure of Merit (Efficiency^2/Background; E^2/B) for tritium counting than commercially available counting vials (1,2). As the use of vial material with a high Figure of Merit is particularly important in low level counting, Teflon vials have been designed, tested and applied to radiocarbon dating, using commercially available liquid scintillation spectrometers.

Carbon-14 Counting

The results of an investigation of ^{14}C counting efficiency of various commercially available 20 ml counting vials and of a 20 ml Teflon vial are shown in Table I. In this study, ^{14}C labelled hexadecane was used and count rates were measured in a Packard 3375 Liquid Scintillation Spectrometer operating at +8°C. Discriminator settings corresponding to a ^{14}C efficiency of about 87% were selected, and remained set for the duration of the experiments. Optimal E^2/B counting ratios were obtained by adjustment of the amplifier gain.

TABLE I

COMPARISON OF ^{14}C COUNTING EFFICIENCY AND E^2/B OF COMMERCIALLY AVAILABLE AND TEFLON COUNTING VIALS

Vial(*)	Efficiency (E) %	Background (B) counts min^{-1}	Figure of Merit (E^2/B)
Low ^{40}K glass	86.3 ± .2	28.7 ± .1	260
Quartz	87.1 ± .2	28.1 ± .1	270
Polyethylene	87.5 ± .2	27.1 ± .1	280
Nylon	87.7 ± .2	25.5 ± .1	300
Teflon	88.0 ± .2	22.5 ± .1	340

*) Each vial contained 1.4×10^{-2} µCi of n-Hexadecane-1-^{14}C and .08 g PPO in 18 ml of O-Xylene (L.R. grade). O-Xylene with the same concentration of PPO was used for the background determinations.

The differences in E^2/B between low ^{40}K Glass, Quartz and Polythene vials are not very significant. Polythene vials (even the high density grade used for this experiment) absorb aromatic hydrocarbons and distort on prolonged use and hence are not suitable for low-level counting experiments. Only the Nylon and Teflon vials show a significant increase in the Figure of Merit which is due to increased ^{14}C counting efficiency and a significant decrease in background. Further tests on Teflon, which had the largest Figure of Merit, showed that it is not permeable to hydrocarbons, and that vials can be re-used if thoroughly washed in benzene and dried under vacuum at +50°C for at least 2 hours. Under these conditions, no memory effect (solvent retention and contamination of subsequent sample) has been observed when alternating ^{14}C labelled and background samples, a factor of particular importance to high precision low-level counting such as radiocarbon dating.

Radiocarbon dating

The organic sample, whose radiocarbon age is to be determined by the liquid scintillation counting method, is generally first converted to carbon dioxide which is further synthesised to acetylene (3,4). The acetylene is then catalytically trimerized to benzene (5). The ^{14}C activity of the sample benzene is then related to the ^{14}C activity of a modern reference standard. The benzene for the sample and standard is prepared following the same synthesis procedures, and both activities are measured on the same equipment, using the same settings and the same counting vials for which the background has been predetermined. For reasons of sample size available for dating, ease of handling of synthesis procedures and taking into consideration the required precision of the ^{14}C age determinations, most workers using the liquid scintillation counting technique have chosen to work with 3 g to 5 g equivalents of elemental carbon, that is 3.7 ml to 6.2 ml of benzene. For this purpose a modified 5 ml volume glass vial is the most efficient (6,7,8).

Design of Teflon vials

Two Teflon vials (Fig 1, A and B) both of 5 ml capacity have been designed and their performances evaluated.

Both are equally suitable for radiocarbon age determinations. Teflon (DuPont PTPF rod SG 2.1-2.3) was used in their manufacture, particular care being taken to obtain a uniform side wall thickness of 0.95 $\pm$ 0.05 mm. The width of both designs (27.5 mm) conforms to standard 20 ml glass vial; the height of design A has been adjusted to place the sample in the centre of photomultiplier tube axis. However as this was not found to contribute to a greater detection efficiency, the overall height of design B is that of a 20 ml standard glass vial and the sample volume is placed at the bottom of the vial to allow the maximum thickness of aluminium shielding cap. Both vials have the same ^{14}C detection efficiency and background in fully (top) shielded liquid scintillation counters.

In Design A (Fig 1), a 20 ml glass counting vial plastic screw cap lined with cork and tin foil was used to seal the vial. The Teflon vial is mounted on an aluminium base by a small lug. It is particularly suitable for bottom loading and top shielded refrigerated liquid scintillation counting systems. At +8°C the solvent loss through the screw cap seal is in the order of 2 mg of benzene per week. For this loss, if so desired, an appropriate count rate correction can be made.

Design B (Fig 1) was developed specifically for ambient temperature top loading counting systems. The Teflon vial, on a thickened base, is sealed by a Viton 'O' ring which is held in an aluminium screw cap, which also acts as additional shielding. The solvent loss at +18°C is about 0.5 mg of benzene per week.

Performance of Teflon vials

A study comparing the efficiency of 5 ml low ^{40}K glass and the Teflon vials described above indicates that the 5 ml Teflon vials have distinct advantages.

A significant reduction in background (10 to 40%) resulted from the use of Teflon vials for all instruments tested (Table II), due to elimination of ^{40}K induced count rate, reduction of Cerenkov radiation and due to the light scattering properties of Teflon which probably bring about a reduction of the cross talk between the opposed photomultiplier tubes.

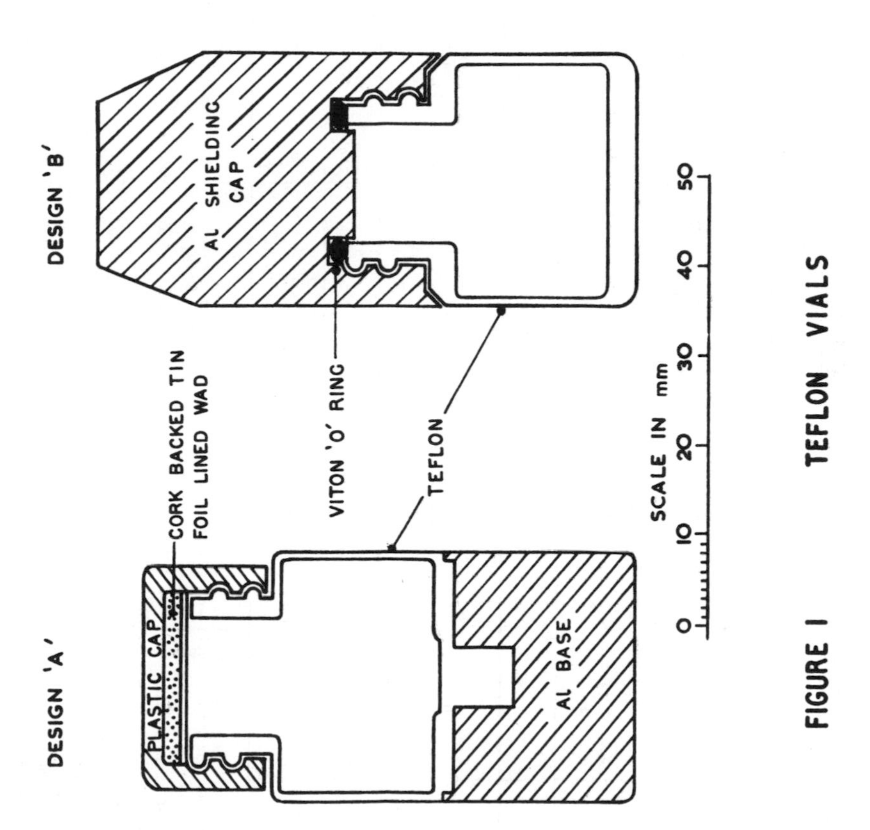

FIGURE I TEFLON VIALS

TABLE II

COMPARISON OF ^{14}C COUNTING EFFICIENCY, F, E^2/B OF SPECIAL DESIGN 5 ml LOW ^{40}K GLASS AND TEFLON VIALS

INSTRUMENT	Sample (a) grams Benzene	5 ml GLASS VIAL					5 ml TEFLON VIAL						COMMENT ON L.S. SYSTEM
		No (b) counts min^{-1}	B (c) counts min^{-1}	F (d)	% E(e)	E^2/B	No counts min^{-1}	B counts min^{-1}	F	% E	E^2/B	VIAL DESIGN	
PACKARD 3375	4.395	49.38	14.8	12.8	90.0	540	47.88	9.5	15.5	87.3	800	'A'	counter set-up for high ^{14}C efficiency and operating at 8°C.
PACKARD 3375	4.395	30.98	4.4	14.8	56.5	730	32.44	3.4	17.6	59.1	1030	'A'	counter set-up for max. E^2/B at minimum photomultiplier EHT setting.
AAEC COUNTER (f)	4.395	29.72	3.3	16.4	54.2	890	32.77	2.0	23.2	59.7	1780	'A'	counter optimised for max. E^2/B and operating at 8°C.
PICKER NUCLEAR LIQUEMAT 220	3.516	30.41	4.8	13.9	69.3	1000	31.07	3.9	15.7	70.8	1290	'A'	Ambient temperature operation (11)
BECKMAN LS-200	3.516	31.95	7.6	11.6	72.8	700	-	-	-	-	-	'B'	Ambient temperature operation
BECKMAN LS-200	3.516	30.77	5.5	13.1	70.1	890	32.66	5.1	14.5	74.4	1090	'B'	6 mm mask on photomultipliers
BECKMAN LS-200	4.395	Theoretical values only to demonstrate merit of full vial counting					40.83	5.1	18.1	74.4	1090	'B'	Not Tested

(a) 4.395g benzene = 5 ml benzene with scintillant PPO added in dry form.
3.516g benzene = 4 ml benzene with scintillant PPO and POPOP dissolved in 1 ml. toluene which is added to sample to make volume to 5 ml.

(b) No = 95% of observed count rate of NBS Oxalic acid contemporary standard in counts min^{-1}. (corrected to the year 1950 and $\delta^{13}C = -19^{o}/oo$ PBD).

(c) B = Background count rate in counts min^{-1} for 5 ml A.R. benzene plus scintillant or for 4 ml benzene plus 1 ml toluene containing scintillants.

(d) F = Radiocarbon dating system figure of merit $F = No/(B)^{\frac{1}{2}}$

(e) % E = Percent efficiency calculated using the data that 95% of the N.B.S. oxalic acid is equivalent to 13.53 ± 0.07 disintegration min^{-1} g^{-1} carbon (corrected to the year 1950 and $\delta^{13}C = -19^{o}/oo$ PBD). (9)

(f) Australian Atomic Energy Commission counter designed for ^{14}C low level counting (10).

Between 0.5 and 2 background counts min^{-1} (depending on the ^{14}C window width setting) can be directly attributed to external and cosmic radiation. All top loading instruments such as the Beckman, Packard and Picker will therefore benefit from the additional shielding being placed above the counting vial. Thus the aluminium cap of Teflon vial design 'B' (Fig 1) reduces the background count rate by about 0.5 counts min^{-1} for a 5 ml vial in top loading liquid scintillation spectrometers.

A further improvement is suggested for Radiocarbon Dating Laboratories. As the use of a 5 ml vial is established, it is more efficient to use all the available vial volume for sample counting and weigh in the required amount of dry scintillant. This replaces the previous commonly used method suggested by Tamers (7) where the scintillant is dissolved in 1 ml of toluene, which is added to 4 ml of sample benzene.

Procedures used to increase Figure of Merit of Commercial Counters

Extensive tests were carried out using two types of liquid scintillation counting systems:

(a) Packard 3375 which has a linear amplifier

and (b) Beckman LS-200 which has a logarithmic amplifier.

Both counters were operated with the lower discriminator set to cut off most of the tritium spectrum (at about 18 keV) which resulted in a loss of not more than 20% of the ^{14}C count and with the upper discriminator set at about 145 keV, towards the end of the ^{14}C spectrum, which gives a further loss of ^{14}C count of not more than 5%. Optimal count rates and balance point operation (8,12) were then achieved by adjusting the amplifier gain settings.

In the Packard 3375, it was found that the E^2/B can be increased by decreasing the photomultiplier high voltage to its minimum setting, resetting the discriminators, and compensating for the loss in gain in the photomultiplier tubes by increasing the amplification. This operation significantly reduces the background, reduces the ^{14}C counting efficiency by about 40% but increases the E^2/B by about 30%. (See Table II).

In the Beckman LS-200, it was found that cross-talk between the photomultiplier tubes is the main contributing factor to the background which remained constant (for equivalent discriminator settings) at various EHT settings. This cross-talk is believed to originate in the periphery of the photomultiplier tubes, and a 6 mm copper shim mask was fixed onto the outside edge, of each photomultiplier. This resulted in a reduction in ^{14}C background by about 30%, a slight reduction in ^{14}C counting efficiency and an increase in the E^2/B of about 25%. (See Table II).

Significance of Standard and Background Count Rates to Radiocarbon Dating

One of the most important factors in radiocarbon dating is the detection of count rates marginally above background from the background count rate. The maximum age that can be determined is governed by the minimum sample count rate that can be distinguished statistically from the mean background count rate.

Callow (13) proposed the 4 sigma detection criterion, and Polach (8) suggested the use of the 3 sigma detection criterion* for counting systems giving stable (statistically reproducible) background determinations.

The age of the sample is calculated from

$$\text{Age (in years)} = \frac{T\frac{1}{2}}{\ln 2} \ln (No/N_s)$$

$$= 8033 \ln (No/N_s) \qquad \ldots (1)$$

where $T\frac{1}{2}$ = the half life of ^{14}C = 5568 years

No = Modern References Standard corrected net count rate in counts min^{-1}

N_s = net sample count rate in counts min^{-1}

---

* The difference in net sample count rate (N_s) and background count rate (B) is not detectable if it is less than three times the square root of the combined squares of the standard deviations of the sample $\sigma^2(N_S)$, and background $\sigma^2(B)$ that is $N_s - B < 3[\sigma^2(N_s) + \sigma^2(B)]^{\frac{1}{2}}$.

If the maximum age corresponds to the lowest detectable sample count rate based on 3 σ criterion and using the same approach as Jansen (14) then

$$N_s \text{ minimum} = 3\left[\sigma^2(N_s) + \sigma^2(B)\right]^{\frac{1}{2}} \quad . \qquad \ldots (2)$$

For count rates close to background, where $N_s \approx B$, equation (2) approximates to

$$N_s \text{ minimum} = 3 \left[2\ B/t\right]^{\frac{1}{2}} \qquad \ldots (3)$$

where t = the counting time in minutes and B = background count rate in counts min^{-1}.

The maximum measurable age (A max) can therefore be expressed as

$$A \text{ max} = 8033 \left[No/N_s \text{ minimum}\right] \qquad \ldots (4)$$

$$= 8033 \ln (t/_{18})^{\frac{1}{2}} + 8033 \ln \left[No/(B)^{\frac{1}{2}}\right] . \qquad \ldots (5)$$

The first term in equation (5) is a constant for a given counting time, t minutes, and is independent of the counting system. It depends on the detection criterion (in this case taken to be 3 σ). The second term depends on counting efficiency, background and activity of the modern standard.

We introduce here a new concept which we call the Radiocarbon Dating System Figure of Merit, F ($F = No/(B)^{\frac{1}{2}}$), which defines the limitation of the counting system for determination of radiocarbon age. When the value of F is substituted into the second term of equation (5) it gives the number of years which can be added to (or subtracted from) the counting time dependent but counter independent first term.

If 4000 minutes is a reasonable maximum counting time, and substituting for t in the first term of equation (5) then

$$\left.\begin{array}{l} A\ max\ (t = 1000\ min) \simeq 16{,}100 \\ A\ max\ (t = 2000\ min) \simeq 18{,}900 \\ A\ max\ (t = 3000\ min) \simeq 20{,}550 \\ A\ max\ (t = 4000\ min) \simeq 21{,}700 \end{array}\right\} + 8033\ \ln\left[No/(B)^{\frac{1}{2}}\right].$$

Substitution of F values from Table II gives the theoretical maximum ages of an assumed 4000 minute counting period as listed in Table III.

TABLE III

MAXIMUM ^{14}C AGES, BASED ON A 4000 MIN COUNTING PERIOD FOR GLASS AND TEFLON COUNTING VIALS

System	Maximum theoretical ^{14}C age	
	5 ml glass vial	5 ml Teflon vial
BECKMAN (Lowest F)	42,400	43,200
AAEC (Highest F)	44,200	47,000

The increase in the Maximum Age Limit of 800 and 2,800 years is due to the improved performance of counting systems when Teflon vials are used.

Conclusion

The significant increase in Figure of Merit of Teflon vials used for ^{14}C counting as indicated in Table I and II makes the use of these vials particularly desirable for low level counting and offers an opportunity for existing liquid scintillation Radiocarbon Dating Laboratories to improve their performance without costly or time consuming modifications.

For commercial liquid scintillation spectrometers it is suggested that Vial Design 'B' be used because of its aluminium shielding cap. While more expensive shielding studies are indicated, they were beyond the scope of this paper.

Acknowledgements

We wish to thank Mr. B.W. Seatonbery and Mr. L.W. Smith AAEC and Mr. J. Gower A.N.U. for assistance with Teflon vial design, count rate efficiency and shielding experiments.

References

1. G.E. Calf, Int. J. Appl. Radiat. Isotopes. 20, 611 (1969)

2. G.E. Calf *in* Organic Scintillators and Liquid Scintillation Counting, p.719 (D.L. Horrocks & C. Peng Ed.) Academic Press (1971)

3. H. Barker, Nature 172, 631 (1953)

4. H.A. Polach and J.J. Stipp, Int. J. Appl. Radiat. Isotopes 18, 359 (1967)

5. J.E. Noakes, J.J. Stipp and D.W. Hood, Geochem. Cosmochim Acta. 27, 797 (1963)

6. L.L. McDowell and M.E. Ryan, Radiocarbon, 7, 174 (1965)

7. M.A. Tamers *in* Proc. VI Int. Conf. Radiocarbon and Tritium Dating, p.53. Pulman, Washington (1965)

8. H.A. Polach, Atomic Energy in Australia 12, 21 (1969)

9. International Commission on Radiation Units and Measurements. Measurement of Low-level Radioactivity ICRU Report 22, p.47 (1972)

10. P.E. Hartley and V.E. Church, A Low Background Liquid Scintillation Counter for ^{14}C, Int. Symp. on Liquid Scintillation Counting, Sydney, 1973.

11. F.J. Pearson, U.S. Geological Survey, Personal communication.

12. J.R. Arnold, Science 119, 155 (1954)

13. W.J. Callow, M.J. Baker and G.I. Hassel, Radiocarbon 7, 156 (1965)

14. H.S. Jansen in Institute of Nuclear Sciences Report No.13, New Zealand p.9 (1960)

Application of Glass Ampoules in the Assay of β-Radioactivity in Small Biological Samples

L.F. Sharry, C.A. Maxwell and A.M. Downes

C.S.I.R.O., Division of Animal Physiology, Ian Clunies Ross Research Laboratory, Prospect, N.S.W., Australia.

ABSTRACT

Some applications of sealed glass ampoules, in the 1-5 ml range, to the radioassay of ^{3}H, ^{14}C and ^{35}S in biological samples were examined. After sample preparation, the ampoules were placed in cylindrical Perspex holders, machined to hold the ampoules in a central position, and radioassayed. The results were compared with those obtained with similar samples prepared in conventional 20 ml glass screw-cap vials.

Counting efficiencies and E^2/B values, especially with ^{3}H, were higher in ampoules than in vials and samples could be measured with similar precision in the two systems. For monitoring quenching, the automatic external standard system could be applied satisfactorily with ampoules even though the ratios were lower than for vials (with unquenched samples: 0.85 for 20 ml vial; 0.42 for 1 ml ampoule).

For the assay of tritiated water (0.2 ml), a system using 2 ml ampoules and a toluene-scintillator - Triton X-100 mixture (balance point counting efficiency 25.6%) was superior to one using 20 ml vials and a toluene-scintillator-ethanol solution (efficiency, 13.3%).

Satisfactory procedures for the assay of β-radioactivity in sheep blood plasma, urine and faeces, and in competitive protein binding assays of steroid hormones have also been developed.

INTRODUCTION

The assay of β-radioactivity in small biological samples is usually carried out with screw-cap glass or polyethylene vials ranging in size from 20 ml to the more recently available 5 ml "mini vial". Small plastic bags have also been used successfully by Gupta (1). In our laboratory, sealed glass ampoules of 1 to 5 ml capacity have been used to assay samples containing [^{14}C]formic acid which, because of its volatility, slowly escapes from screw-cap vials (2). Small ampoules have other advantages, especially in reducing sample costs and problems associated with the storage and disposal of samples (3, 4).

Since samples in the 1-5 ml range are large enough for many other applications of liquid scintillation spectrometry, we have examined the feasibility of using ampoules for assaying ^{3}H, ^{14}C and ^{35}S in various types of biological samples commonly studied in this laboratory.

MATERIALS AND METHODS

Scintillation solutions.- Toluene containing p-terphenyl (3g/ℓ) and dimethylPOPOP (0.1g/ℓ) was the basic solution used in most experiments (Solution A). For aqueous samples Solution A was mixed with the emulsifying agent Triton X-100(5). Solution B comprised two parts of Solution A to one part of Triton X-100 and Solution C seven parts of Solution A to six parts of Triton X-100. Solutions D and E contained 2, 5-diphenyloxazole (PPO) (4g/ℓ) and dimethylPOPOP (0.1g/ℓ) in toluene plus ethanol (D, 45% v/v, E, 40% v/v).

Two reference liquid scintillation solutions were prepared by adding n-[1-^{14}C]hexadecane (1.1 ml; 0.88μCi) and n-[1,2-^{3}H]hexadecane (1.3 ml; 2.18μCi) respectively to Solution A (500 ml). These solutions were dispensed with an automatic Hamilton Precision Liquid Dispenser fitted with a 5 ml gas-tight adjustable syringe. The coefficient of variation due to the dispensing errors ranged from 0.02% (5 ml) to 0.09% (1 ml).

Sample preparation.- Glass ampoules (1, 2, 3 and 5 ml) and vials (20 ml screw-cap) were obtained from Australian Glass Manufacturers Ltd. (Sydney, Australia). After adding the sample and scintillation solution each ampoule was transferred to a revolving brass holder and sealed using a gas-oxygen hand torch. It should be emphasized that we perform this step in a well ventilated hood away from stock scintillation solutions and other flammable materials.

In preliminary tests the ampoules were placed in vials for counting and the effects of water coupling (6) and of positioning the ampoules in the vials were studied. In subsequent experiments each ampoule was placed in a cylindrical Perspex (polymethylmethacrylate) holder (5 cm long, 2.75 cm diameter) in a central well whose depth was such that the ampoules' mid-points were on the same horizontal plane. No water coupling was used with these holders.

In all experiments (except the preliminary one and the progesterone assay) groups of 10 samples were prepared for each comparison of ampoules and vials. For studies of samples in aqueous solutions tritiated water (0.27μCi/ml) and aqueous solutions of L-[^{3}H]leucine (lmCi/ml), L-[^{14}C]lysine (2μCi/ml) and L-[^{35}S]cystine (0.5μCi/ml) were used. ^{14}C-labelled plasma (7nCi/ml) was separated from blood collected four days after administering L-[^{14}C] cystine (lmCi) intravenously to a sheep. Plasma containing L-[^{3}H]leucine (1.6μCi/ml) was used to study the radioassay of tritium in plasma.

To test the applicability of ampoules for the assay of radioactivity in sheep urine, two specimens were selected, one lightly coloured and the other much darker, to represent extremes of colour. To 50 ml of each, L-[^{14}C] lysine or L-[^{3}H]leucine was added giving solutions with specific radioactivities of about 0.05μCi/ml and 0.15μCi/ml respectively.

To measure ^{35}S in faecal samples, L-[^{35}S]cystine (0.2 ml; 0.1μCi) was added to ampoules (5 ml) containing homogenised sheep faeces (50 mg). The samples were dried at 105°C (2 hr), and heated with 1 ml of an oxidizing solution (3 volumes conc. HNO_3 plus 1 volume 60% $HClO_4$ with 10g $Mg(NO_3)_2$ per 100 ml mixture) (7) in an aluminium block on an electric hot plate. The temperature of the block was raised to 260-280°C (during 1 hr) and the samples digested further until only a white crust remained. Then they were cooled, water (1.5 ml) added to dissolve the solid, Solution C (2 ml) added and the ampoules were sealed. To determine whether any ^{35}S was lost during oxidation, similar samples were prepared except that the L-[^{35}S] cystine (in 1.5 ml H_2O) was added after the oxidation step.

To test the application of ampoules for measuring steroids in plasma an assay method for progesterone in sheep plasma (8, 9) was chosen. Dog plasma containing [1,2-^{3}H]corticosterone was the competitive protein binding agent. The assay procedure involved the setting up of a binding curve covering the range 0 to 8 ng of progesterone. The labelled reagent was added to a tube containing the unlabelled progesterone. After incubation at 40°C for 15 minutes the proportion of label in the protein-bound fraction (1.5 ml) was collected from a Sephadex G25 fine column (30 x 8 mm diam.) in either a 20 ml vial or 5 ml ampoule containing 0.1 ml 5N HCl. Solution A (7.5 and 3.5 ml respectively) was added and after thoroughly shaking the samples and allowing them to settle, their counting rates were determined.

Radioassays.- A Model 3375 Packard Tri-Carb Liquid Scintillation Spectrometer, refrigerated at 3°C, was used. The optimum gain setting for balance-point operation (RED channel discriminators 50-1000) was selected for each series of samples using the technique of Neame and Homewood (10). The GREEN channel was operated in the integral mode (100% gain, 20-∞). In each series quenching was determined using the automatic external standard (AES) ratio.

The counting efficiency (%) was determined for the plasma, urine and faeces samples by several methods. For plasma a dried sample was combusted in an oxygen flask system (11). The counting efficiency of the solutions containing the combustion products was calculated after adding radioactive hexadecane as an internal standard ('spiking'). The plasma and urine counting efficiencies were also measured by spiking samples with [^{14}C] or [^{3}H]hexadecane or with tritiated water.

Some faeces samples with no added ^{35}S were oxidized (7) and spiked with [^{14}C]hexadecane. The error in using a ^{14}C spike instead of ^{35}S is small (12). From the curve relating AES ratio and counting efficiency the counting rates of the oxidized faecal samples containing ^{35}S were corrected for quenching.

RESULTS

Unquenched ^{14}C and ^{3}H Samples.- In a preliminary experiment 5 ml ampoules containing 5, 4, 3, 2 and 1 ml of Solution A and containing [^{14}C] or [^{3}H]hexadecane, 10µℓ/ml, were tested with the ampoules placed in 20 ml vials. Other ampoules (3, 2, 1 ml capacity containing 3, 2 and 1 ml respectively) were similarly prepared. The ampoules were compared with vials containing 20 ml of the radioactive solution. Since there were only small differences in balance point gain setting (maximum differences, 0.5% for ^{14}C and 5% for ^{3}H) a mean value (9% for ^{14}C, 55% for ^{3}H) was selected for all ampoules in this experiment. Vials were counted with 10% (^{14}C) or 55% (^{3}H) gain.

The counting efficiency was higher with the ampoules than the vials. For ^{14}C the improvement was about 2% and for ^{3}H about 10%. Water coupling increased the integral counting efficiency for ^{14}C by 0.6% and for ^{3}H by 2-3%. There was no change in counting efficiency with different volumes in 5 ml ampoules. The AES ratio decreased with a decrease in ampoule size. The variability in measuring the AES ratio was greater when the ampoules were not held in a central position.

The reproducibility of the results was studied more carefully with ampoules accurately positioned in Perspex holders. A series of each ampoule size and of the vials were prepared with the corresponding volumes of the standard solutions. Balance point gain settings were 9% for ^{14}C and 55% for ^{3}H. A minimum of 2×10^{5} counts was recorded for each measurement of counting rate.

The balance point and integral counting efficiencies for ^{14}C in ampoules were about 2% greater than the corresponding values for vials (Table 1). There was also an increase (about 0.5%) in efficiency with decreasing ampoule volume. The AES ratio, while decreasing from 0.849 for 20 ml vials to 0.421 for 1 ml ampoules was measured accurately enough (coefficients of variation 1.1 - 1.8%) in the smaller ampoules to enable quenching to be monitored.

The ^{3}H counting efficiencies showed larger changes (Table 1). The efficiencies were higher by 6% (5 ml) to 11% (1 ml) for the ampoules compared with vials.

The background counting rates (Table II) at the balance points were reduced from about 28 to 20 cpm by using ampoules instead of vials. However, the main contribution to the background counting rate was attributable to the ampoule holder and counting system, and not to the scintillation solution itself. When maximum counting rates are required integral counting can be used. With a 20 ml vial the background cpm was trebled and with a 1 ml ampoule doubled when compared with the balance point background values shown in Table II.

The effects of the increased counting efficiency for the low-volume ampoules (Table I) and the reduced background counting rates are reflected in the high E^2/B values, up to 377 for ^{14}C and 165 for ^{3}H, under balance point conditions (Table II). These values for both ^{14}C and ^{3}H in 1 ml ampoules were about 30% higher than E^2/B values for the sealed instrument standards supplied by the manufacturer.

Measurement of Tritiated Water.- The use of tritiated water to measure total body water is now a routine procedure (13). Two 'cocktails' (Solutions E and B) with 5 ml and 2 ml ampoules respectively were compared with the cocktail (Solution D) previously used in this laboratory.

The results (Table III) showed that the counting efficiency for tritium in the 2 ml ampoules was approximately twice the efficiency observed with the 20 ml vials and about 1.6 times the value observed with the 5 ml ampoules. All systems showed stable counting rates over a period of three weeks.

Radioassay of ^{14}C and ^{3}H in Sheep Plasma and Urine.- The system routinely used in this laboratory for assaying β-radioactivity in plasma or urine has been based on a 1 ml sample which is mixed with water (6.5 ml) in a vial to which Solution C (10 ml) is added. Besides forming a stable sample (gel) the dilution of the urine with water considerably reduces its colour quenching. Colour quenching is usually not a problem with plasma. However, a gel system is necessary otherwise the plasma proteins slowly precipitate from the cocktail. Since it is difficult to scale down these systems to the ampoule volume (5 ml) other combinations of the sample components were tested.

On the basis of these tests the following cocktails were adopted: for plasma, 1 ml sample, 1 ml water and 3 ml Solution C; and for urine, 0.5 ml sample plus 4.5 ml Solution B. A small amount of material slowly precipitated from the urine cocktail but this had little effect on the counting efficiency. The results (Table IV) show that higher counting efficiencies and a smaller effect of quenching were obtained for urine when the cocktail adopted for 5 ml ampoules was used instead of the one used for the 20 ml vials. The counting efficiencies with plasma were about the same in the two systems.

Radioassay of ^{35}S in Sheep Faeces.- The results from the digestion of 10 faeces samples containing ^{35}S showed that the counting efficiencies (based on [^{14}C] hexadecane) ranged from 62.5 to 66.8%. This comparatively wide range of values was due to variable quenching. When corrected for quenching the mean efficiency corresponding to the highest AES value (0.200) was 65.8($\pm$ 0.4 S.D.)%. On comparing samples in which the same amount of ^{35}S was added before and after performing the oxidation it was found that the mean recovery of added radioactivity was 101.9($\pm$ 0.9 S.D.)%.

Competitive Protein Binding Assay for Progesterone.- Separate binding curves were prepared for the vials and ampoules. The mean cpm of triplicate samples was graphed against 0, 0.25, 0.5, 1, 2, 4, 8 ng of progesterone. The counting rates obtained with the vials and ampoules were similar (ampoules 95.4($\pm$ 3.3 S.D.)% of vials) and gave substantially the same results (r = 0.9998) for equal masses of added progesterone. When the HCl was omitted, there was a larger difference between the counting rates obtained with the two systems, which differed only in the amount of scintillator added. The radioactivity measured in a sample of the aqueous phase indicated that a larger proportion (about 20%) of tritium remained in the samples which had not been treated with acid.

DISCUSSION

The above results show that a wide variety of biological samples containing β-radioactivity can be assayed as satisfactorily in sealed glass ampoules as in conventional screw-cap vials. Although the size of the sample is obviously restricted by the size of the ampoule, in our experience the procedures described for the assay of radioactive plasma, urine and faeces samples are sensitive enough for most tracer experiments, even with animals as large as sheep. Any loss in sensitivity due to using smaller samples is counteracted, at least to some extent, by the higher counting efficiencies, lower background counting rates, and smaller effects of quenching in ampoules.

The AES ratio was lower for the smaller samples. In the spectrometer used the ratio of the upper portion of the β-spectrum due to Compton electrons produced by the γ-source is compared with the whole spectrum. With the smaller sample volumes a greater proportion of the more energetic γ-rays pass through the scintillation solution without producing electrons. This presumably explains the reduction in the AES ratio. The precision in measuring the AES ratio for 20 ml vials has been studied by Stanley (14) who showed that the major error is caused by irregular vial geometry and vial positioning within the detector. When the same vial was assayed repeatedly with the sample changer operating in the normal cycling mode Stanley found that the coefficient of variation in the measurement of the AES ratio was 0.7%. This may be compared with the value of 1.1% which we observed with a set of ten 20 ml vials.

For many experiments the solubilization or oxidation of the sample can be conveniently performed in the ampoule before adding the scintillation solution. Losses of radioactive material during digestion are less likely with ampoules than with vials. The use of sealed ampoules eliminates losses of solvent or sample on storage or on mixing samples such as those containing Triton X-100 which have to be shaken vigorously to form a gel. With some batches of screw-cap vials in which poorly fitting caps were supplied, such losses have caused considerable inconvenience in this laboratory. The use of ampoules also eliminates another potential source of error - cap luminescence - which is sometimes possible when screw-cap vials are used (15). Another advantage in using ampoules is that toluene vapour does not accumulate in the sample changer of the scintillation counter. Ampoules are easier to store and considerably reduce the problems in disposing of large volumes of scintillation solution.

In analytical applications of liquid scintillation spectrometry, such as radioimmunoassays and allied techniques, much larger numbers of samples are envisaged in future research. For such testing, ampoules would be useful because they can be filled and sealed by fully automatic and readily available machines. With the manual technique of sealing ampoules which we have used, experienced operators can easily prepare and seal more than 100 per hour. We have sealed many thousands of ampoules by the method described, without experiencing any problems, but again we emphasize that appropriate precautions should be taken to avoid a fire.

A difficulty arises when it is necessary to spike samples in sealed ampoules after their counting rates have determined. It is possible to open and re-seal ampoules after adding a spike, but this is not as easy as removing and replacing a screw-cap. An alternative is to prepare duplicates of samples covering the range of quenching being studied and to add the spike before the initial sealing.

Improvements could be made in the counting efficiencies which we described, by using better scintillators and solubilizers. However, our aim has been to reduce costs as much as possible in a laboratory where large numbers of samples (currently about 100,000 per year) are assayed. Ampoules containing 5 ml scintillation solution cost about a fifth as much as the cheapest available screw-cap vial containing 20 ml solution.

ACKNOWLEDGEMENTS

We wish to thank Dr. C.D. Nancarrow and Mr. P.J. Connell for their helpful advice and assistance with the progesterone assay, and Mr. W. Ward for his technical assistance.

REFERENCES

1. G.N. Gupta In Organic Scintillators and Liquid Scintillation Counting, p. 747 (D.L. Horrocks and C. Peng, Eds). New York and London: Academic Press (1971).
2. A.M. Downes In ibid p. 1031.
3. P.S. Rummerfield and I.H. Goldman, Intern. J. Appl. Rad. and Isotopes 23, 353 (1972).
4. G.M. Connell and J.A. Linfoot, Intern. J. Appl. Rad. and Isotopes 24, 239 (1973).
5. M.S. Patterson and R.C. Greene, Anal. Chem. 37, 854 (1965).
6. B.E. Gordon and R.M. Curtis, Anal. Chem. 40, 1486 (1968).
7. H. Jeffay, F.O. Olubajo and W.A. Jewell, Anal. Chem. 32, 306 (1960).
8. J.M. Bassett and N.T. Hinks, J. Endocr. 44, 387 (1969).
9. G.D. Thorburn, J.M. Bassett and I.D. Smith, J. Endocr. 45, 459 (1969).
10. K.D. Neame and C.A. Homewood, Anal. Biochem. 49, 511 (1972).
11. F. Kalberer and J. Rutschmann, Helv. Chim. Acta, 44, 1956 (1961).
12. J.P. Buckley, Intern. J. Appl. Rad. and Isotopes 22, 41 (1971).
13. T.W. Searle, J. agric. Sci. Camb. 74, 357 (1970).

14. P.E. Stanley In Liquid Scintillation Counting, Vol. 2, p. 285 (M. Crook, P. Johnson and B. Scales, Eds.). London, New York, Rheine: Heyden and Son Ltd. (1972).
15. B. Scales In ibid. p. 101.

TABLE I. Liquid scintillation counting of [^{14}C] and [^{3}H]hexadecane in vials and ampoules.

Volume	AES Ratio	Counting Efficiency (%)			
		Balance Point		Integral	
		^{14}C	^{3}H	^{14}C	^{3}H
20 ml*	0.849(1.1)	84.0	42.5	91.8	43.7
5 ml†	0.762(1.1)	85.5	48.4	93.6	49.9
3 ml	0.666(1.2)	85.6	49.9	93.6	51.5
2 ml	0.581(1.1)	85.9	50.3	93.8	51.9
1 ml	0.421(1.8)	86.0	53.2	94.1	55.1

The sample containers were either 20 ml vials (*) or 5, 3, 2 or 1 ml ampoules (†).

The balance point settings were for ^{14}C, 9%, 50-1000 and ^{3}H, 55%, 50-1000.

The results are means for 10 samples with 2 x 10^5 counts recorded for each. The coefficients of variation of the counting efficiency ranged from 0.2 to 0.4 (^{14}C) and 0.8 to 1.8 (^{3}H). The values in brackets are the coefficients of variation for the AES ratio.

TABLE II. Background counting rates and figures of merit (E^2/B) for unquenched ^{14}C and ^{3}H samples at balance point.

	Carbon - 14		Tritium	
	Background cpm	E^2/B	Background cpm	E^2/B
20 ml*	28(17)	257	29(26)	63
5 ml†	20(20)	360	24(21)	100
3 ml	20(20)	362	20(19)	126
2 ml	20(20)	372	19(18)	138
1 ml	20(19)	377	17(16)	165
Standard‡	24	303	20	120

The sample containers were 20 ml vials (*), 5, 3, 2, 1 ml ampoules (†) and the instrument sealed standard (‡).

The balance point settings were for ^{14}C, 9%, 50-1000 and ^{3}H, 55%, 50-1000.

The total number of counts was 10^3 (for each of 10 samples). Values in brackets were the counting rates when vials or holders plus ampoules without added scintillator were used. The counting rates without holders were 16 and 6 cpm respectively for ^{14}C and ^{3}H.

TABLE III. Radioassay of tritiated water.

Volume	Solution	AES Ratio	Counting Efficiency (%)	
			Balance point	Integral
20 ml*	D	0.441(1.3)	13.2(2.2)	14.2(2.1)
5 ml†	E	0.399(3.3)	16.1(4.4)	17.1(4.2)
2 ml†	B	0.372(3.6)	25.6(3.4)	26.7(3.2)

The sample containers were 20 ml vials (*), with 10 ml of Solution D and 5 and 2 ml ampoules (†) with 5 ml of Solution E and 2 ml of Solution B respectively.

The balance point setting was 80%, 50-1000.

The results are means for 10 samples with 2 x 10^4 counts recorded for each. Coefficients of variation are shown in brackets.

TABLE IV. Radioassay of plasma and urine samples.

Sample	Volume	Carbon - 14		Tritium	
		E‡	AES Ratio	E	AES Ratio
Plasma	20 ml*	67.3	0.322(2.8)	5.8	0.271(1.1)
	5 ml†	67.5	0.252(1.1)	5.7	0.198(5.7)
Urine-1	20 ml	56.6	0.228(1.9)	3.7	0.211(4.2)
	5 ml	70.3	0.358(3.4)	13.2	0.371(3.0)
Urine-2	20 ml	40.7	0.086(3.6)	1.6	0.089(4.0)
	5 ml	64.2	0.242(4.9)	9.2	0.246(4.8)

The sample containers were 20 ml vials (*) with 1 ml of plasma or urine, 6.5 ml water and 10 ml Solution C, 5 ml ampoules (†) with 1 ml plasma, 1 ml water and 3 ml Solution C, or 0.5 ml urine and 4.5 ml Solution B.

Urine-1, -2 were the light and dark coloured samples respectively.

‡Counting efficiency (E) was determined at balance point settings for ^{14}C of 40%, 50-1000 and ^{3}H, 100%, 50-1000.

The results are means for 10 samples with 2 x 10^5 counts recorded for each. Coefficients of variation are shown in brackets.

A DIRECT TECHNIQUE FOR COUNTING ^{14}C AND ^{3}H IN TISSUES

S. Apelgot, R. Chemama and M. Frilley

Fondation Curie-Institut du Radium
11, Rue P.et M.Curie et 26, Rue d'Ulm. 75005, PARIS

ABSTRACT - The technique to be described uses liquid scintillation counting to measure ^{14}C and ^{3}H present in tissues without the necessity of destroying the tissue by digestion or combustion. It is based on the property of cellular membranes to be permeated with liquid scintillators, and the extraction of certain cellular components. The use of the enzyme pronase makes it possible to measure the total activity in each sample. By using liquid scintillation which may or may not contain dioxane, it is possible to easily differentiate the fraction of radioactivity within the water soluble and lipid-soluble components. Preliminary calibration curves and the study of pulse height spectra make it possible to define in each case the best conditions for measurement. The calibration curves are reproducible and it is not necessary to estimate the counting efficiency in each case. This rapid technique gives results with high efficiency.

INTRODUCTION - The technique of direct counting, which we have perfected is based on the results obtained by one of us (1) in 1961 and which showed that: A) Microbial cells can be permeated with liquid scintillators; B) The radioactivity (^{3}H or ^{14}C) contained in micro-organisms can be correctly measured without the necessity of destroying them. When demonstrating that it was possible, without any preliminary treatment, to measure the radioactivity of body fluids (urine, whole blood, plasma), we noticed that liquid scintillators extracted various compounds from biological media (2). We then attempted to extend this direct technique to tissues.

MATERIALS

I - Apparatus: a) For radioactivity measurements : automatic liquid scintillation spectrometer from Nuclear Enterprises (G.B.) with a single photomultiplier;
b) ultrasonic generator from Ultrasons (Annemasse, France).

II - Liquid scintillators: NE 220 (Nuclear Enterprises), containing dioxane, which can take up 10% water.

III - Glass fibre papers: Whatman GF/A (refered to as fibre papers in the text).

IV - Standard solutions of ^3H or ^{14}C

V - Labelled experiments: Male or female mice of various strains fed with various labelled compounds; tritiated water; thymidine labelled with ^3H or ^{14}C; pregnenolone labelled with ^3H or ^{14}C.

VI - Enzyme: Pronase (Calbiochem activity 45,000 Proteolytic units/g) as a freshly prepared 1% solution in buffer TRIS-HCl 5.10^{-3}M, EDTA 2.10^{-2}M, pH 7.2.

VII - Hyamine Hydroxide 10-X (Packard, USA)

DEVELOPMENT OF THE TECHNIQUE

At the time of dissection of the control or labelled mice, the samples were laid on fibre paper put into a stoppered vial and weighed. The radioactivity was then counted by the technique to be described or by a classical one (solubilization or combustion). We have worked only with a few organs: liver, mammary gland, muscle, brain, fat, ovaries, and adrenal gland.

I - Principle of the technique

The first attempts showed it was possible to measure the total radioactivity of some samples, by simply immersing them into the liquid scintillator NE 220. In these cases, the weighed samples were immersed in the liquid scintillator with their fibre paper (to avoid losses). When the whole radioactivity was measured in the untreated sample, we observed that the radioactivity was extracted by NE 220. Animal cells, similary to bacterial ones, are therefore permeated with liquid scintillator.

We assumed that the whole radioactivity might be measured in all cases if, by modifying the tissue and cell structures, the energy transfer became possible between the radioisotopes and the liquid scintillator. This assumption was proved correct by experiments in which the samples were treated with the enzyme, pronase.

Our first experiments showed that pronase did not act on an intact sample at room temperature. We therefore adopted the following procedure: after weighing, the sample and its fibre paper were laid on another fibre paper, covered with two others and crushed by rolling with the base of the cylinder and then with a glass stirrer. The four fibre papers surrounding the sample absorbed its fluids and the whole "packet", on which 0.15 ml of pronase solution was deposited was put into a vial. The vial was carefully closed and then left in an incubator at 37°C for three hr for liver, muscle, brain and fat or seven hr for other tissues or organs studied. The samples were then treated by ultrasound for 15 mins at room temperature before being immersed in the liquid scintillator. Working with tritiated water, we checked that losses were negligible, provided the vials were air-tight.

The samples with and without pronase treatment * were then measured. In order to determine in each case the counting efficiency, calibration experiments were carried out.

II - Calibrations

20 µl of standard solutions (^{14}C or ^{3}H) were put into vials containing the liquid scintillator. After counting the activity, we added the samples of organs taken from non-treated mice: these were of various weights and with or without pronase treatment. Preliminary tests enabled us to choose the optimum volume of NE 220 in each case. One ml was used for samples counted intact, and 3 ml for those treated with pronase.

---

*) experiments showed that, when the total radioactivity could not be measured on intact samples the action of pronase was indispensable; in fact, the measurement of samples only crushed, never gave the total activity (3).

The study of the pulse height spectra of NE 220 containing organ pieces either labelled in-vivo, or simply immersed in the presence of ^{3}H or ^{14}C, showed that calibrations had to be made under the conditions we adopted. The simple presence of organs alters the pulse height spectra; this alteration depends only slightly on the nature and weight of the organ and has the same order of magnitude with all samples, whether labelled in-vivo or measured in the presence of radioactivity. The spectra are displaced towards low energies, principally in the case of ^{14}C. Similar results had been obtained with whole blood (2). This led us to measure ^{14}C in the presence of organs with the P setting previously established for the counting of ^{14}C in whole blood; this setting corresponds to the same gain and threshold which was used for ^{3}H, but without an upper discriminator, for high energy pulses remain in such ^{14}C spectra. As observed in the case of whole blood, the modifications of spectra in the presence of organs remain unchanged when the latter are removed from the liquid scintillator (Fig.1). Therefore, the modifications in spectra are due to organic substances extracted by the liquid scintillator. The efficiences of measurement with this calibration technique should give satisfactory results in the case of organs labelled in-vivo.

It was experimentally shown that, when a piece of organ, is immersed into NE 220 in the presence of ^{3}H or ^{14}C, the counted activity decreases with time. It reaches a steady value in a few hours and then remains stable for several days. This drop of activity depends on the nature and weight of the organ and on the added isotope. We think that the decrease of the counted activity is associated with the extraction of various compounds by NE 220 which modify the pulse height spectra. Therefore, the activity measurement must be made at equilibrium, that is 3 to 4 hr after putting the sample into the liquid scintillation solution.

On the contrary, when the sample is treated with pronase, the counted activity most often increases before reaching a constant value 4 to 6 hr later. It was shown experimentally that pronase diminishes the transparency of fibre paper to photons; it recovers slowly, hence, the increase in the counted activity, as time passes. If the sample

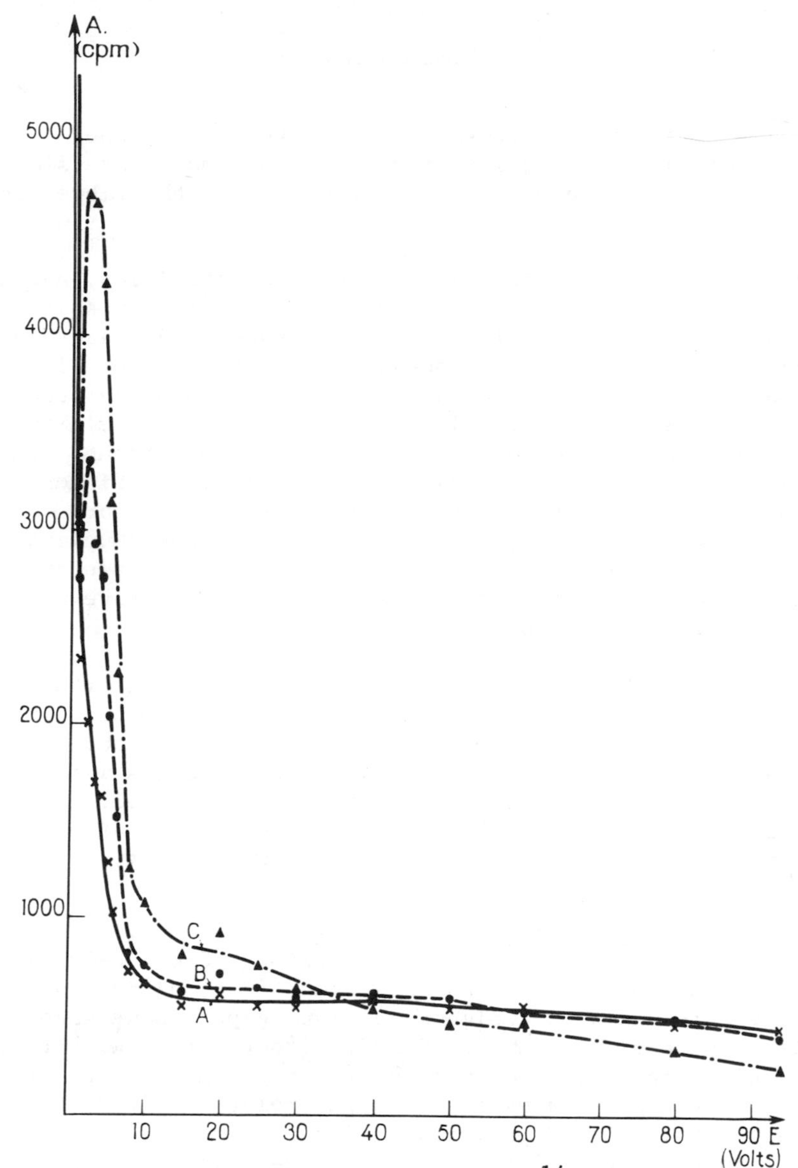

Fig.1 Pulse height spectrum in NE 220, ^{14}C samples (calibrations; P.setting).

The setting of the gain corresponded to that of ^{3}H. ^{14}C standard (48.800 dpm) was added to the following samples: Curve A) Four fibre papers were immersed for 19 hours in 2ml of NE 220, then removed; ^{14}C standard was then added; Curve B) 19,8 mg of spleen crushed between four fibre papers were immersed for 19 hours in scintillator liquid and removed; ^{14}C standard was then added; Curve C) 19,5 mg of spleen crushed between four fibre papers were immersed and left in NE220; ^{14}C standard was then added.

A : counted activity in cpm.

are not treated by ultrasound when taken out of the incubator at 37°C, equilibrium is reached much more slowly. Ultrasound therefore speeds the recovery of the fibre paper transparency.

In this study, we expressed the counted activities, not as counting efficiencies E, but as relative efficiences (or counting yields) R, calculated with respect to measurements made on vials containing the standard radioactive solution, before introducing the organ samples. In this way, the characteristics of the scintillation counter are eliminated and the part played by the samples, is made evident. This study showed (Fig. 2 and 3) that the relative efficiency depends on the state of the sample (taken intact or treated with pronase); it decreases when sample weight increases. This drop is more important in the case of ^{3}H than ^{14}C measured with the P setting. The relative efficiency also depends on the volume of NE 220.

The reproducibility of the calibration curves was shown experimentally, first in the course of time, second with various samples of the same organ taken from various strains of mice. The fluctuation of the results was within 10%. We did not notice any modification of results with various stocks of NE 220. Therefore, the calibration curves (Fig. 2 and 3) can be used systematically.

III - Counting of labelled samples

In the case of mice treated with ^{14}C or ^{3}H-labelled compounds, we have used the experimental conditions defined above. The samples. weighed on fibre paper in weighing flasks were placed in the NE 220, either intact with the weighing paper or treated with pronase. In the case of ^{14}C the P setting was always used. The relative efficiency, R, was determined from previous calibration curves (Fig. 2 and 3); knowing the efficiency (E) of the scintillation counter, we calculated the efficiency (e) of the measurement (e = E x R) which permitted us to express the activity of the sample as dpm/mg.

The experiments showed that for samples taken from animals having received ^{3}H or ^{14}C-labelled compounds, the measured activity, most often increases with time, reaching a limit

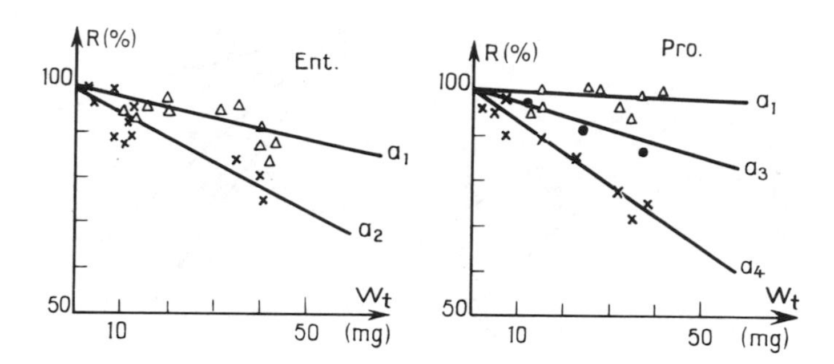

Fig.2 The effect of organs on the activity measurement with ^{14}C (P Setting)

R represented the ratio of the activity counted at equilibrium in the presence of sample, to that counted before the sample was immersed into the scintillator liquid. Measurements made with the samples intact (1 ml NE 220) or treated with pronase (3 ml NE 220), respectively, the curves Ent. and Pro. Curves index 1 : brain, fat, mammary glands and muscle; index 2 : liver spleen, ovaries and adrenal glands; index 3 : liver; index 4 : spleen, ovaries and adrenal glands. Each point represented the mean value of 2 to 4 assays having about the same weight. In the case of C, the effect of the fibre papers and that of pronase on the relative efficiency R is negilible : the curves are extrapolated to R = 100%.

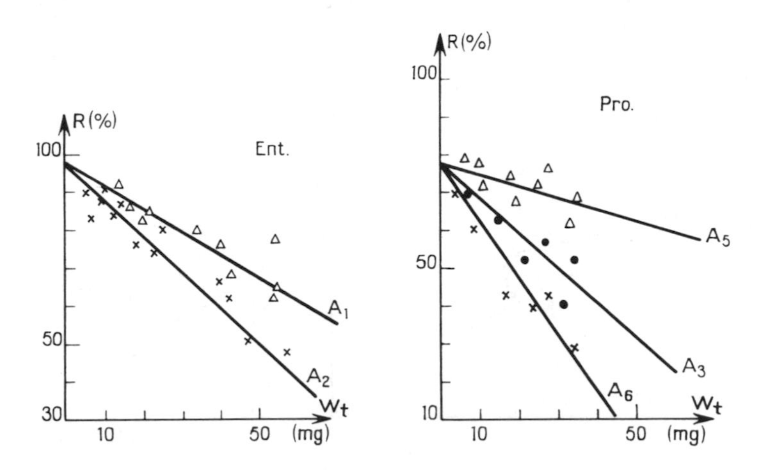

Fig.3 The effects of organs on activity measurements with ^{3}H

R has the same meaning as figure 2.
The measurements made with samples intact (1 ml NE 220) or treated with pronase (3 ml NE 220) respectively, the curves Ent. and Pro. Curves index 1 : brain, fat, mammary glands and muscle; index 2 : spleen, liver, ovaries and adrenal glands; index 3 : liver; index 5 : brain, fat, mammary glands, muscle and avaries; index 6 : spleen and adrenal glands. In the case of ^{3}H, the effect on the relative efficiency R of the fibre papers and pronase is no longer negligible and corresponds to the extrapolation to the origin of the curves.

which remains stable for a few days. The increase in activity and the time necessary to reach this plateau depend on the nature of the radioactive compound and on the state of sample (intact, or treated with pronase); this increase can be quite significant (up to five times the initial activity). Equilibrium is reached five to seven hr after the sample was placed in the liquid scintillator. This kinetics corresponds to the extraction of organic substances by the liquid scintillator, some of which are labelled and alter the pulse height spectra. The extraction of these compounds involves both the increase of counted activity and the decrease in relative efficiency. The experiments also show the preponderance of the former phenomenon, since the measured activity increases. This extraction is confirmed by the following facts: if, after measurement, we remove the radioactive sample with the fibre paper, we find that : a) the pulse height spectra remains modified (Fig.1); b) the radioactivity initially counted, is found, in a significant fraction, in the liquid scintillator (measurement "fibre papers out").

A - ^{14}C

The experiments show that the relationship (activity (dpm) versus weight)is linear for pronase - treated samples, but occurs only rarely for intact samples (Fig. 4). In the latter case, we believe that we have counted all of the "X" compounds which are extracted by the liquid scintillator, and only a fraction of the "Y" compounds, which are not extracted (the fraction located near the surface of the sample). The linear relation between activity (dpm) and the weight of the assay, obtained with the pronase-treated samples (Fig. 4)) seemed to indicate that all of the radioactivity was measured in these conditions. This was demonstrated by comparing the results obtained by this technique with those obtained by a classical method (Table 1).

It should be noted that we attempted to define the optimum reaction conditions of pronase by counting the activities of different fractions of the same organ (taken from a mouse fed with [2-^{14}C] thymidine) treated during different incubation times at 37°C and with different concentrations of the enzyme.

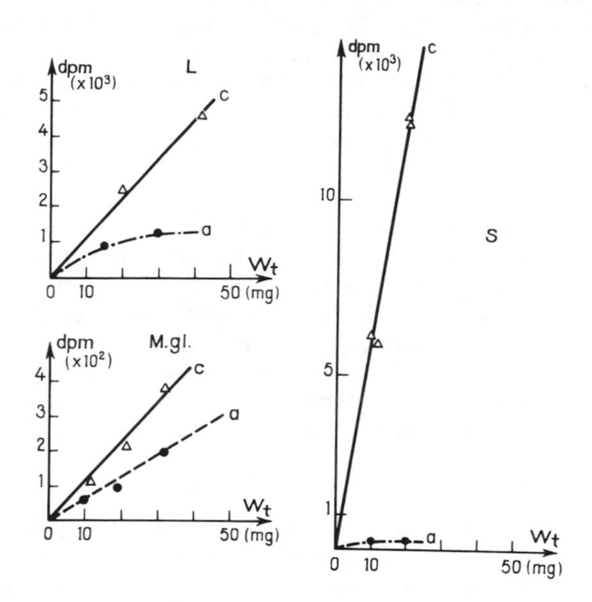

Fig.4 The dependence of activity (dpm) on weight: for ^{14}C
L : liver of a mouse killed two hours after being fed with 200 µg (20 µCi) of [2-^{14}C] thymidine. S and M gl.: respectively, spleen and mammary glands of a mouse killed two hours after being fed with 121 µg (8 µCi) of [2-^{14}C] thymidine.

Curves a (●) : samples counted intact; curves c (△) : samples counted after treatment with pronase.

The "a" curve for mammary gland gives an example of a linear relationship when the total activity is not measured since the "a" curve represented only about half of the "c" curve, in this case.

B - ^{3}H

When we applied the technique established for ^{14}C to ^{3}H, we found that the activity-weight relationships were linear in all cases. However, for the same tissue, the activity measured with untreated samples was sometimes lower than that found with pronase - treated samples (Fig. 5). When the untreated sample is placed in NE 220, we count the total amount of the "X" compounds extracted by the liquid scintillator, but in no case, the non-extractible "Y" compounds, this being due to the weak average β path length of ^{3}H (1 μl in water). A systematic comparison was made between counting results obtained with untreated samples and either those treated with pronase or those treated by a classical method of digestion or combustion. They showed that the results for the pronase-treated samples are always in agreement with those obtained by classical methods, whereas the results from untreated samples are only sometimes in agreement (Table II). These results demonstrated that with ^{3}H, the linearity of the activity-weight relationship is not a sufficient test of the method's validity. However, we must try to interpret why in many cases, it was possible to measure the total ^{3}H on untreated samples. This last result was surprising as it did not occur with ^{14}C (Table I). These results can be explained by the fact that in-vivo the ^{3}H label of organic compounds quickly ends up in a water molecule. Wade and Shaw demonstrated that this transfer exists even when the ^{3}H label of thymidine is in the 2 or methyl position (4). After ingestion of 1 to 10 μCi of this compound, 85-90% of ^{3}H is found in the form of water after 24 hours. A similar result was found in the case of [^{3}H] chlormadinone (5). Our experiments show that water is easily extractable from tissues by NE 220, and this is why, for many types of tissues, the total ^{3}H is counted by simple immersion in liquid scintillator. In the [^{3}H] thymidine experiments (Table II), these are tissues with little mitotic activity (e.g., muscle, brain). In tissues with a greater mitotic activity e.g., spleen, ovaries, adrenal glands, pronase treatment must be used to obtain a measurement of total activity (Table II). However, in order that this non-extractable activity would remain significant compared to the extractable activity we killed the mice after a short period of time (¼ hour). In the case of compounds such as

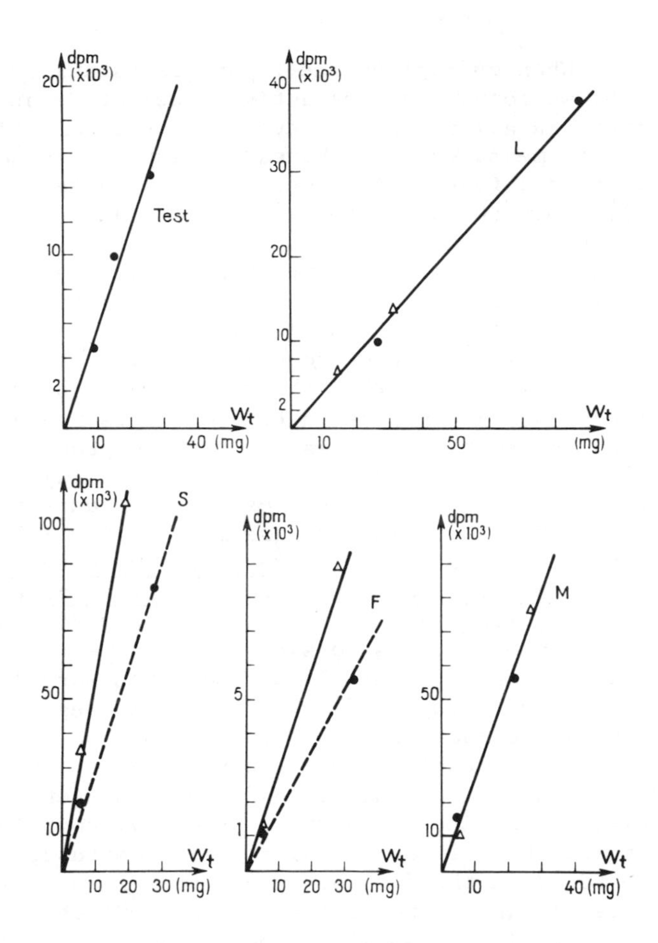

Fig.5 The dependence of activity (dpm) on weight : for ^{3}H
L and Test: respectively liver and testicules of a male mouse killed two hours after being fed with 1,7 μg = 100 μCi of [6-^{3}H] thymidine. S, M and F. : respectively spleen, muscle and fat of a female mouse killed two hours after being fed with 1,7 μg = 100 μCi of [6-^{3}H] thymidine. Points (●) : samples counted intact; points (△) samples counted after treatment with pronase.

TABLE I

Comparison of results of measurements obtained by different techniques for ^{3}H and ^{14}C
Results expressed as dpm/mg tissue

		[methyl-^{14}C] Thymidine			[2-^{14}C] Thymidine		
		Direct Intact	Direct Pronase	Combustion	Direct Intact	Direct Pronase	Combustion
ORGANS	L.	760 ± 95	1.477 ± 230	1.310 ± 81	40 ± 6	71 ± 12	78 ± 5
	S.	129 ± 17	998 ± 156	1.025 ± 64	57 ± 8	417 ± 70	400 ± 27
	M.	120 ± 15	146 ± 25	144 ± 10	20 ± 3	23 ± 4	25 ± 1,5
	B.	441 ± 55	512 ± 84	559 ± 35	21 ± 3,5	20 ± 3,5	21 ± 1,5
	M.Gl	153 ± 20	249 ± 41	285 ± 19	28 ± 4	32 ± 5,5	33 ± 2
	F.	39 ± 5,5	52 ± 9	42 ± 3	19 ± 3	20 ± 3,5	18 ± 1,5
	Ov.		452 ± 84	433 ± 43		109 ± 23	89 ± 9
	Adr.		566 ± 136	584 ± 70		79 ± 20	85 ± 6

B: Brain; Ov.: ovaries: Adr.: adrenal glands
The other letters and abbreviations have the same meaning as in Fig.2,4 and 5.
Mouse killed one hour after being fed with 35 μg (8 μCi) of [methyl- ^{14}C] thymidine or 38.7 μg (8 μCi) of [2-^{14}C] thymidine

TABLE II

Comparison of results of measurements obtained by different techniques for ^{3}H
Results expressed as dpm/mg tissue

ORGANS	[^{3}H] Pregnenolone			[6-^{3}H] Thymidine		
	Direct Intact	Direct Pronase	Hyamine	Direct Intact	Direct Pronase	Combustion
L.	12.500±2000	14.820±3000	15.490±2560	3.144±490	10.970±1725	11.350±850
S.				961±170	1.695± 281	2.090±172
M.	1.555± 250	2.065± 435	1.388± 270	1.004±164	864± 141	926± 70
B.	1.100± 175	1.050± 190	1.050± 204	817±133	735± 124	843± 67
M.Gl	7.005±1050	7.262±1190	6.350± 920	486± 82	656± 111	589± 47
F.	5.720± 860	5.183± 855	5.110± 690	148± 26	250± 44	190± 15
Ov.	--	--	--	--	1.052± 205	834± 92
Adr.	--	--	--	--	1.312± 298	564± 62

The letters and abbreviations have the same meaning as in Table I.
Mice fed with: [^{3}H] pregnenolone : 18,2 μg (560 μCi) and sacrified two hours later;
[6-^{3}H] Thymidine : 0,92 μg (53,5 μCi) " " ¼ hour "

The contradictory result obtained for adrenal glands is probably due to the fact that the two glands were not identical.

tritiated water,[^{3}H] chlormadinone or [^{3}H] pregnenolone, the total activity is recovered for samples counted untreated (Table II).

To avoid loss of ^{3}H corresponding to autolysis of tissues, it is necessary to keep the samples, in liquid nitrogen (-196°C), until pronase is added (3).

DISCUSSION - This study shows that it is possible to measure ^{14}C or ^{3}H which has been incorporated in organs, without using the classical methods of combustion or digestion. The combustion method is precise but requires a special apparatus and presents certain difficulties well known to those who use it. The digestion method gives variable results due to artifacts, of which the two principal ones are, a) the difficulty in obtaining a perfect solubilization of the samples and b) the occurence of spurious luminescence present in liquid scintillators which never completely disappears (6 - 8).

A comparison of the results obtained by the technique described here with those obtained by one of the classical techniques demonstrates that it is possible to directly count the ^{14}C or ^{3}H incorporated in the organs. It also proves the validity of the calibration curves which were established when the organs (or fragments) were added to the liquid scintillator containing one or the other of the radioisotopes. The validity of the curves can be seen following the studies of the pulse height spectra, which have shown that the alterations were of the same order when the sample was either itself radioactive or simply immersed into the liquid scintillator containing radioactivity. We have seen that this result is explained by the fact that the cell membranes are permeated with the liquid scintillator which extracts different substances responsible for the modification of the spectra. These modifications are of the same order as those obtained with whole blood. The reproducibility of the calibration curves shows that it is unnecessary to determine, for each test, the counting efficiency by the internal-standard method, as it is possible to use previously established calibration curves.

The only special equipment necessary for the described

technique is an ultrasonic generator which is used to reduce the time required for the equilibration of the pronase-treated sample in the liquid scintillator. A counter with a single photomultiplier was used in this study. We have shown that the results are similar if one uses a counter with two photomultipliers; thus, with a different geometry the volume of scintillation fluid does not need to be changed. Nuclear Enterprises liquid scintillator NE 220 containing dioxane was also used. It is possible, however, to use liquid scintillators which do not contain this compound. The experiments show that the cell membranes are permeated with the solvent (toluene or xylene) which is common to all liquid scintillators, and that these extract the soluble component from the cells. These two properties are general, and the experiments, show that the activity counted in a labelled sample increases with time, reaching a plateau, whereas that of a control sample, in the presence of a radioactive solution, decreases before reaching a limit; the pulse height spectra are altered, as with NE 220. However, with the liquid scintillator not containing dioxane, it is not possible to treat the samples with pronase, since such media are not miscible with water. The comparison of the activity obtained with two liquid scintillators, of which only one contains dioxane, can give a preliminary indication of the nature (water-soluble and lipid-soluble) of the radioactive components existing in the sample.

In order to count the radioactivity of heavier samples it is necessary to treat several fragments of the same organ. The experiments show that such samplings always give results of the same order when the organ is homogeneous. For precise measurements, the samples must be at equilibrium in the liquid scintillator fluid, that is, 5 to 7 hours after their initial immersion. However, measurements made immediately after immersion give significant approximate pre-results. In the case of ^{14}C, measurements should be made with the "P" setting. Finally, the study shows that samples prepared by this technique remain stable several days in the liquid scintillator at ambient temperature.

ACKNOWLEDGEMENTS - This work was conducted with the technical assistance of G. Tham and M. Guggiari. We would especially like to thank Dr. G. Rudali who kindly provided the animals used in this study, and also the personnel in this laboratory who have been helpful to us. We also thank the Society INTERTECHNIQUE and Dr. Berthelot (C.H.U. of Creteil) who have put at our disposal their combustion apparatus and have allowed us to establish with certainty the validity of the method described. Finally, we wish to thank Drs. L. Montagnier and N. Rebeyrotte for all their suggestions which we have used.

REFERENCES -

1. S. Apelgot and M. Duquesne, J.Chim.Phys.58, 774 (1961)

2. S. Apelgot, R. Chemama and M. Frilley; a) Monatshefte für Chemie 102, 985 (1971); b) Rev.Europ.Et.Clin.& Biol.XVII, 715 (1972)

3. S. Apelgot, R. Chemama and M. Frilley, Bulletin du Cancer, to be published.

4. L.Jr.Wade and E.I. Shaw, Rad.Research, 43, 403 (1970)

5. R. Chemama, S. Apelgot, G. Rudali, M. Frilley and E. Coezy, Bulletin du Cancer, 59, 187 (1972)

6. For a critical restatement of these works, see :
   a) M. Pollay and F.A. Stevens, p.207
   b) J.D. Davidson,V.T. Oliverio and J.I. Peterson p.222
   c) D.A. Kalbhen, p.337 in The Current Status of Liquid Scintillation Counting (Edwin D. Bransome, Ed.) New York and London: Grune and Stratton (1970)

7. D.A. Kalbhen in Liquid Scintillation Counting, p.1 (A. Dyer, Ed.) London, New York and Rheine: Heyden & Son Ltd (1971)

8. S. Apelgot, Results to be published.

QUANTITATIVE DETERMINATION OF HORMONE METABOLITES AND GLYCOGEN BY USING LIQUID SCINTILLATION QUENCHING METHOD

K.R. Laumas and S.A. Rahman

Department of Reproductive Biology
All-India Institute of Medical Sciences
New Delhi-110016, India

ABSTRACT

The colour quenching in liquid scintillation counting has been a major problem and requires an internal or external quenching correction for each coloured sample. However, colour quenching presents a good and a rapid method for the quantitation of those compounds for which appropriate colour could be developed. But this technique has not been applied for the estimation of hormone metabolites. The objective of present work was to use the colour quenching technique and develop it into a method for the estimation of hormone metabolites.

Sealed glass ampules containing ^{3}H-labelled source in toluene base scintillation fluid was placed, precisely centred, in a scintillation vial containing about 5.0 ml of solution of the developed colour to be quantitated. The vial was counted in a liquid scintillation counter (Packard Model 3380) to find out the extent of colour quenching.

The colour produced by the Kober colour reaction for estriol (1) and the phenol - sulphuric acid reaction for glycogen (2) gave linear decrease in the counting rate of the scintillation source. The critical evaluation of the method and its application in quantitative determination of urinary estriol and tissue glycogen, in this study, demonstrate that the sensitiveness, accuracy and precision of the scintillation quenching quantitation equals that of the spectrophotometric method.

INTRODUCTION

In liquid scintillation counting, the colour quenching is usually linear. The quenching results in the decrease in counting rate which is proportional to colour. The linearity in quenching may be exhibited by any colour which absorbs in the region of the spectrum in which the maximum response is shown by the photo-tubes used in the liquid scintillation spectrometer. Making use of this linearity, it has been possible to utilize the liquid scintillation quenching for the quantitation of lipid mass (3).

The use of liquid scintillation quenching technique for the quantitation of colour presents a good and a rapid method for the estimation of those compounds for which appropriate colour could be developed. However, this technique has not been applied for the estimation of hormone metabolites. The objective of the present work was to use the colour quenching technique and develop it into a method for the estimation of hormone metabolites. This rapid technique has been developed to standardise methods for the estimation of urinary estriol during pregnancy and glycogen in rat uterus, liver and muscle.

MATERIALS AND METHODS

Scintillation Source. About 28,000 cpm of ^{3}H-labelled source in 0.5 ml of simple scintillation fluid (toluene base) was pipetted in a small pyrex glass ampule. The glass ampule was sealed. Care was taken to prevent the scintillation medium from being warmed up during sealing. A large number of such glass ampules containing the β-emitting source were prepared. Each ampule was placed in a glass counting vial (20 ml) containing 5.0 ml of toluene. The position of the ampule within the vial was precisely centred with the help of a polyethylene adaptor and counted in a liquid scintillation counter (Packard Model 3380). Ampules containing 27,900 to 28,100 cpm were selected. For the purpose of colour quantitation, the developed colour product (about 5.0 ml) was taken in the counting vial, an ampule containing scintillation source was placed and counted as above .

Quantitation of various colour reactions.

Kober colour reaction for Estriol. Taking various levels of estriol (0 to 40 μg), Kober colour reaction was performed according to Brown's method (1). The optical density of the developed pink colour was taken at 500, 542 and 576 nm (Zeiss PMQ II Spectrophotometer). The corrected optical densities [2 x O.D. at 542 - (O.D. at 500 + 576)] were plotted against the various amounts of estriol (μg). The duplicate samples were transferred to counting vials. The glass ampules containing known scintillation source (28,000 cpm of ^3H) were placed centrally in each vial and counted in the liquid scintillation counter. The rate of colour quenching (cpm of ^3H obtained) for increasing amount of estriol was plotted on a graph paper.

Phenol-Sulphuric acid reaction for Glycogen. Colour reaction for various amounts of standard glycogen (0 to 80 μg) was performed according to Dubois et al (2). The glycogen was taken in 2.0 ml of distilled water, 0.1 ml of 80% phenol solution in distilled water added and immediately followed by the addition of 5.0 ml of conc. Sulphuric acid. It was mixed, allowed to stand in ice water for 30 minutes and optical density taken at 490 nm. The duplicate samples were subjected to scintillation quenching. The optical densities and cpm of ^3H obtained were plotted against the various concentrations of glycogen.

Hydrolysis, extraction and purification of urinary estriol. An aliquot (2.0 ml) of 24-hour male urine or pregnancy urine was taken in Kober tube, conc. hydrochloric acid was added until the pH was 2 or below (pH paper) and 2.5 g of sodium chloride was added. Ethyl acetate (4.0 ml) was added and vortexed for two minutes at room temperature. After centrifugation, 2.0 ml of the solvent phase, representing one ml of urine,was taken, dried under nitrogen and dissolved in ether.

Urine extract contains substances which yield

colour products on chemical reaction. These disturbing chromogens are effectively removed by the method of Brown (1). This involved the washing of ether extract with sodium corbonate solution (pH 10.5), sodium hydroxide solution, sodium bicarbonate solution and finally with distilled water. The ether extract was then partitioned between benzene-light petrolêum and water. The water extract containing estriol was acidified and the estriol re-extracted with ether. The ether extract was finally purified on alumina columns.

Extraction of tissue glycogen. The method of Dubois *et al* (2) was adopted for the extraction of tissue glycogen. To every 100 to 200 mg of tissue (rat uterus, liver and muscle) 1.0 ml of 30% potassium hydroxide was added, and heated in boiling water for 10 minutes. After cooling, 0.5 ml of 2% sodium sulphate and 1.2 ml of 95% alcohol was added and kept for overnight at room temperature. After centrifugation, the supernatant was discarded and the residue was dissolved in known amount of distilled water.

RESULTS.

Evaluation of the assay method.

Sensitivity Test. The linearity of the relationship between estriol and the corrected optical density and the counting rate is presented in Fig. 1. Fig. 2 shows the relationship between various concentrations of glycogen and optical density and the counting rate. The counting rate of the ^3H-radioactive scintillation source decreased by the colour intensity. The sensitivity of colour quenching curve for both estriol and glycogen was good and quite comparable with their respective absorbance. The range could be further extended to at least 60 μg for estriol and 160 μg for glycogen by scintillation colour quenching. Slopes were also identical on repeated observations.

Accuracy Test. The accuracy of a quantitative method is usually studied by means of "recovery experiments" in which determinations are made on the material being analysed before and after the addition of known amounts of the substance under investigation. Since both the urinary

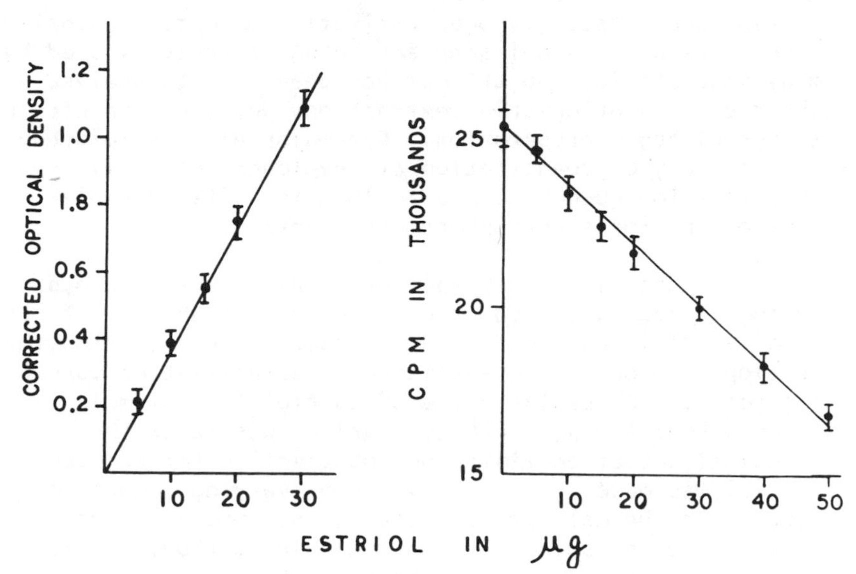

Fig. 1 Estriol Standard Curve : The dose response curve by the spectrophotometric method (542 mμ) and the scintillation quenching method.

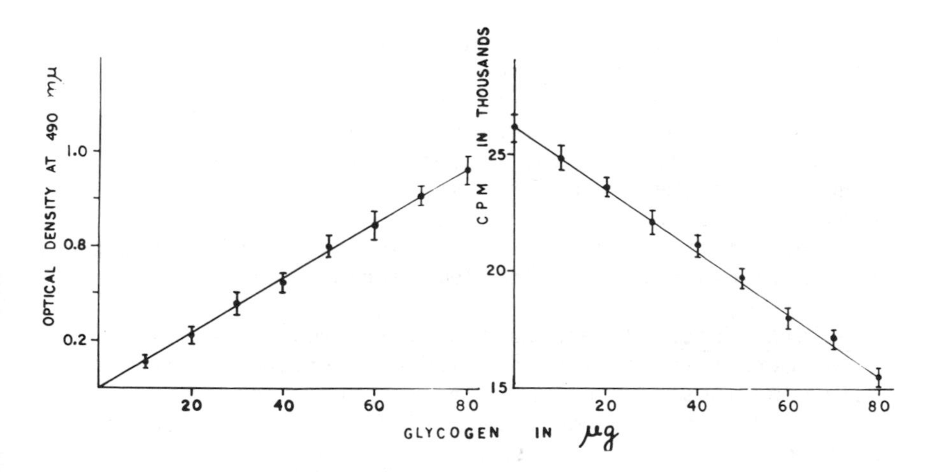

Fig. 2 Glycogen Standard Curve ; The dose response curve by the spectrophotometric method (490 mμ) and the scintillation quenching method.

estriol and tissue glycogen estimation by optical density method are well established and being extensively used by many laboratories, no attempt has been made to analyse the recovery during urine extraction, purification etc. or tissue glycogen preparation. Our major aim was to study the accuracy of quantitation of developed colour by scintillation quenching and to look into its possible replacement for spectrophotometric method.

To each 1.0 ml of male urine extract (urine hydrolysed, extracted and purified as described above) 5 to 25 µg of estriol was added. After Kober colour reaction, the developed colour was quantitated by scintillation quenching method. Since the value of estriol in 1 ml male urine (about 7.0 µg in 24-hour urine) was below the practical detection limit, no substraction for endogenous estriol was made. After addition of varying amounts of estriol to the male urine extract, the amount of estriol was measured by scintillation quenching method. It is presented in Fig.3. It may be observed that the scintillation quenching and the concentration of the estriol added had a linear relationship. The method was quite accurate and reproducible for the quantitation of estriol.

Similarly, 10 to 50 µg of glycogen was added to aliquots from a tissue glycogen preparation. The colour was developed and quantitated by scintillation quenching. The amount of glycogen already present in each aliquot (11.8 ± 1.01 µg) was substracted from each observation. Fig. 4 shows the amount of glycogen measured for every addition of standard glycogen before colour reaction. Accuracy of the scintillation quenching method in quantitation of glycogen was also demonstrable.

Precision Test. An estimate of precision of a chemical assay method is usually obtained by carrying out multiple determinations on the same sample. Precision is usually expressed as the standard deviation of replicate determinations.

Multiple determinations of the urinary estriol in 4 samples from varying weeks of pregnancy by both spectrophotometric and scintillation quenching methods have been presented in Table I. The standard deviations

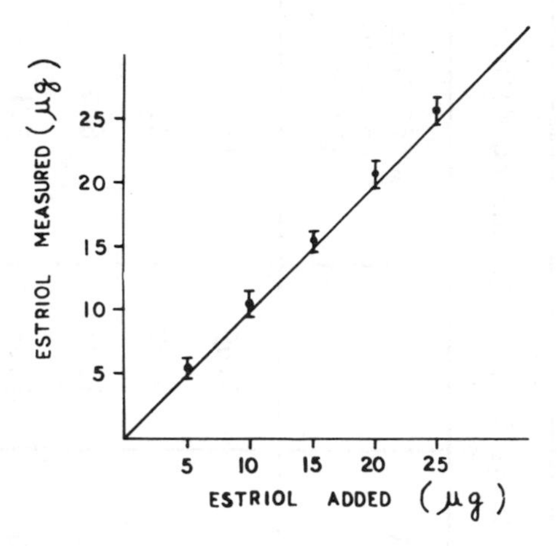

Fig. 3 Accuracy of the urinary estriol quantitation by scintillation quenching method.

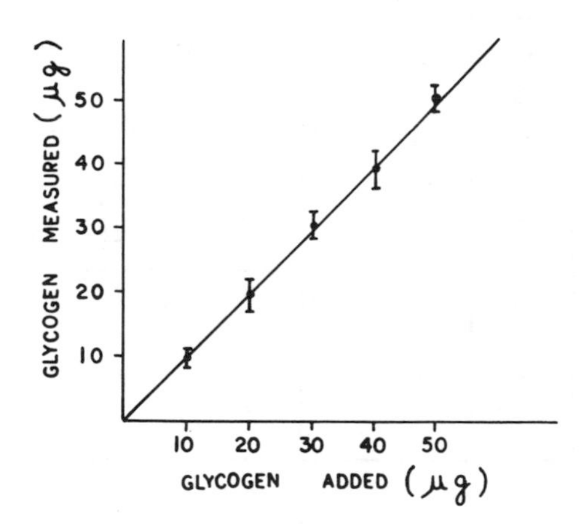

Fig. 4 Accuracy of the tissue glycogen quantitation by scintillation quenching method.

TABLE- I

The Precision of Urinary Estriol quantitation by classical optical density method and scintillation quenching method

Pregnant women	Optical Density Method			Scintillation Quenching Method		
	MEAN (μg)	S.D.	Coefficient of variation (%)	MEAN (μg)	S.D.	Coefficient of variation (%)
S.S.	4.7	0.49	10.4	5.3	0.59	11.1
S.D.	7.1	0.69	9.7	7.2	0.90	12.5
G.A.	15.4	0.91	5.9	16.9	1.02	6.0
K.R.	20.0	0.94	4.7	20.6	1.05	5.1

and the coefficient of variations (%) for the two methods may be observed to be quite comparable. The coefficient of variation for the new method ranged between 5.1 to 12.5 per cent. This may indicate a high degree of precision.

Table II shows the multiple estimations of glycogen in various tissue glycogen preparations by both methods. It may be observed that the precision of the glycogen quantitation by scintillation quenching method was also comparable to the spectrophotometric method.

Urinary estriol estimation in pregnant women. Large number of 24-hour urine from women at various weeks of pregnancy were analysed for estriol. The values obtained by scintillation quenching method has been compared with those obtained by spectrophotometric method (Fig. 5). It may be observed that estriol (μg/ml of urine) quantitated by scintillation quenching method was not different from that obtained by spectrophotometric determination.

Tissue glycogen estimation. Tissue glycogen was prepared from rat uterus, liver and muscle. Also, to evaluate the influence of different concentrations of glycogen, varying aliquots from these tissue preparations were quantitated for glycogen by both methods. Ranging between 10 to 80 μg of glycogen were quantitated by scintillation quenching method and the values did not differ from that of the spectrophotometric method (Fig. 6).

DISCUSSION

The scintillation quenching method for the quantitation of colour has been evaluated and applied for the estimation of urinary estriol and tissue glycogen. The method is as sensitive as that of the spectrophotometric procedure. Measurement of varying concentrations of estriol and glycogen demonstrated that the quantitation by scintillation quenching method is quite accurate. Repeated estimations on a number of samples indicated that the precision of the quenching method equals that of the spectrophotometric method.

TABLE II

The Precision of Tissue Glycogen quantitation by classical optical density method and scintillation quenching method

Tissue preparations	Optical Density Method			Scintillation Quenching Method		
	MEAN (μg)	S.D.	Coefficient of variation (%)	MEAN (μg)	S.D.	Coefficient of variation (%)
1. Rat liver	10.6	1.26	11.9	10.3	1.32	12.8
2. Rat muscle	12.5	0.85	6.8	11.8	1.30	11.0
3. Rat muscle	18.8	1.03	5.5	19.1	1.33	7.0
4. Rat muscle	21.7	1.54	7.1	21.1	1.44	6.8
5. Rat liver	34.6	3.38	9.8	34.1	3.62	10.6

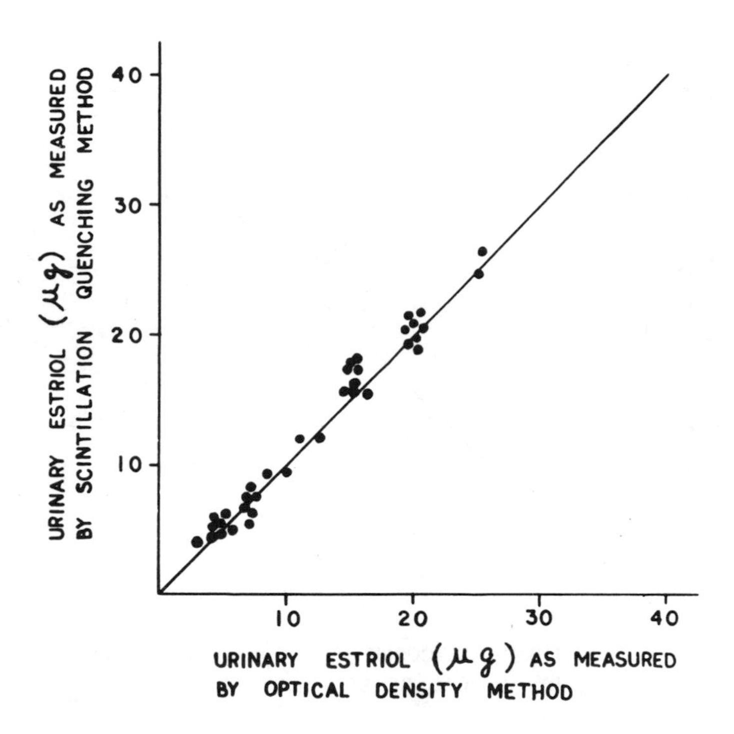

Fig. 5 Quantitation of Urinary Estriol: The accuracy of scintillation quenching method as compared with that of the classical spectrophotometric method.

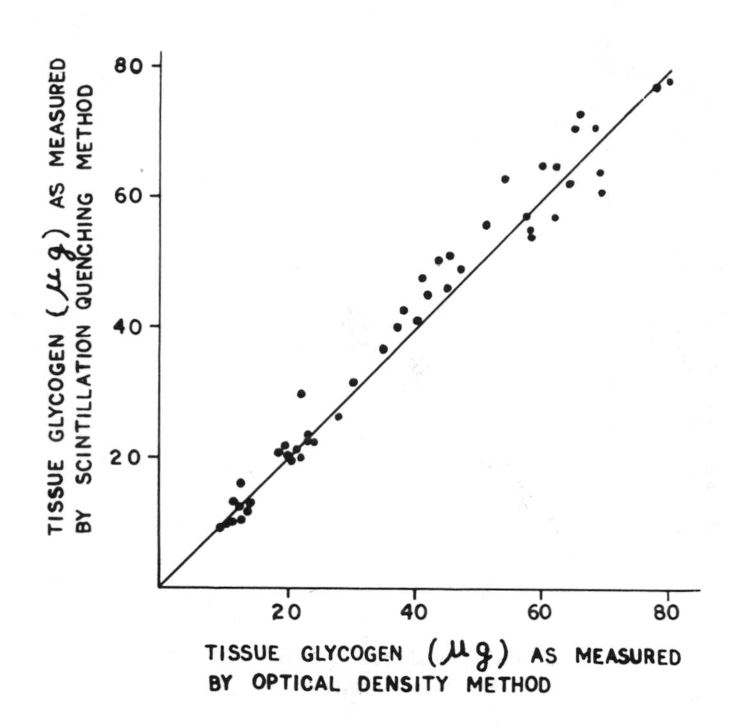

Fig. 6 Quantitation of Tissue Glycogen : The accuracy of scintillation quenching method as compared with that of the classical spectrophotometric method.

The evaluation of scintillation quenching for the quantitation of coloured products and its application in estimating the urinary estriol in pregnant women and the tissue glycogen may suggest that the scintillation quenching method is a good alternative to that of the spectrophotometric procedure.

REFERENCES

1. J.B. Brown, Biochem. J. 60, 185 (1955).

2. M. Dubois, K. Gilles, J.K. Hamilton, P.A. Robers and F. Smith, Nature (Lond.) 168, 167 (1951)

3. F. Snyder and A. Moehl, Anal. Biochem. 28, 503 (1969)

A COMPARISON OF PHOSPHOR SOLUTIONS FOR COUNTING AQUEOUS SAMPLES OF STEROID HORMONES

Bruce A. Scoggins, Aldona Butkus and John P. Coghlan

Howard Florey Institute of Experimental Physiology and Medicine, University of Melbourne, Parkville, 3052, Australia.

Abstract

The counting efficiency of phosphor solutions suitable for counting aqueous samples of non-polar and polar [^{3}H] steroids have been compared. Results show that non-polar steroids such as [^{3}H]-progesterone can be most efficiently counted as a heterogeneous system in toluene phosphor. Relative counting efficiency (sample without water = 100%) increased when water was present and was independent of water content up to 3 ml in 10 ml phosphor. Shell Sol A (methyl benzene fractions) phosphor also gave a counting efficiency independent of water content but the relative efficiency was 10-15% lower than for toluene phosphor. Emulsion systems, triton/toluene and Insta-Gel and a Dioxan based phosphor gave low counting efficiencies dependent on water content.

For the more polar steroids, e.g. [^{3}H]-cortisol, the most efficient phosphor solution was toluene phosphor with the aqueous phase 50% ammonium sulphate rather than water alone. This phosphor gave counting efficiencies independent of water content whereas all other phosphors had lower efficiencies very dependent on the amount of water present.

On the basis of cost, toluene (Shell, commercial grade) phosphor was cheap \$0.73/L and as efficient as toluene (Fluka, purum) phosphor \$2.0/L. Apart from Shell Sol A all other phosphors were more expensive and much less efficient.

The mechanism of the increase in relative counting efficiency seen with aqueous toluene phosphor is discussed.

Introduction

In radioimmunoassay and saturation analysis procedures used for the measurement of a wide variety of steroid hormones in biological fluids, tritium labelled steroids are widely used both as the tracer and as a recovery indicator for extraction and purification losses. Liquid Scintillation counting of aqueous samples of the tritium labelled steroid is often necessary if additional time consuming methods of preparing the samples in a non-aqueous form are to be avoided. The aim of the present study was to evaluate the efficiency and cost of a variety of phosphor solutions suitable for counting aqueous samples of steroid hormones of different polarity.

Materials and Methods

Tritiated steroids, [1,2-^{3}H]-aldosterone, [1,2-^{3}H]-cortisol [1,2-^{3}H]-corticosterone, [1,2-^{3}H]-deoxycorticosterone, [6,7-^{3}H]-17β-oestradiol, [6,7-^{3}H]-oestrone, [1,2-^{3}H]-testosterone and [1,2-^{3}H]-progesterone were obtained from either the Radiochemical Centre, Amersham, U.K. or New England Nuclear Corporation, U.S.A. All were of specific activity 30-50 Ci/mM and purified by paper partition chromatography before use. [^{3}H]-steroids were stored in ethanol at -4°C. Non-radioactive steroids (Ikapharm) were stored as a 1 mg/ml solution in ethanol.

Solvents used in preparation of phosphor solutions, toluene (Fluka, purum) and dioxan (Fluka, puriss), were used without purification. Toluene (Shell, commercial grade) and Shell Sol A (methylated benzene fractions) were filtered prior to use. Triton X 100, Insta-Gel, PPO and dimethyl POPOP were obtained from Packard Instrument Co. All phosphor solutions, except Insta-Gel and the dioxan based phosphor, contained 4g/L PPO and 40 mg/L dimethyl POPOP. The dioxan based phosphor contained 100 ml ethoxyethanol, 20 ml ethylene glycol,

8g PPO, 600 mg dimethyl POPOP, 150 g napthalene and dioxan to make 1 litre (1). All samples were prepared in glass counting vials (Packard Instrument Co.) and counted in a Packard Model 314 EX, 3330 or 3375 liquid scintillation spectrometer. Samples were prepared in triplicate by addition of non-radioactive steroid (50 μg) and at least 50,000 dpm of [^{3}H] - steroid to each counting vial. After removing the ethanol by drying, 0.5 to 3 ml of water was added followed by 10 ml of phosphor solution. Samples were counted for at least 3 x 10 mins. after samples had been thoroughly mixed and allowed to equilibrate at the temperature of the counting chamber for 10-12 hr.

Results

Samples of all [^{3}H] - steroids to be compared were prepared in: a) toluene (Fluka) phosphor, b) dioxan based phosphor, c) triton X-100/toluene phosphor (2/1 v/v). Samples were counted in a Packard 314 EX at -10°C at 25% tritium efficiency and in a Packard 3375 at 0°C or 20°C at 49% tritium efficiency. Results for the toluene (Fluka) phosphor are shown on Table 1. All results are shown as relative efficiency, the sample without water being expressed at 100%.

TABLE I

Relative efficiency of samples counted in toluene (Fluka) phosphor containing 0.5-3.0 ml water. All samples counted in a Packard 3375 at 0°C at a tritium efficiency of 49%. Sample without water expressed as 100%.

STEROID	WATER VOLUME (ml)				
	0.5	1.0	1.5	2.0	3.0
[^{3}H]- Aldosterone	74.7	60.1	50.4	43.3	-
[^{3}H]- Corticosterone	101.3	100.7	98.9	99.3	-
[^{3}H]- Cortisol	68.4	52.2	50.9	34.3	27.1
[^{3}H]- DOC	104.7	104.7	105.1	105.8	-
[3H]- Oestradiol	100.6	101.4	101.8	102.0	-
[^{3}H]- Oestrone	99.4	101.1	101.7	101.8	-
[^{3}H]- Progesterone	102.7	103.2	102.6	101.9	103.9
[^{3}H]- Testosterone	104.0	104.3	105.1	105.3	-

Similar results were obtained when the above examples were counted at either -10°C in a Packard 314 EX at a tritium efficiency of 25% when the aqueous layer was frozen or at 6°C in a Packard 3330 at a tritium efficiency of 65%.

Table I shows that counting efficiency of the most polar of the steroids, [^{3}H]-aldosterone and [^{3}H]-cortisol, fell with increasing water content of the samples. However, for the other non-polar steroids, counting efficiency did not change with water content and in most cases was greater than if no water was present. The degree of increase in relative efficiency appeared to be a function of a particular steroid. In the remainder of the studies to be reported, [^{3}H]-cortisol and [^{3}H]-progesterone have been used as examples of the polar and non-polar steroid types.

A comparison of the relative efficiency (100% is for the water free toluene (Fluka) phosphor sample) of three different phosphor solutions for aqueous samples of [^{3}H]-cortisol and [^{3}H]-progesterone are shown on Figure 1. All samples were counted at 0°C in a Packard 3375 at a tritium efficiency of 49%.

For [^{3}H]-progesterone, the dioxan based phosphor and the triton/toluene phosphor both gave much lower counting efficiencies when compared with the toluene phosphor. Efficiencies were much lower even without water present and showed a considerable decrease with increasing water content. With the toluene phosphor, relative counting efficiency was very high and independent of water content (Table I).

On the other hand, for [^{3}H]-cortisol the results were similar for all three phosphor solutions. All showed a marked reduction in counting efficiency with increasing water content. Figure 1 also shows that recounting the dioxan phosphor at 20°C to prevent freezing increased efficiencies for both [^{3}H]-progesterone and [^{3}H]-cortisol compared with those obtained at 0°C.

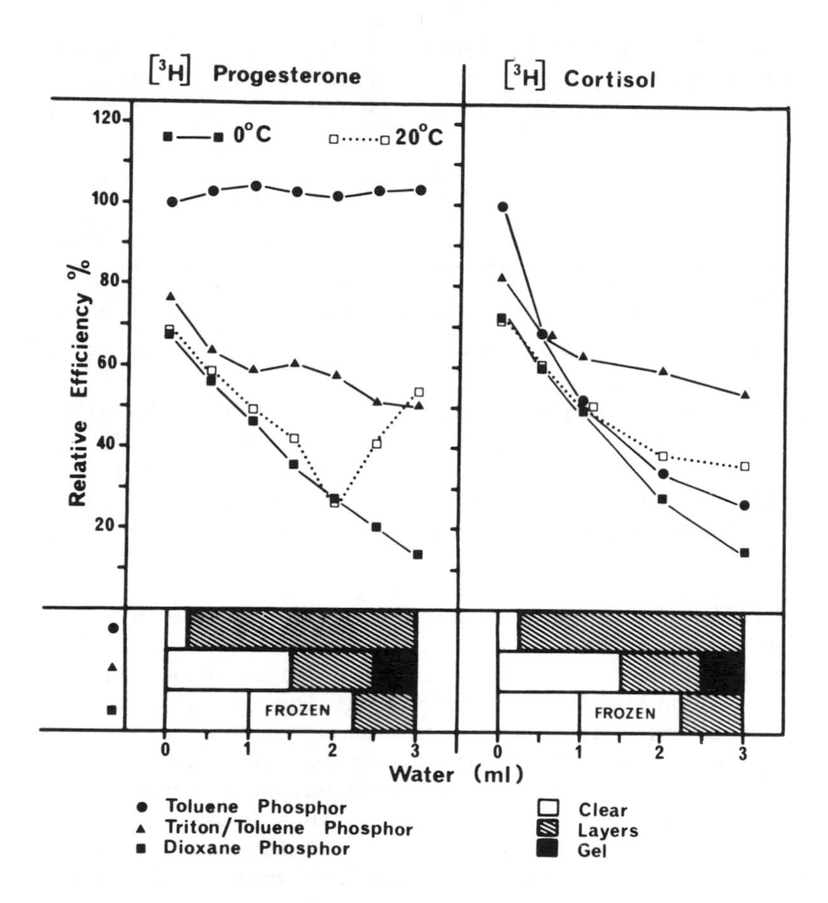

Figure 1. A comparison of the relative efficiency of toluene phosphor, triton X-100/toluene phosphor and dioxan based phosphor for counting aqueous samples of [^{3}H] -cortisol. Samples were counted in a Packard 3375 at 0°C and 20°C at a tritium efficiency of 49%.

TABLE II

Relative efficiency of [^{3}H]-progesterone samples counted in toluene phosphor containing 0-10 ml water. Samples counted in a Packard 3375 at 0°C at 49% tritium efficiency.

	WATER VOLUME (ml)						
	0	2.0	3.0	4.0	5.0	7.0	10.0
RELATIVE EFFICIENCY	100	104.2	103.6	104.2	103.5	101.6	96.7

The effect of increasing the water content up to 10 ml in 10 ml of toluene phosphor for [^{3}H]-progesterone is shown on Table II. The fall in efficiency observed with the larger water volumes may be due to poor positioning of the toluene phase of the phosphor in the counting chamber.

Availability of low cost commercial grade toluene (Shell) and a methyl benzene based solvent (Shell Sol A) have enabled these to be compared with the purum grade toluene (Fluka). Results for samples containing either [^{3}H]-progesterone or [^{3}H]-cortisol are shown on Figure 2. A comparison is also made with Insta-Gel (Packard) and with samples containing 50% ammonium sulphate rather than water. Samples were counted at 6°C in a Packard 3330 with a tritium efficiency of 65%. Relative efficiency is expressed as 100% for the water free toluene (Fluka) phosphor.

[^{3}H]-progesterone had similar counting efficiencies and increase in relative counting efficiency in the presence of water in both the Fluka and Shell toluene phosphor solutions. However, when counted in Shell Sol A,the relative efficiency was much lower and averaged 12% less than with toluene phosphor. Insta-Gel caused considerable loss in counting efficiency even in the absence of water and efficiency was dependent on the water content of the samples. Instability of the solution was observed in

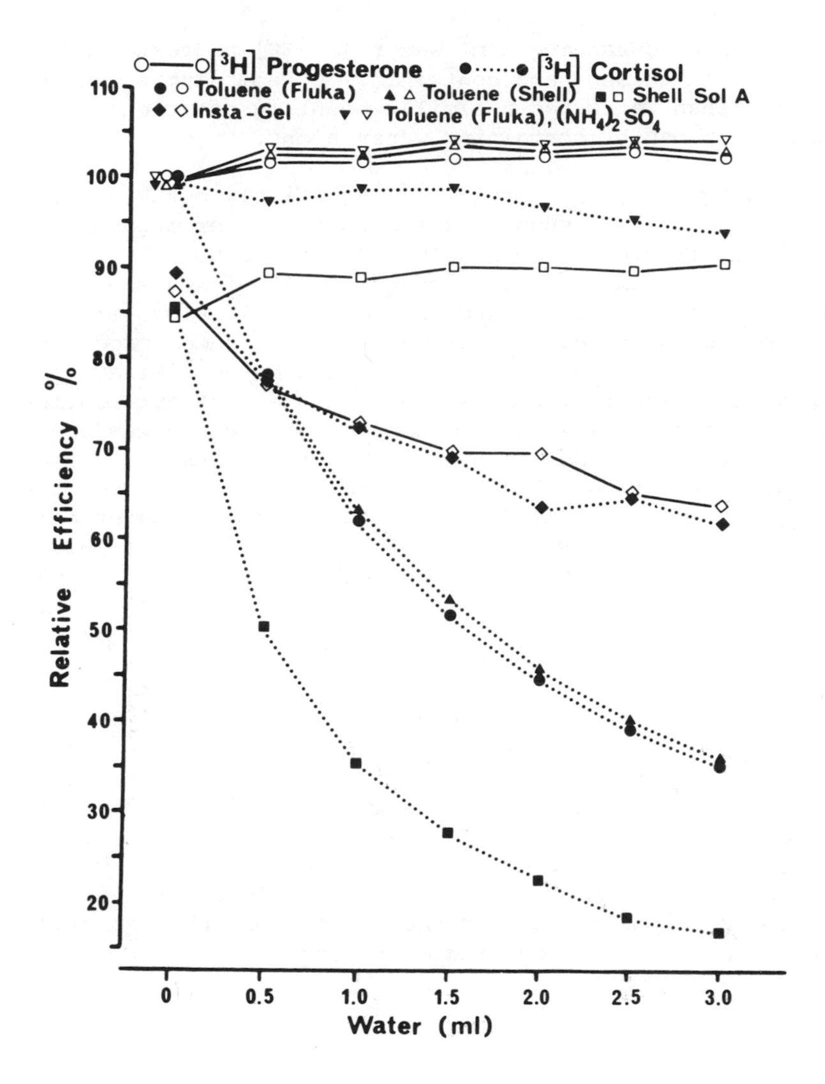

Figure 2. A comparison of the relative efficiency of various phosphor solutions used to count aqueous samples of [^{3}H]-progesterone and [^{3}H]-cortisol. Samples were counted in a Packard 3330 at 6°C at a tritium efficiency of 65%.

the sample containing 2 ml water. [^{3}H]-progesterone counted in toluene and containing 50% ammonium sulphate rather than water had a similar counting efficiency to toluene phosphor containing water alone.

For the more polar steroid, [^{3}H]-cortisol, both the Fluka and Shell toluene phosphors showed the water dependent fall in efficiency previously observed (Figure 1). Shell Sol A had a relative counting efficiency 20-25% lower than for toluene and showed a similar loss in efficiency with addition of water. Insta-Gel produced a relative efficiency - water content curve similar to that seen with [^{3}H]-progesterone. The most interesting result observed was when 50% ammonium sulphate was added to the [^{3}H]-cortisol samples rather than water. Relative counting efficiency increased markedly over that seen with water alone and was independent of water content with up to 1.5 ml water. Efficiency then fell slowly with increasing water content to 95% with 3.0 ml of water.

Table III combines and summarises the results for each of the phosphor solutions examined.

Results are expressed as relative counting efficiency, 100% is for toluene (Fluka) sample without water. Table III shows that for non-polar steroids, e.g. [^{3}H]-progesterone, there is an increase in counting efficiency in toluene phosphor and in Shell Sol A when water is added. Once water has been added efficiency is independent of water content. For polar steroids, e.g. [^{3}H]-cortisol, 50% ammonium sulphate added to the toluene phosphor solution increases relative counting efficiency.

TABLE III

Relative counting efficiency for various phosphor solutions used to count aqueous samples of [^{3}H]-progesterone (A) and [^{3}H]-cortisol (B)

(A) [^{3}H]-PROGESTERONE

PHOSPHOR	WATER VOLUME (ml)			
	0	1.0	2.0	3.0
1. Toluene (Fluka)	100	103	102	104
2. Toluene (Shell	100	103	103	104
3. Shell Sol A	85	89	91	91
4. Dioxan based	68	46	27	14
5. Triton/Toluene	77	59	58	50
6. Insta-Gel	87	73	70	64
7. Toluene & Amm.Sulphate	100	103	104	105

(B) [^{3}H]-CORTISOL

PHOSPHOR	WATER VOLUME (ml)			
	0	1.0	2.0	3.0
1. Toluene (Fluka)	100	62	45	35
2. Toluene (Shell)	100	64	46	36
3. Shell Sol A	85	35	23	17
4. Dioxan based	73	50	28	15
5. Triton/Toluene	82	63	60	54
6. Insta-Gel	89	73	64	62
7. Toluene & Amm.Sulphate	99	99	97	95

Another factor examined in this study was the cost of the various phosphor solutions. Table IV shows the cost of both the solvent component and the total phosphor solution. Brays Solution (2) is shown for comparison. Prices were those available in Australia and have been converted to U.S.$.

TABLE IV

PHOSPHOR SOLUTIONS FOR AQUEOUS SOLUTIONS*

SOLVENT	COST/L $	PHOSPHOR COST/L	RELATIVE COST
1. Toluene (Fluka)	1.40	2.00	2.7
2. Toluene (Shell)	0.13	0.73	1.0
3. Dioxan (Fluka) Blend (1)	6.27	10.57	14.5
4. Shell Sol A	0.14	0.74	1.0
5. Triton/Toluene (2/1)	1.40	1.79	2.4
6. Insta-Gel	-	8.75	12.0
7. Brays Solution (2)	-	6.34	8.7

* All solutions except (3) and (6) contain 4g/L PPO ($0.85/L) and 0.04 g/L dimethyl POPOP ($0.05/L). All costs based on Australian $ converted to U.S.$.

Also, they are the lowest available for the particular product of a stated quality. The relative cost shown on Table IV is based on the cheapest toluene (Shell) phosphor. When the cost factor is taken into account with the counting efficiency data previously presented in Figures 1 and 2 and Table III it is clear that for non-polar steroids in aqueous solution the Shell toluene solution is the phosphor of choice. It is cheap and gives high counting efficiency independent of water content. For the more polar steroids addition of ammonium sulphate to the aqueous phase of toluene (Shell) phosphor provides a cheap scintillation solution giving a high relative counting efficiency independent of the amount of water present.

The reproducibility of counting a $[^3H]$-progesterone sample in 2 ml of water in toluene phosphor has been examined. Ten replicate samples containing 3.5 x 10^5 dpm were prepared and each counted 7 times for 10 mins in a Packard 3330. The mean between sample variability was 0.2% and that between individual counts of the same sample 0.6%. A similar check on reproducibility was carried out for $[^3H]$-cortisol in samples containing 2 ml of 50% ammonium sulphate. Variability was similar to that found for $[^3H]$-progesterone.

To further investigate the increase in counting efficiency caused by the presence of water in the toluene phosphor, the effect of adding water to phosphor containing $[^3H]$-toluene was examined (Table V).

An increase in efficiency was observed with as little as 0.2 ml of water in the 10 ml of $[^3H]$-toluene phosphor. A small increase in relative counting efficiency was also observed for both $[^3H]$-progesterone (102.5%) and $[^3H]$-cortisol (101.3%) if water saturated toluene rather than anhydrous toluene was used to prepare the phosphor solution.

TABLE V

The effect of water on the relative counting efficiency of $[^3H]$-toluene phosphor. Samples were counted in a Packard 3375 at 0^oC at a tritium efficiency of 49%.

	WATER VOLUME (ml)				
	0	0.2	1.0	2.0	3.0
RELATIVE COUNTING EFFICIENCY	100	103	105	107	106

Discussion

The most interesting finding of these studies was the increase in relative counting efficiency observed for the non-polar steroids when they were counted in aqueous toluene phosphor. Further, this increase in efficiency appeared to be relatively independent of water content up to 30%. The high counting efficiency of non-polar steroids can probably be explained on the basis of the high extraction of these steroids from the aqueous phase into the toluene. However, this does not explain the increase over 100% seen in many of the experiments reported in this paper. This will be discussed in detail later. It is not possible to explain the difference in behaviour between the individual non-polar steroids reported on Table I without more detailed studies. The lower counting efficiency, dependent on water content, observed for the more polar steroids is presumably due to their poor partitioning from the water into the toluene phase. Addition of ammonium sulphate to the aqueous phase increases the extraction of the [^{3}H]-cortisol into the toluene and results in greater counting efficiency. A number of explanations of the 4-5% increase in relative counting efficiency seen when water is added to the toluene phosphor have been considered.

Firstly, it is unlikely that adsorption of the [^{3}H] steroid to the wall of the vial, which is removed by addition of water, has occurred since at least 50 µg of non-radioactive steroid is added to each vial as carrier. Secondly, the samples without water could be quenched and when water is added the quenching material partitions into the water phase with a resultant increase in efficiency. Although this could explain the observed results if a narrow counting window was used it is unlikely to have been responsible when the spectrometer was set to give maximum tritium efficiency. Thirdly, if the counter had been not set up properly at balance point for tritium,addition of water may have quenched the samples and shifted the spectra back into the window resulting in an apparent increase in counting efficiency. This would also be unlikely for a counter set up for maximum tritium efficiency. The most likely explanation is that it is due to an optical effect within the counting vial; possibly due to internal reflection at the toluene-water interface. Some of the photons which are normally lost through the base of the glass vial are reflected and eventually are observed by the photomultiplier tubes. A similar effect has been observed by Gordon and Curtis (3); they reported an increase in efficiency in samples containing optically diffusing white materials or surfaces. The enhanced efficiency they suggested resulted from a reduction in the amount of light lost through total internal reflection.

Although it is not easy to measure actual counting efficiency in such a heterogenous counting system as is described in this paper, this is not so important in saturation analysis where standards and samples are counted in the same way.

In summary, a cheap, efficient phosphor solution suitable for counting aqueous samples of [^{3}H]-steroid hormones is described. A feature of the toluene phosphor system for non-polar steroids is the independence of counting efficiency on water content.

Acknowledgements

This work was supported in part by Grants in Aid to the Howard Florey Institute from the National Health and Medical Research Council of Australia, U.S. Public Health Service, Grant No.HL-11580-06, Laura Bushell Trust and National Heart Foundation of Australia.

References

1. D. R. White, Intern. J. Appl. Radiat & Isotopes 19, 49 (1968).

2. G. A. Bray, Anal Biochem. 1, 279 (1960).

3. B. E. Gordon and R. M. Curtis, Anal Chem. 40, 1486 (1968)

SIMULTANEOUS MEASUREMENT OF FINENESS AND YELLOWNESS OF WOOL SAMPLES

A.M. Downes

C.S.I.R.O. Division of Animal Physiology, Prospect, N.S.W. Australia

F.E.L. ten Haaf

Philips Nuclear Applications Laboratory, Eindhoven, Netherlands

N. Jackson

C.S.I.R.O. Division of Animal Genetics, Epping, N.S.W. Australia

ABSTRACT

The effects of fibre diameter and yellowness on the liquid scintillation counting of wool containing [^{14}C]formic acid were studied. The main effect of an increase in diameter was to reduce the scintillation spectrum approximately uniformly over the entire energy range whereas the main effect of yellowness (caused by pre-heating the wool at 153°C) was to shift the spectrum towards the lower energies.

Fibre diameter (D) and yellowness index (Y; calculated from measurements of reflectance of light in the red, green and blue regions of the spectrum) were regressed simultaneously on a number of linear and quadratic functions of relative integral counting efficiency (E) and channels ratio (R), using conventional least squares methods. This confirmed that D is closely related to E ($r = -0.992$) but is also slightly correlated with R($r = 0.184$), and Y is closely related to R($r = -0.897$) and is slightly correlated with E ($r = -0.232$).

With the calibration equations thus derived, D and Y may be measured simultaneously by liquid scintillation spectrometry with 95% confidence limits of $\pm$ 0.45 μm and $\pm$ 4.5 units respectively.

INTRODUCTION

Objective measurements of the fineness (mean fibre diameter) and yellowness of wool samples are becoming increasingly important in the wool industry. Fineness is the main determinant of the quality, and hence the price, of different lots of wool; yellowness restricts the range of colours which may be used in dyeing wool and so may also influence price.

Until recently most measurements of fineness were made after the wool had been processed to form the approximately parallel array of fibres known as "tops", but there is a rapidly increasing demand for the testing of raw wool in order to specify wools objectively before sale. Fineness is usually measured with equipment which operates on the air-flow principle (1,2) : the resistance to flow of a given mass of wool varies with total surface area and hence with fineness. A yellowness index may be calculated from measurements of light reflectance at three wave lengths (3,4).

Another method of measuring fineness has been developed in which a liquid scintillation solution containing [^{14}C]formic acid is added to wool samples and the relative counting efficiencies are subsequently determined after the formic acid has been absorbed by the wool (5,6,7). The counting efficiency is related to diameter, due to the self-absorption of β-radiation within the fibres. The method has been calibrated with a set of standard tops of known fineness, and has been shown to produce results with sufficient accuracy for most commercial and research applications (8), even though this calibration includes small errors due to variable quenching. The most likely cause of quenching in raw wool or tops is yellowness. Some wools acquire a natural yellowness during their growth, while others become yellow during processing due to factors such as heat, ultraviolet light, or alkaline washing (9).

We have examined the possibility that both fineness and yellowness of wool samples could be measured simultaneously by liquid scintillation spectrometry. The use of colour quenching in the study of yellowing of wool fabric has been described (10). We have also obtained further evidence on the conditions in which the same fineness calibration may be used with different liquid scintillation spectrometers.

MATERIALS AND METHODS

Wool Samples. - Two groups were studied. Group I comprised six samples of raw wool from individual sheep (7), and six sub-samples of each of these wools which had been heated at 153°C for up to 8 hr to produce a range of yellowness. Each sub-sample (3g) was heated in a "drying pistol" in which the wool was held in a glass cylinder surrounded by the vapour of boiling anisole. Group II comprised eight standard tops (Interwoollabs, Brussels) whose mean diameters were based on measurements in about 70 laboratories, and 19 other tops exhibiting a wide range of yellowness.

The fineness of each sample was measured by the air-flow (1) and sonic (2) methods after calibrating the instruments with the standard tops. The heating described above produced no detectable change in diameter. Yellowness index, expressed as a percentage, was determined with a "Meeco Colormaster Model V" Colorimeter. Samples were prepared by carding 2.5g of clean wool, which was mounted on a white card behind PT300 cellophane and compressed laterally between glass plates. The colorimeter was calibrated to adjust for the effect of the glass and cellophane on the reflectance readings. Readings at three wave lengths were duplicated with the sample rotated 180° between readings and yellowness index was calculated as described by King (4).

Preparation of Samples for Scintillation Counting. - Each test specimen of wool (75mg; 88% dry matter) was placed in a glass ampoule (3ml) to which was added 3ml of toluene containing 3g ButylPBD/ℓ, 0.2g dimethylPOPOP/ℓ, 50 μl Triton X-100/ℓ, and 50μl [^{14}C]formic acid/ℓ. About 0.2 μCi ^{14}C was added to each ampoule in most experiments. Ampoules containing no wool but with extra Triton X-100 (60μl) (7) and with 3ml of the above scintillation solution were prepared with each batch to provide solution standards. After adding the solution, the ampoules were immediately sealed and heated at 70°C for 5 hr. For assay, each ampoule was placed inside a 20ml glass vial which was then filled with water to improve light collection (6).

A small proportion of the radioactive material was not absorbed by the wool but remained in solution. In order to determine the true counting rates attributable to the ^{14}C within the wool, some ampoules in each batch were opened and a measured volume of the solution withdrawn to enable the proportion of the added ^{14}C remaining in solution to be measured.

Radioassays. - Two Packard Model 3375 Liquid Scintillation Spectrometers and a Philips Model PW4510 Liquid Scintillation Analyser were used. The sample changers were held at 3°(3375s) and 16°C(PW4510). One channel in each instrument was operated in the integral mode, with a counting efficiency of about 94% for an unquenched ^{14}C standard solution (6). A second channel was set with its lower level raised to exclude about 40% of the pulses from an ampoule containing unheated wool from Group I. The upper levels were not used as the background in both channels was negligible compared to the sample count rates. The ratio of the counting rate in the second channel to that in the first was used to monitor quenching. These values, multiplied by 1000, were used as the "channels ratios". The integral counting rates were corrected by subtracting the counting rate due to ^{14}C labelled compounds left in solution and the corrected counting rates of the wool samples were divided by the corrected mean rate of the solution standards to give relative counting efficiencies.

The spectra of representative samples were determined in a multichannel analyser whose detector, preamplifiers, high voltage, coincidence circuitry and sum amplifier were those of a standard Philips PW4510 instrument. The conventional measuring channels, however, were replaced by a special amplifier/discriminator unit. This contained a linear pulse amplifier and 24 discriminator circuits, connected to various stages of the amplifier. The bias levels of the discriminators were arranged so that the instrument behaved as a 23 channel pulse-height analyser, all channels having the same relative width of 1.414 : 1. A pulse-height range from 2 millivolts to 5.793 volts was covered in this way. Pulses below 2 millivolts were registered in channel 0, and those larger than 5.793 volts were registered in channel 24.

Statistical Methods and Calibrations. - Using conventional least squares methods, fibre diameter (D) and yellowness index (Y) were regressed simultaneously on a number of linear and quadratic functions of relative integral counting efficiency (E) and channels ratio (R). The most complex model fitted was in the form of two simultaneous quadratic equations in which, for the ith sample,

$$D_i = a + bE_i + c\,E_i^2 + d\,R_i + e\,R_i^2 + \text{residual} \quad)$$

$Y_i = a' + b'E_i + c'E_i^2 + d'R_i + e'R_i^2 + \text{residual}$) ...1

and a, b, ...e, a'....e' = regression coefficients.

A number of less complex models (sub-sets of equations 1) were also fitted. The model with the smallest residual mean square was chosen as the calibration equation.

This technique of calibration corresponds to the "inverse" rather than the "classical" method of Krutchkoff (11), who showed that the former is superior. The distinction is of little importance in the present case because, although D_i and Y_i are the controlled variables they are not estimated without error. The inverse method had the additional advantage that the resultant equations estimated D_i and Y_i directly, and confidence limits for the estimates could be obtained from conventional multiple regression formulae (12), while the classical method required solving simultaneous quadratic equations.

RESULTS

Yellowness Index. - When first measured by the colorimetric method, the yellowness indices of the unheated wools in Group I ranged from 19.9 to 23.8%, the coarser wools having the higher values, while the heated wools gave values of up to 61.4%. During the following six months the indices of the unheated wools increased slightly (to 20.6-24.4%) while those of the heated wools decreased by up to 7 units.

Eight representative wool samples (2.5g) with yellowness indices ranging from 22 to 38% were immersed in a liquid scintillation solution (100ml each) of the same composition as described above except that unlabelled formic acid was used. After heating at 70°C for 5 hr (7) the wool was washed with ethanol and light petroleum, conditioned, and yellowness index measured again. All values decreased by about 1.6 units, and the correlation between values before and after immersion was 0.988.

The yellowness indices of the wools in Group II ranged from 21 to 50%. This covers the range of natural yellowness observed in Merino wool produced in Australia (Jackson, unpublished).

Liquid Scintillation Spectrometry. - The effects on the scintillation spectrum due to the presence of wools varying in fineness and yellowness were first studied. The major effect of increasing diameter was to reduce the spectrum approximately uniformly over the whole energy range

(Fig. 1) whereas the major effect of yellowness was to shift the spectrum towards the lower energies (Fig. 2). Thus, it is legitimate to use a channels ratio method to correct the relative integral counting efficiency, and hence diameter, for the effect of yellowness.

An examination of the relationships between D, Y, E, and R was carried out with six test specimens of each wool sample, prepared by the procedure described and assayed twice with a counting time of two minutes per sample. The measurements of Y by means of the colorimetric method were made within two weeks of conducting the radioassays.

The proportion of added ^{14}C remaining in solution varied slightly, being larger with the finer wools and with the wools which had been pre-heated for the longest times. However, as the range of values (0.7-1.4%) was small, the mean (1.04%) was used as the correction for all samples.

The data for the wools of Group I were subjected to regression analysis to provide calibration equations. The correlations confirmed the results in Figures 1 and 2 - diameter is closely related to integral counting efficiency ($r = -0.992$) but is also slightly correlated with channels ratio ($r = 0.184$); yellowness index is closely related to channels ratio ($r = -0.897$) and is slightly correlated with integral counting efficiency ($r = -0.232$).

The "goodness of fit" of model 1 and its various sub-models is shown by the residual mean squares in the analyses of variance (Table I). It is clear that the full model is not necessary for either fibre diameter or yellowness index, but a different sub-model is appropriate for each. For diameter the equation with the smallest residual ($0.04284\ \mu m^2$) was:

$$\hat{D}_i = 202.36 - 4.381\ E + 0.02445\ E^2 + 0.00614\ R \quad \ldots\ldots\ 2$$

where $\hat{D}_i$ = the estimated mean fibre diameter of the ith sample.

For yellowness index the equation with the smallest residual ($3.962\%^2$) was:

$$\hat{Y}_i = -88.2 - 0.560\ E + 0.818R - 0.000947R^2 \quad \ldots\ldots\ 3$$

where $\hat{Y}_i$ = the estimated yellowness index of the ith sample.

For fibre diameter the correlation between D_i and $\hat{D}_i$ (multiple correlation coefficient) for the calibration

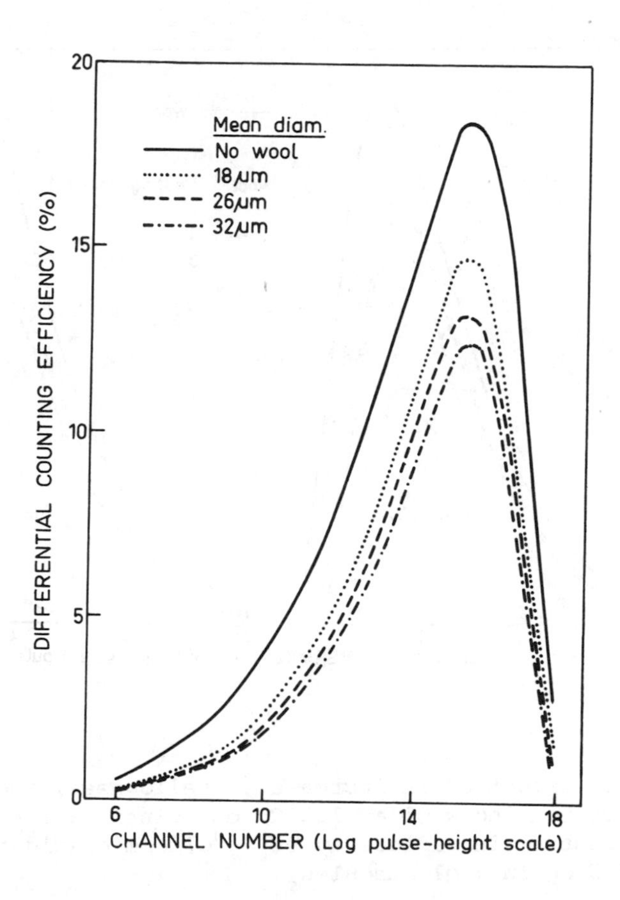

Figure 1. Effect of an increase in fibre diameter on scintillation spectrum in radioassay of [^{14}C]formic acid in wool (75 mg in 3 ml ampoules). The yellowness indices of these wools were in the range 20-23%.

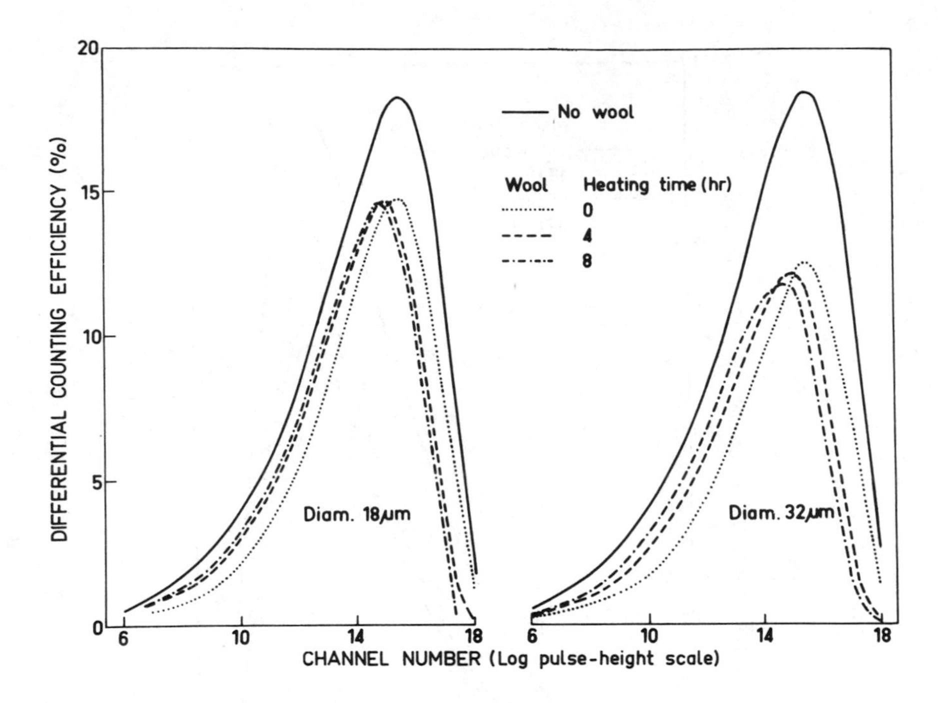

Figure 2. Effect of an increase in yellowness, due to pre-heating wool at 153°C, on scintillation spectrum in radioassay of [^{14}C]formic acid in wool (75 mg in 3 ml ampoules).

set of data was 0.995 (Fig. 3). For yellowness index the corresponding correlation between Y_i and $\hat{Y}_i$ was 0.916 (Fig. 4).

The precision of an estimate of diameter ($\hat{D}_i$) and yellowness ($\hat{Y}_i$) from a set of E_i and R_i values, using equations 2 and 3, is given in Table II. The estimates are slightly more precise at intermediate E_i and R_i values. The confidence intervals in Table II are appropriate when the same number of ampoules (six, each assayed twice) as in the calibration experiment are averaged to obtain E_i and R_i before using equations 2 and 3. If a multiple of that number were used, say N times as many, an approximate confidence interval would be that in Table II divided by $\sqrt{N}$. An exact confidence interval could be obtained (as were those in Table II) from formulae given by Draper and Smith (12).

An excellent correlation (r = 0.999) between observed and predicted values was obtained when equation 2 was used to calculate $\hat{D}_i$ for the wools of Group II (Fig. 3) which were not used in deriving the calibration equations. For the standard tops the maximum difference was 0.4µm for the sample of nominal diameter 34.9µm. For the other tops the maximum difference was 0.5µm for diameters below 35µm and 1.5µm for those above 35µm. The larger differences for the coarse wool were probably due to differential effects of medullation (13) on measurements by the two methods.

In general there was also a good correlation between the observed and predicted yellowness indices for the wool tops. However, the values of some of the yellower tops, as calculated from equation 3, were several units higher than the observed values (Fig. 4), showing that the calibration made only with heat-yellowed wools did not apply strictly to all of the tops.

In several experiments carried out over a period of a year, ampoules containing sub-samples of the wools from Group I were prepared using the procedure described. Throughout this period the samples were assayed on two of the counters (3375s) and the channels ratios were measured with the gain and discriminator settings originally chosen. The relative integral counting efficiencies remained constant, but the channels ratios varied significantly unless the instruments were carefully normalized on each occasion (Packard Instrument Manual). However, the ranking of samples on the basis of channels ratio remained the same (r = 0.98-0.99) when the results from different experiments or instruments were compared.

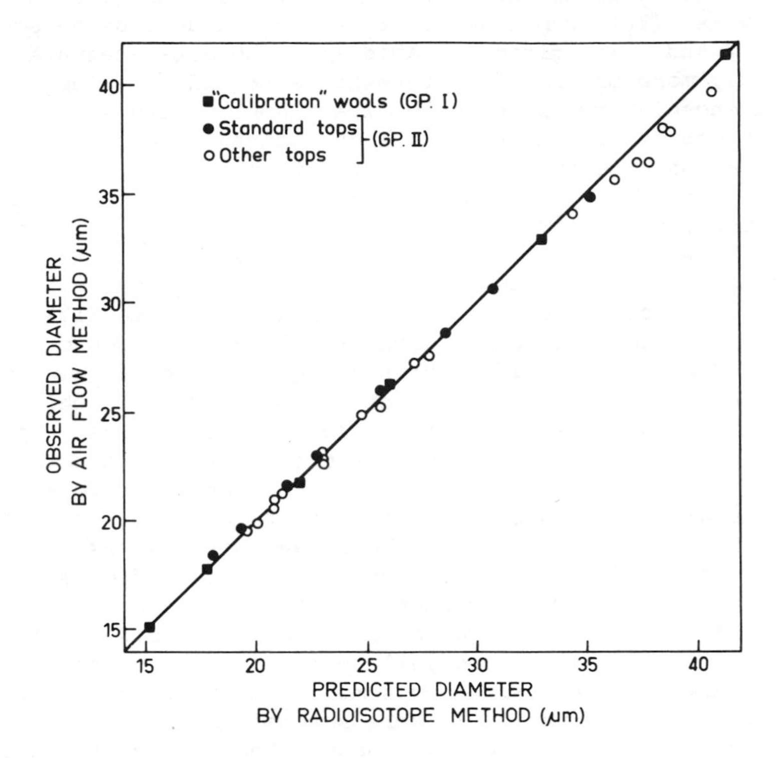

Figure 3. Comparison of mean fibre diameters predicted from calibration equation 2 (based on results for wools of Group I) with values obtained by air-flow (standard tops) and sonic (other wools) methods.

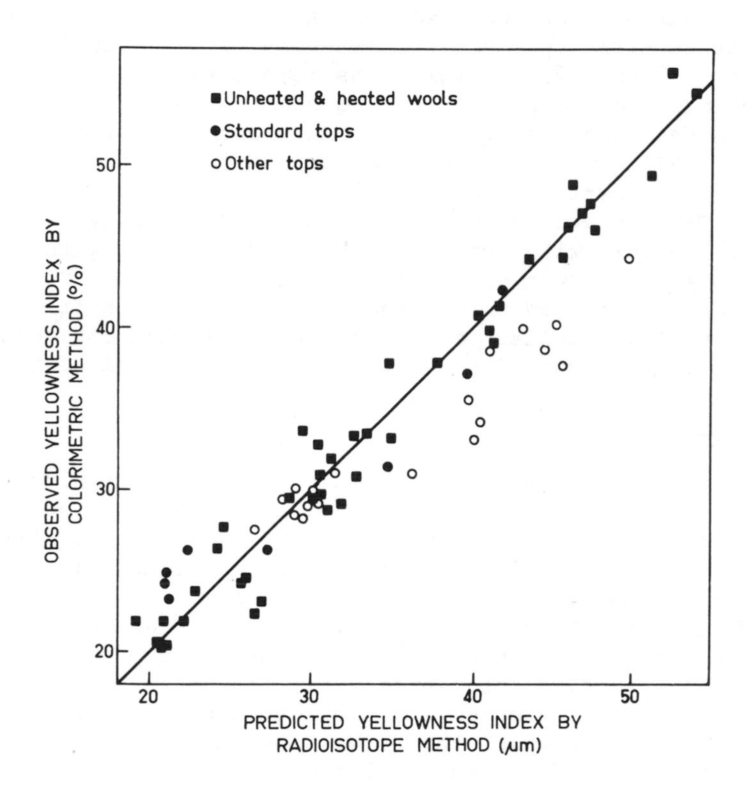

Figure 4. Comparison of yellowness index values predicted from calibration equation 3 (which was based on results for wools of Group I) with those obtained by measurements of reflectance in the colorimeter.

The integral counting efficiency of ^{14}C in solution was slightly different in the three instruments studied. Because of this the relative counting efficiencies of the wool samples differed slightly, but significantly, in the three instruments, being lowest in the instrument with the highest sensitivity. However, when the lower discriminator setting of the most sensitive instrument was raised so that the counting efficiency of the solution standards was the same in each instrument, the relative counting efficiencies of the [^{14}C]formic acid-wool samples were also the same, within experimental error, in the three instruments (Table III). This is due to the fact that proportionally more counts are lost with the solution standards than with the wools when the lower discriminator is raised. Apparently small differences in the average photocathode sensitivity of the photomultiplier pair in different instruments can be adequately compensated by varying the lower level of the measuring channel.

DISCUSSION

We draw three main conclusions from the results. Firstly, the fineness of wool samples can be measured by liquid scintillation spectrometry with good accuracy irrespective of yellowness, at least over the range of yellowness index observed in Australian Merino wools and in the standard tops. Secondly, a simultaneously measured channels ratio may be used to predict yellowness index provided other causes of quenching (pigmented wool, vegetable matter) are excluded from the samples and provided a standardized procedure is used in which the water content of the wool and the composition of the scintillation solution are kept constant. Thirdly, the evidence suggests that the same calibration equations can be used with different liquid scintillation spectrometers.

In measuring channels ratios, care is needed in checking for drifts in the liquid scintillation spectrometer. A set of unquenched ^{14}C standards and of standards containing various amounts of an appropriate yellow dye would enable the desired counting conditions to be set in a given instrument from time to time or in different instruments. Wool is not suitable as a standard for colour measurements because of the changes which can occur in its yellowness.

We have mainly studied the quenching effect produced by pre-heating wool. Electron microscopic evidence (14) shows that wool can withstand the amount of heat which we applied, without undergoing any detectable changes in its physical shape and characteristics except for the discoloration produced. Further research is needed on the degree of correlation of quenching with yellowness caused by other agents and on the correspondence between the measuring conditions, with respect to colour, of the scintillation counter and the colorimeter. Several points can be mentioned:

(i) Type of illuminant. The colorimeter used is a tristimulus instrument - that is, it illuminates the sample in turn with broad wavelength bands in the red, green, and blue regions. The yellowness index used in this study is a combination of the reflectances of the sample using all three illuminants - in fact it is (red-blue)/green. In the scintillation ampoule the "illuminant" is the scintillations of the primary scintillator as modified by the secondary scintillator or wavelength shifter - that is, there is only one illuminant instead of three and its spectrum is in the blue region. The channels ratio is therefore closely related to blue reflectance. However, for a wide sample of Australian raw wool (Jackson, unpublished), blue reflectance is very highly correlated with yellowness index, so that channels ratio can be used as a predictor of yellowness for raw wool, but not as a general chromatic measure such as might be required for pigmented raw wools or dyed tops.

(ii) Type of observer. The phototubes of the scintillation counter also differ from that of the "Colormaster", being sensitive mainly to pulse energies in the range corresponding to the blue region of the spectrum.

(iii) Geometry of sample viewing conditions. The "Colormaster" colorimeter illuminates the sample surface at 45 degrees and views the scattered light at 90 degrees. The sample is usually mounted in air. The sample in the scintillation counter ampoule is illuminated from all points within the ampoule and is viewed from a fixed position. The sample is immersed in toluene (refractive index 1.496). Since the refractive index of toluene is much closer to that of wool (1.548) than to that of air, the quenching effect could be much more a measure of transmission than reflection. The arrangement and packing of fibres is known to be important in the colorimetric measurements, but have not yet been investigated for the scintillation technique.

The scintillation counting method described in this

paper was developed primarily for measuring the fineness of small wool samples. A more accurate prediction of yellowness index could probably be obtained by changing the procedure, for example by using a much larger sample (and a larger ampoule), and accepting a larger quench correction and hence a less accurate prediction of diameter. Alternatively a separate technique, based on liquid scintillation counting of some other ^{14}C-labelled compound or another radionuclide in the presence of wool, could perhaps be developed for the prediction of yellowness index alone. However, in view of the much greater importance of diameter, the former approach of simultaneous measurement of both properties remains more attractive.

Although the prediction of yellowness index by this method is less accurate than that obtained with the colorimeter, the results so far obtained show that it is good enough for use in genetic selection of sheep or for allocating different lots of wool to one of several yellowness classes. At present these are the two main applications of yellowness measurement in the wool industry.

ACKNOWLEDGEMENTS

We wish to thank Mr W.H. Clarke for the measurements of fineness by the sonic method; and Miss B. Royal, Miss E. McKay, and Mrs L. Phillips for assistance with the yellowness index measurements, statistical calculations, and radioassays respectively.

REFERENCES

1. S.L. Anderson, J. Text. Inst. 45, 312 (1954).
2. A.E. Stearn, J. Text. Inst. 61, 485 (1970).
3. R.S. Hunter, J. Opt. Soc. Amer. 32, 509 (1942).
4. M.G. King, J. Text. Inst. 61, 513 (1970).
5. A.M. Downes and A.R. Till, Text. Res. J. 38, 523 (1968).
6. A.M. Downes, "Organic Scintillators and Liquid Scintillation Counting", 1031 (1971). Eds D.L. Horrocks and C.T. Peng. Academic Press Inc.,New York and London.
7. A.M. Downes, Appl. Polymer Symp. 18, 857 (1971).
8. A.M. Downes and C.H. Gray, in preparation.
9. F.O. Howitt, J. Text. Inst. 55, 136 (1964).
10. A.M. Downes, A.R. Till and F.G. Lennox, Text. Res. J. 35, 772 (1965).
11. R.G. Krutchkoff, Technometrics 9, 425 (1967).
12. N.R. Draper and H. Smith, 121 (1966). Applied Regression Analysis. Wiley, New York.

13. W. von Bergen, Wool Handbook Vol. 1. p.137 Interscience Publishers (1963).
14. R.S. Asquith, M.S. Otterburn, J.A. Swift, J. Text. Inst. 63, 544 (1972).

TABLE I

Analyses of variance of fibre diameter and yellowness index for Model 1 and various sub-models, fitted to the calibration set of data

Fibre Diameter			Yellowness Index		
Source†	D.F.	M.S.	Source†	D.F.	M.S.
R(E)	1	3363.9**	R(R)	1	3363.9**
resid.	40	1.372	resid.	40	20.393
$R(E^2\|E)$	1	50.501**	$R(E\|R)$	1	575.18**
resid.	39	0.1121	resid.	39	6.167
$R(R\|E, E^2)$	1	2.743**	$R(R^2\|R,E)$	1	89.798**
resid.	38	0.04284	resid.	38	3.962
$R(R^2\|E,E^2,R)$	1	0.0404 N.S.	$R(E^2\|R,R^2E)$	1	0.116 N.S.
resid.	37	0.04290	resid.	37	4.066

† The notation is exemplified by the following:

R(E) = reduction due to regression on E

$R(E^2|R,R^2,E)$ = reduction due to regression on E^2 given regressions on R, R^2, and E.

** Significant at 1% level.

N.S. Not significant.

TABLE II

Precision of predicted fibre diameter ($\hat{D}_i$) and yellowness index (Y_i) expressed as the 95% confidence interval for an estimate of $\hat{D}_i$ or $\hat{Y}_i$ from a single set of E_i and R_i values (mean of 6 ampoules) using equations 2 and 3.

E_i	R_i	$\hat{D}_i$	Precision for $\hat{D}_i$	$\hat{Y}_i$	Precision for $\hat{Y}_i$
55	465	38.6	± 0.47	56.8	± 4.9
55	565	39.2	± 0.44	41.2	± 4.3
55	615	39.5	± 0.44	26.3	± 4.3
65	465	24.0	± 0.46	51.2	± 4.9
65	565	24.6	± 0.43	35.6	± 4.1
65	615	24.9	± 0.44	20.7	± 4.2
75	465	14.3	± 0.47	45.6	± 5.0
75	565	14.9	± 0.44	30.0	± 4.2
75	615	15.2	± 0.45	15.1	± 4.3

TABLE III

Comparison of integral counting rates on three liquid scintillation counters

Solution Standards	Mean integral counting rate* (counts/min)			
	A 808520	B 795780	C 793920	D 795880
Wool Samples Mean diameter (μm)	**Relative counting efficiency* (%)**			
	A	B	C	D
15.1	75.0	75.5	75.5	75.4
17.8	71.5	72.0	72.1	72.0
21.8	67.9	68.2	68.3	68.1
26.2	63.9	64.3	64.3	64.3
32.9	58.9	59.3	59.4	59.3
41.4	53.8	54.0	53.9	53.8

* A = Philips Model PW4510; attenuator 0.0, discriminators, 40-∞

B = Philips Model PW4510; attenuator 0.0, discriminators, 320-∞

C,D = Two Packard Model 3375; gain 100%, discriminators, 20-∞

In this experiment each ampoule contained about 0.38 μCi ^{14}C.

DATA PROCESSING FOR A MULTI-USER SYSTEM WITH A SMALL ONLINE COMPUTER.

P.A. Gresham and T.P. Carter,
Department of Experimental Pathology,
University of Birmingham Medical School,
Birmingham B15 2TJ. England.

Abstract. The online processing of the output from liquid scintillation counters with small computers (e.g. 8K memory) raises a series of problems, usually arising from the limitations in program size and in the relatively slow methods of changing programs in store.

A Searle PDS/3 8K computer, both with and without a dual magnetic tape cassette unit, has been used to operate online to several counters. The basic programs derive and store standard curves and apply them as necessary to compute disintegration rates or handle data from radio-immunoassay procedures. User identification is provided, and instructions are given by dialogue programs.

The procedures and times for changing programs are considered; also the required automatic system of checks on the processing procedures and the results, such as graphical displays of standard curves, curve limits, etc. Details of some erroneous results and of limitations imposed by small programs are discussed.

Introduction. The Faculty of Medicine and Dentistry at the University of Birmingham operates a centralised counting service for radioisotopes: over 50 research projects and several postgraduate courses produce samples for liquid scintillation counting. In our experience, the production of samples for counting in most biological and medical work can be termed 'routine', since the sample preparation procedures are well defined for each project and many of the problems have been minimised; thus although care is still needed, the procedures are repetitive and routine. Numerous samples can be rapidly produced and counted automatically on multi-sample machines set for sample or external standard channels ratio determination of quenching. All too often the bottleneck in the procedure occurs at the stage's of data handling and processing, particularly with variably quenched or multi-labelled samples. Consequently, automatic data processors are becoming routine adjuncts to liquid scintillation counting equipment, with the primary objectives of reducing the processing time for the results and the number of man hours expended, and increasing the accuracy and usefulness of the final data by reducing arithmetic errors and utilising more sophisticated processing techniques.

The increasingly common use of small computers (e.g. 8K memory) means that large, complex programs cannot be handled, and small programs have to be interchanged frequently. The fact that many users, with a wide variety of projects, utilise the machines means that programs have to be fairly general in application and flexible, rather than designed for individuals. The provision of a service facility means that most users know little (and do not wish to know more) about the data processing: the final result becomes the only important feature of the process. This makes it imperative to have a comprehensive series of checks of the processing and the data, with an automatic series of visual warnings.

This paper is therefore concerned with practical aspects of data processing as we have experienced and applied them, rather than with advanced aspects of programming, which have been dealt with by many other workers (e.g. 1,2,3).

Materials and Methods. The systems used in these studies were: 1) Philips LSA 00 with 4 counting channels, 2 external standard channels and built-in data analyser. 2) Nuclear Chicago (Searle) Isocap 300 refrigerated counter with PDS/3 8K computer and Teletype. 3) Two Nuclear Chicago (Searle) Isocap 300 refrigerated counters with PDS/3 8K computer, Sykes dual magnetic cassette tape unit and three Teletype outputs.

The programs used were developed by Nuclear Chicago (G.D. Searle, Ltd.) for use in the processing of liquid scintillation results and in radioimmunoassay and competitive protein binding experiments. The programs are designed for multi-user applications, and 12 independent sets of quench correction curves for each counter can be held and recalled automatically.

Results and Discussion. The simplest forms of on-line data analyser used to produce disintegrations per minute (dpm) require the user to predetermine and manually enter the quench correction parameters; the Philips LSA 00 offers this type of data processing. The accuracy is often rather restricted and built-in checks are minimal or missing, but if these limitations are taken into consideration the simple data analysers can be extremely useful.

Computers offer a much more comprehensive system. Easy and rapid access to a large computer is often the best arrangement, since large and sophisticated programmes for a wide range of users can be accommodated, but in many cases this is not feasible and a small computer with limitations in program size and changing has to be utilised.

A computer for use in a multi-user service in a biological or medical environment requires programs, as a basic minimum, for producing dpm and analysing radio-immunoassay results. These programs must handle the results from STANDARD samples to produce, print and store the required parameters of the standard curves, and check the validity and suitability of the curves. The relevant parameters then have to be applied to the results from EXPERIMENTAL samples to produce the required figures (e.g. dpm), after which a basic statistical analysis must be carried out, together with a series of quality control checks. The multi-user application requires a positive IDENTIFICATION to be made (and printed) of the user, the experiment and the set of standard parameters used in the calculation. The instructions to the computer on how to handle each set of standards and experimental samples, and information for identification, are usually given by means of a DIALOGUE. The computer asks a series of questions via the Teletype, and the user types the answers: in a computer with a small core memory, the dialogues may have their own set of programs, to be called up independently of the programs they control.

In a large computer, the programs can all be held in core or on disc and called up rapidly. A major problem with a small computer, particularly handling a variety of results from several counters, is the time taken to change programs. Read-in time for a program occupying 4K of store (e.g. our programs for generating standard curves, or those for handling experimental results) is approximately 15 min using a Teletype paper tape reader, while our radioimmunoassay program takes 25 min. Use of a fast tape reader reduces this time considerably, but manual handling of paper tape is eradicated only by use of a backing store. Our programs are now held on magnetic tape cassettes, and called up via the Teletype keyboard. Call-up time for the dialogues and for dpm programs is 25 to 35 sec, and for the radioimmunoassay system is 55 sec. When the computer is on-line to the counters, it is programmed to call up the requisite programs automatically. The 'dialogue' to give instructions to the computer on how to handle results from a set of samples takes up to one minute for dpm estimations and about 3 minutes for radioimmunoassay procedures, on each set of samples. Three minutes may seem a long time, but the information given (number, replicates and dilutions of standards and unknown, user identification, etc) enables the computer to process completely the data from standard and experimental samples in almost any radioimmunoassay or competitive protein binding assay.

The aspects of time taken in changing programs, and in dialogues, is of particular significance when a small computer is used by many research workers with a variety of problems. A more important problem in the multi-user context is the provision of visual checks on the data and its processing. When one or two people are involved in the development of programs for the computer they rapidly become aware of the limitations of their processing system. Workers in biological or medical fields who use scintillation counters as just one method in their work are usually (not always) relatively uninterested in how a program works and its limitations: as long as a printed result is obtained the user is happy, despite the fact that computers are notoriously lacking in intelligence. The critical ability of the user is often considerably reduced. For these reasons it is even more necessary than usual to have a comprehensive, visual series of checks on processing and the results.

With the possibility of using up to 12 standard curves for each counter, and with many users utilising the counting service, it is important as the first step in eliminating errors to have a printed record of the experimenter, the experiment, and the set of parameters applied to the results: our system uses a fairly standard display as follows:

USER ID. GRESHAM, DIGOXIN. CURVE ID. 3H-SCR

Since many groups of users prepare their own standards, which may then be utilised by other groups, it is necessary to print the basic data for each group of standards, preferably with details of the closeness of the standard curve to each standard point. Thus the range and distribution of the standards is displayed as well as the 'goodness of fit' of the curve. A typical print-out for a set of aldosterone standards in a radioimmunoassay experiment is shown in Table I.

TABLE I. COMPUTER PRINT OUT OF STANDARDS DATA FOR RADIOIMMUNOASSAY EXPERIMENT

ACTUAL DOSE	COMPUTED DOSE	DIFFER-ENCE	PERCENT BOUND
8.0	7.6	0.4	4.68
4.0	4.2	0.2	5.76
2.0	2.1	0.1	7.40
1.0	0.8	0.2	10.72
0.5	0.6	0.1	12.44

The computed curve must be displayed graphically: this is demonstrated clearly by the computed quench correction curve shown in Figure 1, for a set of ^{14}C standards; the curve passes very close to each standard, yet is obviously invalid. An interesting point is that these standards were purchased from a large manufacturer of scintillation counters. The distribution of quenched standards made by our users is usually very much better (but is occasionally much worse!). Another point of interest is that a fairly generally accepted 'rule-of-thumb' is that the power, n, of the fitted curve should not exceed one-half the number of standards, and the curve in Figure 1 obeys this rule: it is a 4th power (quartic) curve and there are 9 standards. It is obvious that rules-of-thumb are potentially dangerous.

This type of procedure is important in all computerised systems - it is indispensible in a multiuser system. The time taken for a print out of these results derived from a set of standards using a Teletype is only 3 to 5 minutes.

However, the print-out of a standard curve can lead to a false sense of security, particularly if a line is strongly curved: an example is shown in Fig 2 where the curve as printed out (Fig 2a) gives an impression of a good fit. Inclusion of the B-zero value, and reprogramming of the computer to align the points in a less strongly curved array, showed the poorness of fit at low concentrations (Fig 2b). A strongly curving or particularly steep or shallow line can mask numerous inaccuracies, particularly in radioimmunoassay procedures where points near the curve limits often have a relatively low level of accuracy. In general, the printed curve should be as nearly rectilinear as possible, and of gradient fairly close to +1 or -1. This can usually be programmed.

The use of a computer makes it a simple process to do a basic statistical analysis of the data: thus when dpm are being calculated, a figure for the best attainable level of accuracy is given. The Nuclear Chicago system prints

DPM Standard Deviation
(absolute) (%)

This gives the accuracy of counting assuming perfect standards, curve fitting, counter and samples. In other words, it gives the level of accuracy that <u>cannot</u> be

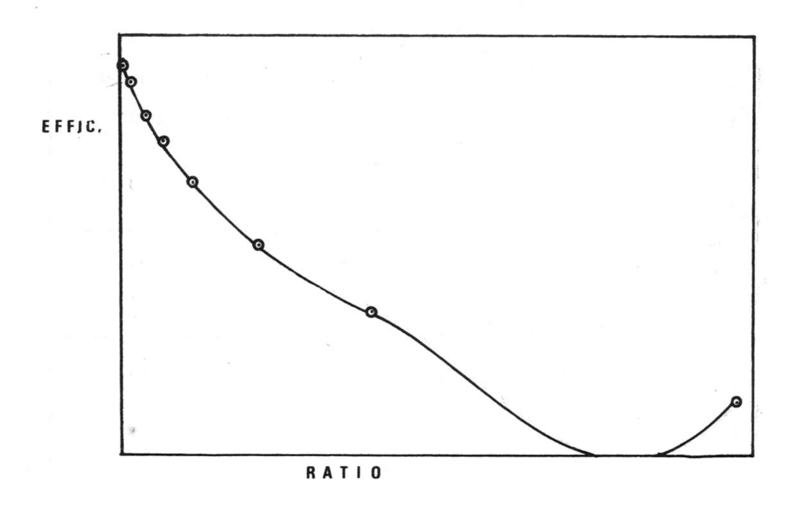

Figure 1. ^{14}C quench correction curve as derived by computer program.

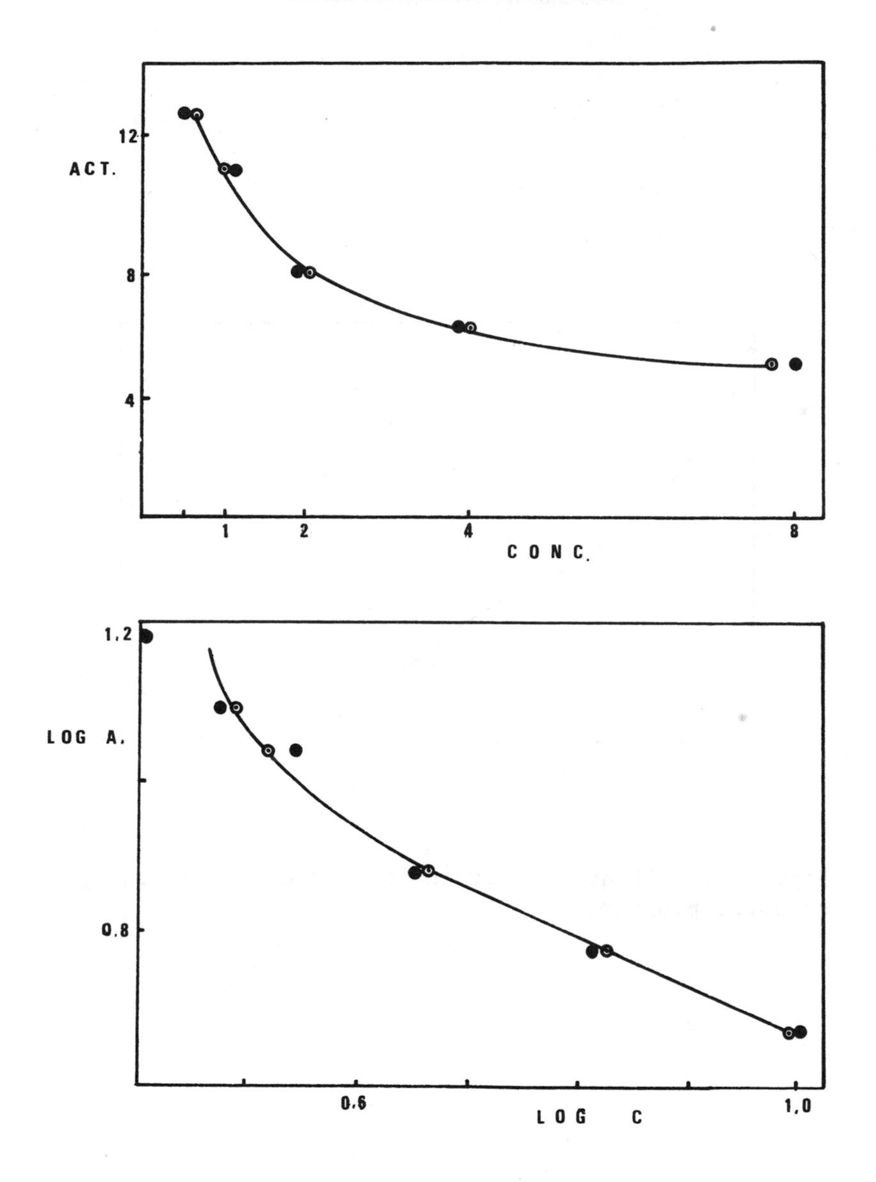

Figure 2. Digoxin radioimmunoassay standard curve. (2a) Linear axes of digoxin concentration and bound activity. (2b) Logarithmic transformation with B_0 value included.

bettered without counting for a longer period. It shows the limit of accuracy but is often taken as representing actual accuracy. Needless to say, the true level of accuracy is usually many times lower.

When the results are printed, a warning must be displayed if the experimental samples fall outside the range of the standards: i.e. if the limits of the quench correction curve are exceeded. The warning printed is:

RESULTS ARE EXTRAPOLATED

We have experienced some highly unusual results when the samples fall outside the range of the standards: in one case we had a group of tritiated samples which the computer calculated as having a counting efficiency of between 700% and 3000%! At that time, no warning messages were printed and a certain amount of confusion ensued. Equally serious are the cases where efficiency of counting is being estimated by the sample channels ratio technique, and some low activity samples are included. The poor statistics of counting can lead to low estimates of counting efficiency, resulting in a higher (and, at first sight, reasonable) estimate of dpm. For this reason it is important to print out channels ratio <u>and</u> efficiency figures.

The use of slightly larger computers allows checks to be made for the presence of luminescence (4), the validity of the set of quenching parameters applied (5), and correct functioning of the counting system (4,5). Application of these type of programs in a small computer would be extremely useful.

The emphasis on a built-in quality control system in a data processing system for liquid scintillation counting is particularly important in a multi-user system, since the programs have to be fairly flexible and general; and also because an expert in (for example) immunology cannot also be expected to be equally expert in all aspects of sample preparation, scintillation mixtures, figures of merit, computer programming, etc., as well as the finer points of radioimmunoassay, gas-liquid chromatography and all the other procedures in common use. This is one reason for running a centralised counting service.

In addition, when a computer is introduced into a scintillation counting system on a routine basis with programs provided, many users apparently lose their critical abilities when surveying their results. Many research workers in the Medical Faculty had a remarkable facility for scanning columns of printed counts and picking out any erroneous results, due to factors such as chemiluminescence. When a computer is introduced into a system, it becomes a 'magic' black box and printed results are accepted as absolute truth. The oft-repeated warning is ignored, that if rubbish is inserted at one end of a computer, one is liable to obtain rubbish accurate to six-significant figures at the other end.

Overall, it can be seen that the use of a small computer in a liquid scintillation counting system offers almost all the advantages which are given by a large computer, and rather more disadvantages which are mainly caused by restricted program capacity and the consequent need to change programs (often by relatively tedious methods). There are significant advantages in using a computer, particularly in the increased accuracy and power of calculation, and the considerable saving in man-hours spent processing data, but in a multi-user service these advantages are tempered by the need to cater for research workers who are not really interested in the functioning of computers (except in so far as they produce a result) or in scintillation counting (except as a convenient technique). However, in our experience, virtually all the problems originate basically from poor preparation of standard or experimental samples, or experimental design: a computer rarely initiates problems, but merely tends to amplify other problems.

References.

1. J.L. Spratt, Intern.J. Appl. Radiation Isotopes 16, 439 (1965).

2. J.L. Spratt and G.L. Lage, Intern. J. Appl. Radiation Isotopes 18, 247 (1967).

3. M. Strolin-Benedetti, P. Strolin and B. Glasson, Computers Biomed. Res. 2, 461 (1969).

4. D.S. Glass, in Organic Scintillators and Liquid Scintillation Counting. p.803 (D.L. Horrocks and Chin Tzu Peng, Ed.). New York and London: Academic Press (1971).

5. J.F. Lang in Organic Scintillators and Liquid Scintillation Counting, p.823 (D.L. Horrocks and Chin Tzu Peng, Ed.). New York and London: Academic Press (1971).

PLUTONIUM ANALYSIS BY LIQUID SCINTILLATION TECHNIQUES

John J. Fardy

Chemical Technology Division, Australian Atomic Energy Commission, Private Mail Bag, Sutherland, N.S.W. 2232, Australia

Abstract. Liquid scintillation counting procedures developed during studies of the aqueous complexes of plutonium and having high accuracy and simple sample preparation are discussed. Alpha counting techniques for the assay of plutonium will not discriminate against the natural ^{241}Am content. However, the ^{241}Pu parent is the only beta emitter present in high purity standard plutonium metal and measurement of the beta activity with a commercial liquid scintillation counter effectively differentiates plutonium from americium. This paper examines a variety of scintillator solutions, both commercial and laboratory-prepared mixtures as well as extractive scintillators, for the determination of ^{241}Pu in acid media. Liquid scintillation spectra were repeatedly measured over a 24 hour period, counting efficiencies determined, the maximum solubility of the aqueous phase in each scintillator solution recorded, and the extent of plutonium adsorption on the counting vials examined for a variety of solution conditions including plutonium concentration, valency of the plutonium ions, acidity and acid type. Based on these results suitable counting media are suggested.

Introduction. While liquid scintillation counting is normally associated with the measurement of the weak beta emitters, tritium and ^{14}C, in organic samples, more inorganic chemists are using this technique for counting other beta nuclides and alpha emitters. The determination of the alpha activity associated with the actinides using this technique has been well described in a recent paper by Ihle (1). However, no liquid scintillation methods for the radiometric assay of plutonium were listed in a recent hand book on plutonium (2). The determination of ^{241}Pu in a mixture of plutonium isotopes was described in 1958 (3) and the liquid scintillation determination of plutonium in biological samples was described recently (4).

While plutonium metal can be purchased as a high purity standard it is not normally isotopically pure. The only important impurity is ^{241}Am from the ^{241}Pu parent. Alpha counting will not discriminate against the ^{241}Am content and many investigators neglect its presence. This study demonstrates the use of beta activity measurements of ^{241}Pu with a liquid scintillation spectrometer to determine the plutonium content of acidic samples. Unfortunately, the low beta energy (18 keV maximum) makes the measurement of ^{241}Pu difficult and this paper examines some of these problems.

Materials and Methods. All chemicals used in these studies were of reagent grade.

A concentrated plutonium solution was prepared from high purity plutonium metal (NBS, Washington) and various plutonium solutions in the +3 and +4 state were prepared as described by Fardy and Pearson (5). These were added to perchloric acid, and mixtures of perchloric/sulphuric, perchloric/hydrochloric and perchloric/nitric acids in which the ionic strengths and total acidity were maintained constant at 1 or 4$\underline{M}$ and the concentration of added acid (H_2SO_4, HCl, HNO_3) maintained constant at 0.5$\underline{M}$. Since the isotopic composition of the plutonium metal was known the theoretical beta and alpha activity of each solution could be calculated and used as a basis for counting efficiency measurements.

Liquid scintillation counting was performed with an Ansitron liquid spectrometer with a freezer temperature of -5°C. Discriminator circuits permitted manual plotting of the energy spectrum. Alternatively the signal was fed to a 256 Channel Analyser (RCL) and the spectra recorded either as a digital output or on a chart recorder. The operating conditions were selected to obtain maximum counting rate for the beta activity of the ^{241}Pu but with the low energy discriminator adjusted so that the preamplifier noise was eliminated. Figure 1 shows typical beta and alpha spectra of a high purity plutonium sample. The alpha particles from ^{239}Pu, ^{240}Pu and ^{242}Pu yield pulses at an energy level of a corresponding 0.7 MeV beta ray.

A number of scintillator mixtures were examined. Two commercially available liquid scintillators, NE-220(NE) and Insta-Gel (IG), were used but later modified by the addition of 4g tri-n-octylphosphine oxide (TOPO) per 100 ml of scintillator (NE/T and IG/T). Another scintillator mixture (O), comprising 5g of 2,5-diphenyloxazol (PPO) and 50 mg of p-bis- 2-(5-phenyloxazol) -benzene (POPOP) per litre of 38.5 vol % p-xylene, 38.5 vol % p-dioxane and 23.0 vol % absolute ethanol, and used as a general scintillator at Oak Ridge National Laboratory (ORNL), was also studied. Other alternatives of this mixture involved the addition of TOPO (O/T) and/or naphthalene (O/N and O/N/T). Finally an extractive scintillator recommended by McDowell (6) which was made 0.01 - 0.2$\underline{M}$ in an organic soluble extractant was tested.

Each scintillator was compared by adding 100 to 500 μl of an acidic solution of plutonium, to 10 ml of the scintillator in a scintillator vial, and measuring the beta and alpha activity as a function of time (0,4 and 24 hours). Similarly, the energy spectra were recorded over the same time interval. At the conclusion of the 24 hour period the adsorption of the plutonium on the walls of the counting vials was measured. Each sample was removed from the counter, the solution discarded, the vial shaken with 10 ml of the scintillator for one minute, this removed and a fresh aliquot of scintillator added prior to recounting.

Insta-Gel was tested as an emulsion scintillator. The scintillator (10 ml) contained in counting vials were stood in a bath of hot water, 100-500 μl aliquots of Pu III in 0.5$\underline{M}$ $HClO_4$/0.5$\underline{M}$ H_2SO_4 added to the respective bottles and sufficient demineralised water or acid mixture introduced to give a total aqueous phase of 7 ml. The vials were shaken while being cooled to room temperature with tap water. Each vial was counted after 30 minutes in the freezer of the liquid spectrometer.

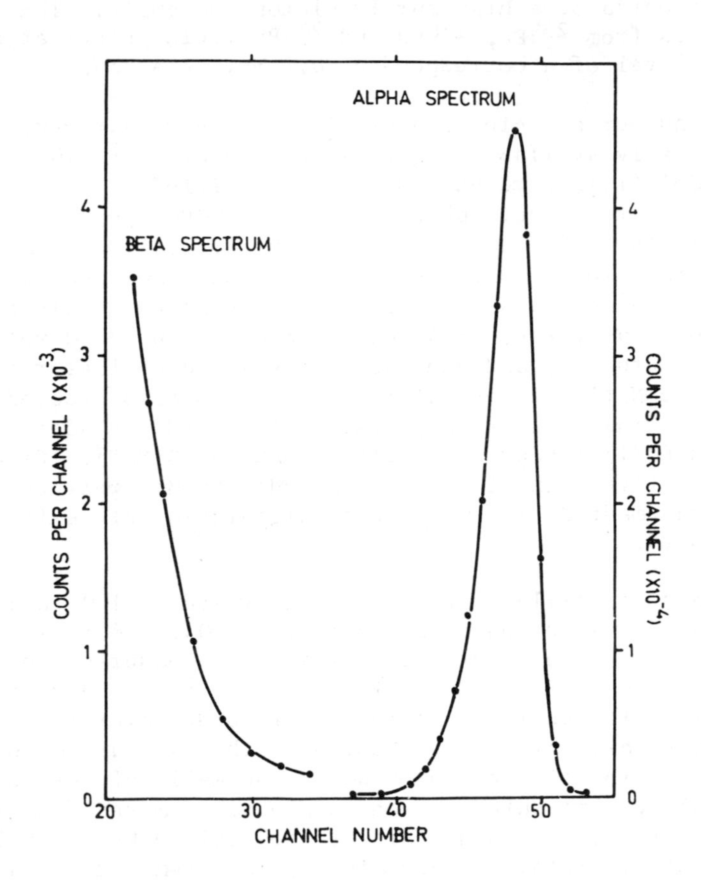

Figure 1 Beta and alpha liquid scintillation spectra of NBS plutonium

Finally, the maximum volume of water and mineral acid mixture that can be added to 10 ml of each solvent system, without causing cloudiness or separation, was determined at room temperature and -5°C.

Results and Discussion. Initial studies showed that the most marked variation in the scintillation counting occurred when the concentration of the plutonium solutions was low. Therefore, all scouting tests were restricted to 100 μl aliquots of 1.125×10^{-3}M of PuIII solution (5×10^3 - 10^4 cpm).

Figure 2 shows a variation observed in some liquid scintillation spectra over 24 hours. It illustrates a marked reduction in the alpha spectrum accompanied by a shift to lower energy of the peak. This change produced a marked shift in the beta spectrum causing an increase of 20% in the total beta activity. The decrease observed in alpha activity opposed this increase. These changes normally occurred through a combination of colour quenching by water and acid, and adsorption of the plutonium on the surface of the counting vial. Another change was observed spasmodically in ORNL type scintillators and which is the opposite of that observed in Figure 2. Initially there was poor resolution of the alpha spectrum while the minimum between the alpha and beta peaks showed an unusually high count rate. However, after a time interval of 24 hours, the plutonium spectrum returned to normal. This change was characterised by an increase of the total alpha activity but a decrease in the beta activity. We have no explanation for this reversal.

Two other forms of spectral changes were observed during the course of this study. A similar reduction in the size of both peaks, which resulted in a decrease of both beta and alpha activity, was attributed to a loss of plutonium from solution either as a precipitate or by surface adsorption. Liquid scintillation counting in 0.5M $HClO_4$/0.5M H_2SO_4 solutions usually yielded this phenomenon. The other change involved an overall increase in the size of the peaks. Under the initial extraction conditions the plutonium was not transferred quantitatively from the aqueous phase to the solvent. On standing, the plutonium

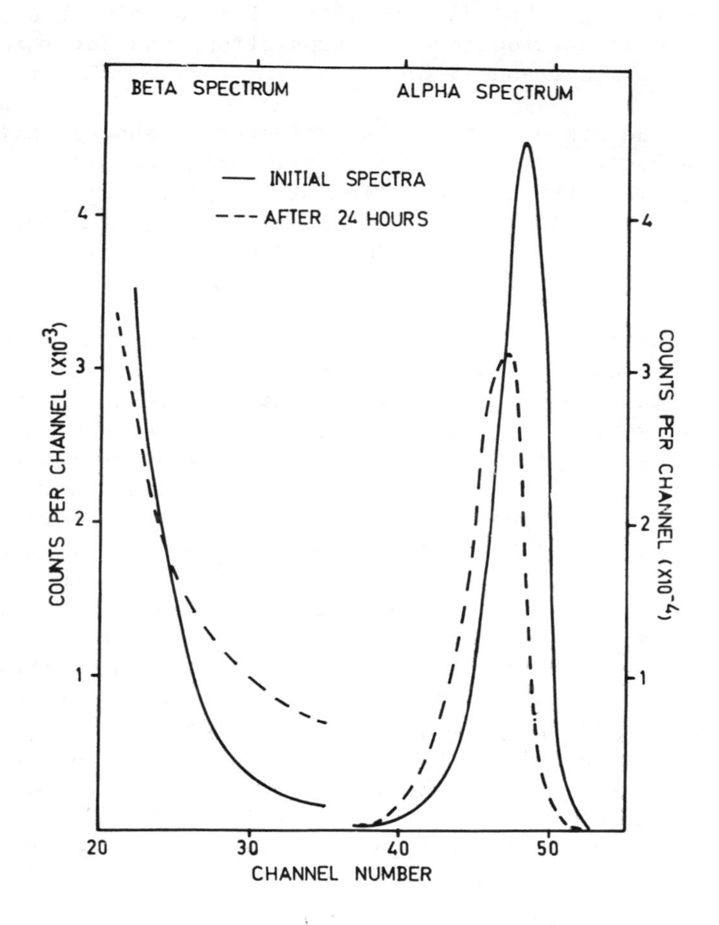

Figure 2 The effect of ageing on the liquid scintillation spectra of plutonium in the scintillator, NE-220

continued to diffuse slowly into the scintillator.

Tables I - IV summarise the results for the comparative tests on seven scintillator mixtures which gave homogeneous solutions of the plutonium samples.

Table I compares the efficiency of each scintillator for the detection of beta and alpha activity. These are marginally dependent on the acid mixture and most dependent on the scintillator type. Only NE-220 produced coloured solutions but the intensity was reduced when TOPO was added, and led to a large increase in its efficiency for beta detection. Generally $HClO_4/HNO_3$ was the preferred acid medium while the presence of H_2SO_4 decreased the efficiency. However, TOPO reduced most acid effects. The more energetic alpha readiation was less susceptible to changes in the aqueous or scintillator media but the small changes observed parallel those obtained for beta detection. Based on these figures the preferred scintillators were placed in the order NE-220/TOPO > Insta-Gel/TOPO > Insta-Gel.

The change in activity recorded over 24 hours for each scintillator mixture is summarised in Table II. The magnitude and sign of these changes can be correlated with spectral changes discussed earlier in this paper. Since the beta activity was about one fifth of the alpha activity the largest changes occurred in the measurement of ^{241}Pu. These figures again showed that NE-220/TOPO was the best counting medium followed by ORNL/TOPO and Insta-Gel/TOPO. The presence of H_2SO_4 caused instability in the latter scintillator which contrasted its extreme stability in other acid mixtures. This follows our initial observation that a 100 μl sample of $HClO_4/H_2SO_4$ produced a slight cloudiness but on further additions of this acid solution the scintillator cleared.

Table III lists the adsorption of plutonium on the surface of each counting vial as a percentage of the original activity. Since the beta activity was more susceptible to the nature of the scintillator all surface adsorption calculations were based on the measured alpha activity. These results explain the difficulty found when counting plutonium in $HClO_4/H_2SO_4$ solutions. Plutonium losses by adsorption were highest from this acid mixture especially when using the commercial scintillants, NE-220 and Insta-Gel. The addition of TOPO stabilised the plutonium in NE-220 but could not prevent significant adsorption losses from Insta-Gel. Past experiences in the solvent extraction of actinides from sulphate solutions (7) suggested that the addition of a primary amine, Primene JMT (JMT), would correct this situation. Tests with Insta-Gel/TOPO made 0.01M in JMT confirmed this. Adsorption was reduced to less than 4%. The scintillant mixture, ORNL/TOPO, was the best choice, when adsorption losses are required to be minimised, followed closely by NE-220/TOPO and ORNL/N/TOPO. The low rating of scintillator, Insta-Gel/TOPO, was due to its adverse reaction in sulphate systems but the addition of JMT upgraded this scintillator to the level of NE-220/TOPO.

The quantity of aqueous phase that can be maintained as a single phase for each of the scintillators is summarised in Table IV. Both Insta-Gel and Insta-Gel/TOPO mixtures were far superior to the remaining cocktails. An increase in acidity from 1 to 4M did not decrease this capacity. The scintillators not containing TOPO showed a low tolerance for the aqueous phase as its acidity increased. However, TOPO stabilised these scintillants towards highly acid solutions and in combination with some scintillators, particularly NE-220, significantly increased their capacity. For NE-220/TOPO this figure increased from 0.86 ml for 1M $HClO_4$ to 1.50 ml for 4M $HClO_4$. The figures in Table IV were reduced by 0.05 - 0.1 ml when remeasured at $-5^{\circ}C$.

Based on the results of the above studies, the scintillator NE-220/TOPO was the obvious choice as the best solvent system for incorporating various acid solution of plutonium into a homogeneous solution for the liquid scintillation counting of ^{241}Pu. Our second choice for continued study was the mixture Insta-Gel/TOPO. Its high capacity for various aqueous media, coupled with its relative high efficiency for the detection of ^{241}Pu suggested it as a useful alternative, despite its poor performance in sulphate systems (which can be corrected by the addition of JMT).

Both scintillators were re-examined after the volume of 0.5M $HClO_4$/0.5M H_2SO_4 added to 10 ml of the scintillator was increased from 100 μl to 500 μl. While the efficiency for ^{241}Pu detection decreased from 19.0% to 9.8% in NE-220/TOPO, the quenching in Insta-Gel/TOPO was less intense and the efficiency decreased from 16.8% to 12.6%. Even more surprising was the reversal in the stability of the activities in these scintillators measured as a function of time. Over 24 hours the total beta count registered in NE/TOPO changed by 0.77% for a 100 μl aliquot size but decreased by 30% for a 500 μl sample. The reverse trend occurred in Insta-Gel/TOPO. The 16.8% loss in the beta activity for a 100 μl sample was totally eliminated when a 500 μl aliquot was used. Changes in the alpha activity were unaffected by aliquot size when using NE-220/TOPO. However, the stability in Insta-Gel/TOPO was similar to that described for the beta activity.

Increasing the total acidity of the plutonium sample from 1 to 4 M reduced the efficiency for ^{241}Pu detection to 15% in both scintillators, yielded no appreciable change in the total alpha efficiency and gave no noticeable change in the measured activities over 24 hours. This suggested that TOPO stabilised the scintillator against acid effects and the changes wrought by increasing aliquot size were largely due to the water.

The valence of the plutonium solutions was increased from +3 to +4 and the above tests repeated. This change did not alter either the efficiencies or stabilities from that recorded for Pu III solutions. The same problems plagued the $HClO_4/H_2SO_4$ system.

We have investigated the use of Insta-Gel as an emulsion scintillator for the detection of ^{241}Pu. The stabilised Insta-Gel/TOPO was also included in this study. The Insta-Gel was superior to Insta-Gel/TOPO. The presence of TOPO led to the formation of clouded or intensely white gels when acid or water was added. This severely quenched the ^{241}Pu activity and to a lesser degree the alpha activity. When TOPO was omitted the alpha activity could be determined with greater than 95% efficiency, independent of water or acid addition. In contrast, the counting efficiency for ^{241}Pu was reduced from 13.2% to 7.4% when water was added, and to 4.8% with the addition of acid. In all instances the variation in the total beta activity was less than 3% over a 72 hour period.

Finally, we examined some extractive scintillators. Our laboratory experiences with the solvent extraction of plutonium and other actinides (8, 9) suggested a study of the organic extractants, TOPO, di-(2-ethylhexyl) phosphoric acid (HDEHP), JMT and tri-n-octylamine (TNOA). Initial studies were made with 10 ml of the scintillator containing $0.1\underline{M}$ extractant and shaken for one minute with 100 μl of Pu III contained in various acid mixtures whose total acidity was one molar. The alpha activity was used to follow the degree of extraction. In no instance did a single extractive scintillator quantitatively extract plutonium from all acid solutions. Both TOPO and HDEHP extracted all the plutonium and give 21% counting efficiencies for ^{241}Pu from $0.5\underline{M}$ $HClO_4$/$0.5\underline{M}$ HNO_3. In $0.5\underline{M}$ $HClO_4$/ $0.5\underline{M}$ H_2SO_4 solutions JMT quantitatively removed the plutonium but colour quenching reduced the counting efficiency to 13.7% for beta activity. Plutonium was least extracted from $1\underline{M}$ $HClO_4$ but recounting the solution after standing for a day showed a significant increase in the activity of the TOPO scintillator. This suggested that one minute mixing was insufficient for the extraction to reach equilibrium. Several modifications to the above extractive systems were tried. The best results were obtained by adding 100 μl of the Pu III solution to 5 ml $0.05\underline{M}$ HNO_3 and shaking for five minutes with 10 ml of the extractive scintillator containing $0.2\underline{M}$ TOPO. The hydrolytic reactions of Pu IV precluded its use in such a system.

This study indicates that the scintillator mixture, NE-220/TOPO, is the best counting medium for general use with plutonium in various acid media. Insta-Gel/TOPO is a useful alternative especially when aliquots of 500 μl or more of the plutonium solutions are necessary to obtain reasonable counting statistics. If sulphuric acid is present in the plutonium solutions then this scintillator must be counted within 4 hours of adding the aqueous phase unless it is stabilised with JMT. Suitable extractive scintillators can yield counting efficiencies as high as 21% for ^{241}Pu. Emulsion scintillants are practical for alpha measurements on plutonium samples but the necessity of adding large amounts of aqueous solution to stabilise the emulsion detracts from its use for ^{241}Pu detection.

Acknowledgements. The author is indebted to Mrs. J.M. Pearson and Mr. C.H. Randall for their excellent technical assistance.

References.

1. H.R. Ihle, M. Katayannis and A.P. Murrenhoff in Proceedings of a Symposium on Standardisation of Radionuclides, p.485. Vienna, IAEA (1967).

2. R.J. Brouns in Plutonium Handbook, Vol.2, p.707 (O.J. Wick, Ed) New York : Gordon and Breach (1967)

3. D.L. Horrocks and M.H. Studier, Anal. Chem. 30, 1747 (1958).

4. R.F. Keough and G.J. Powers, Anal. Chem. 42, 419 (1970).

5. J.J.Fardy and J.M. Pearson, J. Inorg. Nucl. Chem. 35, 2513 (1973)

6. W.J. McDowell in Proceeding of International Conference on Organic Scintillators and Liquid Scintillation Counting, p.937 (D.L. Horrocks and Chin-Tzu Peng, Ed). New York : Academic Press (1970).

7. J.J. Fardy, D.G. Pinchbeck and M.S. Farrell, AAEC Report AAEC/E189 (1968).

8. J.J. Fardy and J.M. Chilton, J. Inorg. Nucl. Chem., 31, 3247 (1969)

9. J.J. Fardy and J.M. Pearson, J.Inorg. Nucl. Chem., (In Press)

Table I. Efficiency (%) for Beta and Alpha Detection

Solvent	$HClO_4$		$HClO_4$/ H_2SO_4		$HClO_4$/ HCl		$HClO_4$/ HNO_3	
	β	α	β	α	β	α	β	α
NE	9.2	90	8.8	82	10.2	97	11.9	94
NE/T	18.0	98	17.8	96	18.0	98	18.6	99
O/T	7.7	98	7.5	94	7.6	97	7.6	95
O/N	9.5	90	9.8	89	10.0	95	10.6	92
O/N/T	10.7	97	10.6	94	12.9	89	11.4	97
IG	16.0	98	13.2	91	16.8	96	16.1	95
IG/T	16.0	99	16.2	96	16.0	99	16.0	98

Table II. Change in Beta and Alpha Activity (%) over 24 Hours

Solvent	$HClO_4$		$HClO_4$/ H_2SO_4		$HClO_4$/ HCl		$HClO_4$/ HNO_3	
	β	α	β	α	β	α	β	α
NE	-12	-4.1	20	-22	-17	-0.2	-20	-7.8
NE/T	0.1	0.1	-0.8	-2.7	-0.8	0.1	0.1	-0.5
O/T	-0.5	-0.1	3.2	-1.9	-0.9	1.1	-6.8	2.4
O/N	10	-5.7	4.1	-4.5	4.2	-2.3	3.0	-4.2
O/N/T	-0.3	0.3	0.7	-3.9	-25	9.4	-0.3	0.5
IG	-1.5	-2.0	-5.6	-33	-5.4	-4.5	-2.7	-1.1
IG/T	-0.8	-0.4	-15	-6.7	0.4	0.1	-0.7	0.1

Table III. Surface Adsorption of Plutonium (%)

Solvent	$HClO_4$	$HClO_4/H_2SO_4$	$HClO_4/HCl$	$HClO_4/HNO_3$
NE	8.2	25.5	1.1	9.1
NE/T	0.7	3.3	0.3	0.3
O/T	0.1	2.2	0.1	0.1
O/N	9.4	7.5	2.9	8.6
O/N/T	0.1	4.5	0.1	0.2
IG	3.1	37.7	8.7	2.4
IG/T	0.5	19.1	0.4	0.3

Table IV. Maximum Volume of Aqueous Phase (ml) Miscible with Solvents at Room Temperature

Solvent	H_2O	$HClO_4$	$HClO_4/H_2SO_4$	$HClO_4/HCl$	$HClO_4/HNO_3$
NE	1.09	0.67	0.27	0.64	0.29
NE/T	0.88	0.86	0.50	0.70	0.48
O/T	0.82	0.83	0.75	0.79	0.76
O/N	0.80	0.65	0.65	0.63	0.60
O/N/T	0.70	0.68	0.63	0.60	0.60
IG	2.00	1.75	2.00	1.85	1.95
IG/T	1.95	1.70	1.90	1.80	1.90

RADIOIMMUNOASSAY MEASUREMENT OF THE SECRETION AND EXCRETION RATES OF ALDOSTERONE* IN THE GUINEA PIG

Richard H. Underwood and Nancy Dodeja
Endocrine-Metabolic Unit, Peter Bent Brigham Hospital;
Department of Medicine, Harvard Medical School, Boston, Ma.

ABSTRACT

Following the subcutaneous injection of [^{3}H]-aldosterone into guinea pigs, the pattern of tritiated metabolites of the hormone appearing in the urine has been studied. About 60% of the tritiated dose appeared in the urine, within 3 days of administration, indicating the urinary to be the major route of excretion of metabolites of the hormone. Fifty-one per cent of the tritiated dose appeared in the 0-24 urine, and 4.6% of the administered tritium could be extracted from this urine at neutral pH, 6% after pH 1.0 hydrolysis for 24 hr at R.T., 2.8% after glucuronidase hydrolysis and 26% with butanol. About 1.9% (corrected for recovery) of the administered dose was shown to be present in the neutral extract, specifically as aldosterone. No aldosterone could be found in the 24-48 and 48-72 hr urines.

Aldosterone secretion rates were determined using a radioisotope dilution procedure. 0-24 hr urine collections were made following [^{3}H]-aldosterone administration to the animals, and the urine was extracted at neutral pH with methylene chloride. The aldosterone in the neutral extract was isolated and purified by celite partition column chromatography, and its mass was measured by the radioimmunoassay method of Underwood and Williams (J. Lab. Clin. Med. 79:848, 1972), and the tritium assayed by liquid scintillation counting. The secretion rate was calculated from the specific activity of the excreted aldosterone.

Excretion rates were determined from the amount of aldosterone measured in the neutral extract of a 0-24 hr urine. The value of the mass assayed was corrected for losses involved in the process of isolation and purifica-

*Aldosterone = 11β,21-dihydroxy-3,20-dioxo-4-pregnen-18-al-11,18-hemiacetal.

tion by adding [^{3}H]-aldosterone to the urine immediately following collection to act as an internal recovery indicator. For nine animals, the mean value for the secretion rate was 2.9 ± 0.9 (S.D.), range 2.1-4.5 μg/day. The mean value for the excretion rate was 67 ± 34 (S.D.), range 31-136 ng/day. These results show that the metabolism of aldosterone in the guinea pig is very different from that in man, and on a body weight basis, the guinea pig secretes three times more aldosterone than the human.

INTRODUCTION

A previous study by Finkelstein (1) on the metabolism of [^{3}H]-aldosterone in the guinea pig revealed that, unlike in man (2,3), the major urinary metabolites of the hormone were not present as β-glucuronidase or acid hydrolysable conjugates, and were very water soluble and not readily extracted with organic solvents. Thus, a study was undertaken with a view to isolating and measuring a metabolite of aldosterone so that the secretion rate of the hormone could be determined, and with a particular objective of extracting, isolating, and identifying the polar water soluble conjugates. This paper describes the measurement of free aldosterone excreted in the urine and its use to determine the secretion rate of the hormone.

MATERIALS AND METHODS

1. Materials

[1,2-^{3}H]-aldosterone, 52 Ci/mM, and [4-^{14}C]-aldosterone, 46 mCi/mM (New England Nuclear Corporation, 575 Albany Street, Boston, Mass.). Purified by column partition chromatography using the Bush 5 solvent system. Aldosterone and aldosterone-18,21-diacetate* (Ikapharm, Ramat-Gan, Israel), were used without further purification. Methanol, cyclohexane, benzene (spectroquality-Matheson, Coleman & Bell, Curtin Scientific Co., Olympia Park, 3 Normac Rd., Woburn, Mass.). Skellysolve C (Skelly Oil Co., Kansas City, Missouri). Celite 545 - acid washed (Fisher Scientific Co., 461 Riverside Ave., Medford, Ma.).

*Aldosterone - 18,21 diacetate = 11β-hydroxy-18,21 diacetoxy-3,20 dioxo-4-pregnen-11,18-hemiacetal.

Was further washed with acid, then with water until neutral. PCS™ Solubilizer (Amersham/Searle Corp., 2676 So. Clearbrook Drive, Arlington Heights, Illinois).

2. Paper chromatography

Chromatography was carried out at 31° constant temperature. Solvent systems used were: Bush 5 (benzene: methanol:water, 1:4:1); Bush 3 (skellysolve:benzene: methanol:water, 2:1:4:1) and P1 (cyclohexane:benzene: methanol:water, 1:1:7:3, Kliman, B. and Peterson, R.E., J. Biol. Chem. 235:1040, 1960).

3. Urine collection

Ken-Kal random bred male and females guinea pigs were housed in metabolic cages and maintained on an unrestricted diet of regular laboratory guinea pig lab chow (0.63% Na, 1.44% K) and water for seven days. Urine was collected over the third day, for 24 hours. At the beginning of the 4th day, 10 μCi [^{3}H]-aldosterone in 0.4 ml physiological saline containing 20% ethanol was injected subcutaneously into every animal and urine collected in 0-24, 24-48, 48-72 hr batches.

4. Assay of radioactivity

Aliquots of urine were pipetted into Wheaton glass liquid scintillation vials and 10 ml PCS were added. In the case of urine extracts or paper chromatogram eluates, aliquots were first taken to dryness (vac., 55°) before adding the PCS. Every vial was counted for a total of 4000 counts minimum in a Nuclear Chicago Unilux II liquid scintillation counter. When both tritium and [^{14}C] were present, the contributaions from [^{3}H] and [^{14}C] were calculated according to the method of Okita et al. (4). The counting error was 2%.

5. Determination of the aldosterone secretion rate (ASR) from the specific activity of [^{3}H]-aldosterone extracted from urine at neutral pH.

(a) Extraction of urine

1/4 vol of the 0-24 hr total urine volume collected from each animal following the administration of [^{3}H]-aldosterone was adjusted to pH 7.0, made up to a total volume of 20 ml with water and extracted with 1 x 1 vol, 2 x 1/2 vol methylene chloride. The combined methylene chloride extract was washed with 5 ml water, then taken to dryness (vac., 55°).

(b) Column partition chromatography

The column and solvent system used was that described by Flood et al. (2). Aldosterone appears in fractions 10-12 from this column. The dried extract was dissolved in 5 ml mobile phase and added to the top of the column. The column was developed with mobile phase and the eluate collected in 5 ml fractions. An aliquot from every fraction (1-25) was taken to dryness and assayed for tritium. For every one of the nine extracts chromatographed, a tritium peak was found between fractions 9-13 and the three peak fractions for a particular extract were combined and taken to dryness.

(c) Determination of the radiochemical purity of the urinary [^{3}H]-aldosterone

Aliquots of the [^{3}H]-aldosterone from the column for every sample were combined (185,000 dpm [^{3}H] total) and 182,000 dpm [^{14}C]-aldosterone and 50 μg standard aldosterone to act as carrier were added and the sample submitted to successive paper chromatographies in the systems Bush 5, Bush 3, and PI. After chromatography in Bush 5 and before chromatography in Bush 3, the eluate was taken to dryness in a 25 ml r.b. flask and acetylated with 0.3 ml pyridine and 0.15 ml acetic anhydride for 16 hr at R.T. Excess pyridine and acetic anhydride were removed (vac., 55°), 1 ml ethanol was added and the mixture again evaporated. The carrier aldosterone and aldosterone diacetate were detected on the paper chromatograms by absorption of uv light in the short wavelength(240 mμ) region. An aliquot of the eluate from every chromatography was assayed for [^{3}H] and [^{14}C].

(d) Measurement of the specific activity (s.a.) of the purified urinary aldosterone.

The [^{3}H]-aldosterone from the column was chromatographed on paper in the Bush 5 system and an aliquot of the eluate from the paper chromatograms was assayed for radioactivity and duplicate aliquots were taken for radioimmunoassay. The complete procedure for chromatography and assay is described in the method of Underwood and Williams (3). The specific activity in dpm/μg was calculated.

(e) Calculation of the secretion rate

The ASR in μg/day was calculated from the formula:

$$\text{ASR} = \frac{[^3\text{H}]\text{-aldosterone injected (dpm)}}{\text{s.a. urinary aldosterone (dpm } \mu\text{g)}}$$

6. Determination of aldosterone excretion rate (AER)

4475 dpm $[^3H]$-aldosterone to act as internal indicator were added to 1/10 vol of the 24-hr urine collected over the third day. The urine was extracted with methylene chloride and the extract processed and submitted to column chromatography by the procedure described above for measuring the secretion rate. The specific activity (dpm/ng) of the purified aldosterone was calculated and the AER (ng/day) determined from the formula:

$$\text{AER} = 10\text{x} \frac{[^3\text{H}]\text{-aldosterone internal indicator added (dpm)}}{\text{s.a. urinary aldosterone (dpm/ng)}}$$

7. Recovery of the injected tritiated dose in urine.

A 0.3 ml aliquot from every 0-24, 24-48, 48-72 hr urine was assayed for radioactivity.

8. $[^3H]$-aldosterone metabolites extracted from the 0-24 hr urine.

(i) At pH 7.0 and then after acid hydrolysis (pH 1.0, 24 hr, 25°). 1/10 vol of the urine was adjusted to pH 6.5-7.0 and extracted with 1 x 1 vol, 2 x 1/2 vol methylene chloride. The combined extract was washed with 2 ml water and an aliquot assayed for tritium.

(ii) After β-glucuronidase hydrolysis. 1/10 vol of the urine was extracted with 1 x 1 vol, 2 x 1/2 vol methylene chloride and the extract discarded. The urine was then adjusted to pH 5.0, 0.5 ml (2500 F.U.) "ketodase" (Warner-Chilcott, N.J.) added, and incubated at 38° for 48 hr. After the incubation, the urine was extracted with 1 x 1 vol, 2 x 1/2 vol methylene chloride, the extract washed with 2 ml water and then an aliquot assayed for tritium.

RESULTS AND DISCUSSION

The results on the radiochemical purity determination of the isolated [^{3}H]-aldosterone are shown in Table I. The percentage of the administered tritiated dose of aldosterone recovered in the crude urines and as metabolites extracted from the urine is shown in Tables II and III, respectively. Table IV shows the values obtained for the secretion rate in micrograms and the excretion rate in nanograms per day of aldosterone in the guinea pig.

The urinary excretion of 60% of the injected tritiated dose of aldosterone indicates the urine to be the major pathway for excretion of metabolites of the hormone in this animal. The percentage recovery of the injected tritiated dose from the urine by extraction with methylene chloride first at pH 7.0, then after acid hydrolysis, and also after βglucuronidase hydrolysis, 4.6, 6, and 2.8%, respectively, compared favorably with those from the study of Finkelstein (1), i.e., 2.5, 2.5, and 4%, respectively. More than half of the metabolites appearing in the 0-24 urine and representing 26% of the tritiated dose were extracted into butanol after first extracting the urine with methylene chloride. No aldosterone was reported as being found in any of the urine extracts in the previous study (1). However, we found 1.9% of the administered tritiated dose appearing in the urine as free aldosterone which was extracted from the urine at neutral pH. This figure is fully corrected for losses involved in the process of isolating and purifying the tritiated aldosterone prior to specific measurement. In the test for radiochemical purity of the isolated [^{3}H]-aldosterone, the constant [^{3}H/^{14}C] rates over three chromatographies confirmed radiochemical specificity at the stage of assaying both the mass and tritium of the [^{3}H]-aldosterone, (i.e., after chromatography in the Bush 5 system).

The mean value for the secretion rate was 2.9 ± 0.9 (S.D.) μg/day, range 2.1-4.5 and for the excretion rate 67 ± 34 (S.D.) ng/day, range 31-136.

The metabolism of aldosterone in the guinea pig is very different from that in man. In terms of percentage recovery of administered tritiated dose, 4.6% was extracted at neutral pH into methylene chloride and fractionation of this extract revealed aldosterone present to the extent of

1.9%. When the urine was first extracted at neutral pH, then hydrolyzed at pH 1.0, 24 hr at 25° about 6% was then extracted into methylene chloride. However, when this extract was fractionated and specifically assayed for aldosterone, none of the hormone was found to be present. In contrast, in man, less than 0.2% is extracted from the urine at pH 7.0, and none of this tritium has been shown to be associated specifically with aldosterone. 14% is extracted after pH 1.0 hydrolysis and nearly all (> 95%) of the tritium in this extract is specifically associated with aldosterone.

When the daily secretion rate is calculated on a per kilogram body weight basis, then the guinea pig (500 gm, ASR 2.9 μg/day) compared with the human (80 kg, ASR 150 μg/day) secretes three times the amount of aldosterone.

ACKNOWLEDGEMENTS

This work was supported in part by the John A. Hartford Foundation, Grant 9893.

REFERENCES

1. J.W. Finkelstein. Steroids 8, 435, (1966)

2. C. Flood, D.S. Layne, S. Ramcharan, E. Rossipal, J.F. Tait, and S.A.S. Tait. Acta Endocrinol. 36, 237, (1961).

3. R.H. Underwood, C.A. Flood, S.A.S. Tait, and J.F. Tait. J. Clin. Endocrinol. & Metab. 21, 1092, (1961).

4. G.T. Okita, J.J. Kabara, F. Richardson, and G.V. Leroy. Nucleonics 15, 111, (1957).

TABLE I

Radiochemical purity of urinary free [^{3}H]-aldosterone. Ratios of [$^{3}H/^{14}C$] after successive paper chromatographies in Bush 5, Bush 3, P1 systems.

Solvent	$^{3}H/^{14}C$
Bush 5	0.29
Bush 3	0.29*
P1	0.27*

*As aldosterone diacetate

TABLE II

Per cent recovery of injected [^{3}H]-dose in urine.

Animal	Urine collection, hr			
	0-24	24-48	48-72	0-72
1	64.4	6.2	1.1	71.7
2	47.9	6.7	1.9	56.5
3	51.6	7.1	1.2	59.9
4	41.4	8.2	3.0	52.6
5	25.1	4.5	1.4	31.0
6	52.3	5.6	1.7	59.6
7	65.2	5.4	1.3	71.9
8	59.4	5.3	1.6	66.3
9	54.9	7.0	1.7	63.6
Mean	51.1	6.2	1.6	59.0

TABLE III

Per cent recovery of injected [^{3}H]-dose in extracts from 0-24 hr urine: (1) at pH 7.0; (2) after acid hydrolysis (pH 1.0, 25°, 24 hr); (3) after β-glucuronidase hydrolysis; and (4) with butanol.

Animal	Extract			
	1	2	3	4
1	4.3	5.7	2.0	Combined urines 1-9
2	3.7	4.5	2.0	
3	3.6	4.2	2.4	
4	4.1	3.8	1.5	
5	4.0	8.7	4.9	
6	5.8	6.4	3.1	
7	5.8	7.9	3.2	
8	5.3	6.9	3.5	
9	4.9	8.0	2.4	
Mean	4.5	6.2	2.8	25.8

TABLE IV

Aldosterone secretion and excretion rates

Animal	Excretion (ng/day)	Secretion µg/day)
1	64	2.8
2	63	2.1
3	85	2.2
4	86	2.3
5	42	2.4
6	83	2.9
7	136	3.1
8	31	4.2
9	47	4.5
Range	31-136	2.1-4.5
Mean	67	2.9
S.D.	±34	±0.9

COMMENTS ON ASPECTS OF ABSOLUTE ACTIVITY MEASUREMENTS OBTAINED BY LIQUID SCINTILLATION COUNTING

G. C. LOWENTHAL

Australian Atomic Energy Commission
Research Establishment
Lucas Heights. N.S.W. 2232.

These comments are concerned principally with reported work on pure β emitters especially those with end point energies below about 200 keV. The overwhelming majority of radioactivity measurements of these β emitters are made by means of liquid scintillation counting but this technique has not been very successful for determining absolute activities, the more accurate results being obtained by gas proportional counting. Two of the reasons for the relatively less accurate performance of liquid scintillation counters are considered in more detail: the problems created by the marked rise in pulse rates at low pulse heights and the relative inefficiency with which scintillations due to low energy electrons are converted into detectable signals. Accurate results have been realised by extrapolating plots of countrates taken as function of discriminator levels but only for β emitters with end point energies above about 200 keV. Besides, extrapolation procedures are as yet unsupported by accepted models of the liquid scintillation detection process. However, more recently introduced methods are making it possible to count the activities even of ^{14}C and ^{3}H with accuracies comparable to those attained in internal gas proportional counters, i.e. about $\pm$ 1 per cent.

1. INTRODUCTION

The activity A of a radionuclide is expressed as $A = \Delta N/\Delta t$ where ΔN is the number of nuclear transformations in the sample which occur within the time interval Δt. Radioactivity measurements are described as absolute if every nuclear transformation is accounted for. It is then a necessary though not a sufficient requirement for absolute counting that overall counting efficiencies in a given apparatus are known for each sample and that differences between a measured countrate and the rate, which would be obtained if the overall counting efficiency were 100 per cent, can be determined uniquely and with the necessary degree of accuracy.

To limit the scope of this discussion only pure β emitters will be considered. Actually, the bulk of liquid scintillation counting (LS counting) is in any case made with pure β emitters namely ^{14}C and ^{3}H which between them account for the overwhelming majority of LS measurements. Their number has been estimated as well in excess of 10 million per year. Beta spectra extend to zero energy so that source self-absorption effects are perennial problems with β measurements. When radionuclides are dissolved in a scintillator solution, source self-absorption is negligibly small because neither carrier nor other inactive materials, in practice mainly organic materials, prevent the interaction of emitted electrons with the scintillant.

However, there are other problems which users of LS counting techniques must face (1,2). These include the following:

(i) the various quenching effects and changes in these effects as functions of time and of many other variables;

(ii) problems connected with keeping radionuclides in solution and generally with maintaining the homogeneity of solutions;

(iii) problems due to the marked rise in the pulse rate at low pulse heights; and

(iv) the relative inefficiency with which scintillations caused by low energy electrons (roughly $E < 10$ keV) are converted into detectable signals.

Problems which affect other counting techniques as much as LS counting will be disregarded.

The extent to which the difficulties listed under (i) to (iv) affect LS counting results depends strongly on the physical and chemical characteristics of a radionuclide but very much less strongly on whether the aim is a true disintegration rate or only a relative countrate. For the purpose of this brief review, attention can be focussed only on points (iii) and (iv) which are in fact of somewhat greater relevance for absolute than for non-absolute counting (2).

2. EXTRAPOLATION METHODS

When LS counting is employed for absolute measurements one often plots countrates as functions of discriminator settings resulting in a linear plot which is extrapolated to obtain the countrate at zero bias which is accepted as the disintegration rate of the source (3). In what follows this procedure will be called the linear extrapolation method (Figure 1).

Largely for reasons stated in (iii) and (iv) above LS counting requires the imposition of a relatively high energy threshold generally well above 1 keV (it is 10 keV in Figure 1). This leads necessarily to the exclusion of the low energy end of the β spectrum and so requires some form of extrapolation procedure or an equivalent to close the resulting gap. The loss in counting efficiencies which must be allowed for is clearly the greater the lower the end point energy of the β spectrum. For example a 2 keV threshold will cause a loss of around 20 per cent of ^{3}H betas but only around 2 per cent of ^{14}C betas (Figure 2).

An important function of the imposed thresholds is to exclude small pulses which are not part of a measurable background but which do not correspond with disintegrations either. Most of the small pulses are due to

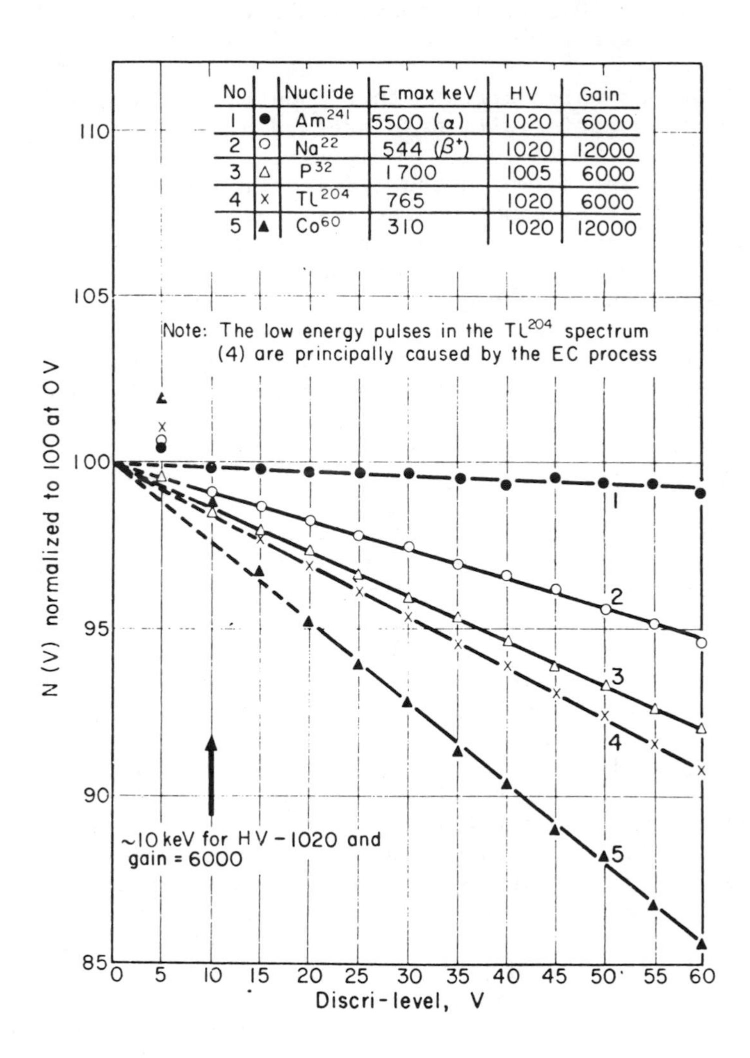

Figure 1 Examples of integral spectra for a number of radionuclides (from Ref. 3, Figure 3).

chemiluminescence and to after pulsing (1,2), effects which can be expected to be the higher, the higher the end point energy of the β spectrum. However, they are also potentially serious sources of error when counting low energy events.

When the linear extrapolation method is used it is often assumed that it is only these spurious counts easily seen at low bias settings (Figure 1) which cause the plotted points to be above a straight line defined by results obtained at the higher bias settings, but this is an arbitrary assumption (2,4). A linear plot is in itself no proof of a correct trace of disintegration events. Many β spectra including those of ^{3}H and ^{14}C are markedly non-linear in the region approaching zero electron energies (Figure 2) so that a linear extrapolation could be expected only if it could be shown on general grounds that the pulse height distribution remains a horizontal line down to zero energies. This has not been done so far.

As a rule it is possible to adjust the gain of a system to obtain a reasonably close to linear plot for extrapolation to zero bias (Figure 3). However, even if the shape of the pulse spectrum is actually horizontal or very nearly so this could be due to compensating errors (see below), bearing in mind that a horizontal pulse spectrum is not predicted by any established model. The validity of extrapolations is in doubt not only for linear but equally for exponential extrapolations which have been used for ^{3}H activity measurements (2,5).

Notwithstanding this, extensive experiments have shown that provided the end point energies of the β spectra exceed about 200 keV and that one takes account of known sources of errors, including those listed earlier, the results of activity measurements made with the linear extrapolation method agree with results obtained by established absolute methods, i.e. 4π gas flow proportional counting, and the uncertainties in these results are also of the same order, generally 1 to 2.5 per cent for a 99 per cent confidence interval (c.i. 99) (3).

It is still necessary to refer to the use of LS detectors with 4π β-γ coincidence counting, a method

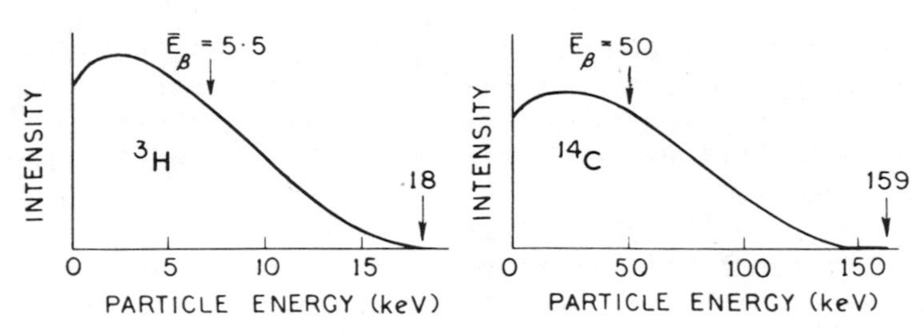

Figure 2 Computed β spectra for ^{3}H and ^{14}C.

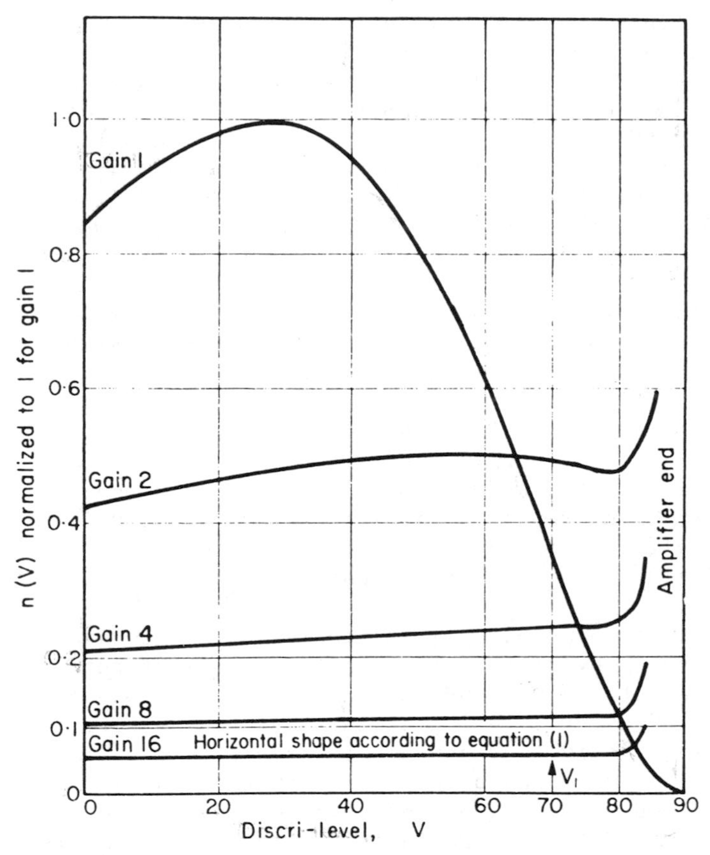

Figure 3 Differential pulse height spectra obtained with a LS counter at different amplifications (from Ref. 3 Figure 2).

applicable e.g. to ^{60}Co where β decay is followed virtually instantaneously be γ emissions. With this method the overall counting efficiencies of β-γ emitters can be unambiguously measured and it is this fact which established 4π β-γ coincidence counting, wherever it can be properly applied, as the most accurate procedure for absolute activity measurements (6). Since 4π detection is achieved with LS counting as well as with gas counting the LS method can be and has been used for 4π β-γ coincidence counting.

Moreover, and this is the relevant point for the present purpose, it has been possible to use β-γ coincidence counting as an aid for measuring the activities of pure β emitters. The procedure is known as the efficiency tracing method (6). A β-γ and a pure β emitter, e.g. ^{60}Co ($E_{\beta max}$ = 310 keV) and ^{63}NI ($E_{\beta max}$ = 67 keV) are mixed homogeneously and in known proportions. Sources are prepared from this mixture which differ in their counting efficiencies and these sources are counted to obtain their total β countrates (^{60}Co + ^{63}Ni) as function of the measured counting efficiencies for the betas of ^{60}Co. The data is plotted (Figure 4) and the plot is extrapolated to 100% counting efficiency for ^{60}Co. The plot is a quadratic and not a linear function and this was predicted by theory based on simplifying assumptions (7) and so still subject to significant residual uncertainties.

The pure β activity, ^{63}Ni in this example, can be calculated from the corresponding total countrate value since the ^{60}Co activity in the source is known. Non-linear extrapolations of plots of results of efficiency tracing measurements yield far more reliable pure β activity measurements than do estimates of counting efficiencies when only 4πβ counting is used.

Unfortunately, the use of LS counting in β-γ coincidence measurements has never so far been really successful. To obtain acceptable results it has been necessary to introduce arbitrary assumptions and the accuracies were always below the levels attainable with 4π gas counting of thin sources (8-10). However, there are many cases when solutions of radionuclides contain too high a solute concentration for making adequately thin sources and it is in these cases that LS counting can be the

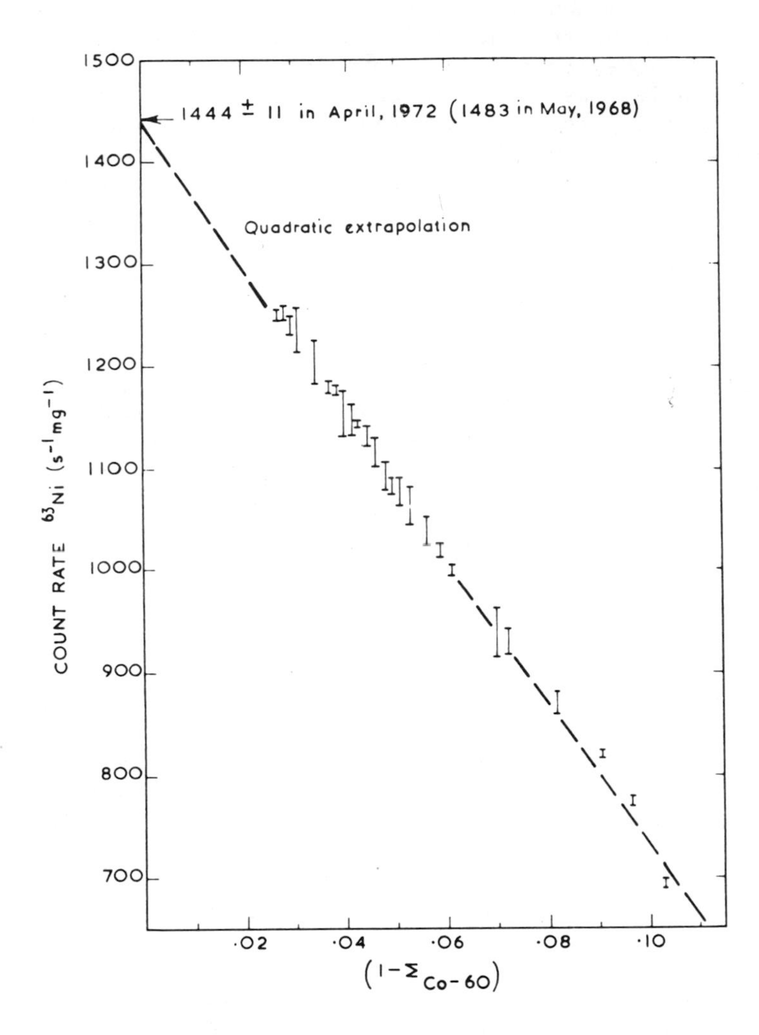

Figure 4 Plot of efficiency tracing measurements in a 4π (proportional gas flow) β-γ coincidence apparatus where ^{60}Co was employed for the efficiency tracing of ^{63}Ni. The ^{63}Ni countrate marked along the ordinate was obtained by subtracting in each case the known ^{60}Co contribution from the measured 4πβ countrate.

relatively more accurate technique but the emphasis is on the term relative.

3. THE PROBLEM OF ZERO RESPONSE PROBABILITY

When LS counting is used to measure the activities of β emitters with end point energies around 150 keV or less the production of spurious pulses is much reduced. It is the non-detection of scintillations (point iv above) also described as the zero response probability, which is then likely to be the dominant factor in causing uncertainties in the results (1,2,4).

The relative inefficiency of conversions of weak light pulses into detectable signals is reflected in the poor energy resolution of LS detectors. The resolution shown in Figure 5 is fairly typical for contemporary general purpose counters equipped with PM tubes with bialkali cathodes. It is seen that to discriminate against the 25.5 keV Ag K edge the level must be set to correspond not to just above this energy but to at least 36 keV. Thus, poor energy resolution introduces uncertainties into the setting of discriminator thresholds and so also into the results of activity measurements which involve extrapolations where one of the variables is a function of the discriminator setting.

A substantial improvement in the performance of LS counters was made possible some 5 years ago with the introduction of photomultiplers (e.g. RCA type 8850) whose first dynode is coated with gallium phosphide which gives this dynode a 5 to 6 times greater amplification factor than it has in other tubes with bialkali cathodes. The detection efficiency for ^{3}H which had already risen from 20% in 1960 to 60% in the late sixties (1) became 80 to 82% when using just a single tube and close to 90% when using two tubes operated in the pulse summation mode.
This detection efficiency corresponds to an effective average cut-off energy close to 0.7 keV.

With this and other recently developed instrumentation the activity of ^{3}H can now be measured with an estimated accuracy close to 2.5% (c.i. 99) and the activity of ^{14}C to within 1.0% (11). However, the equipment used to achieve these accuracies is costly and the required

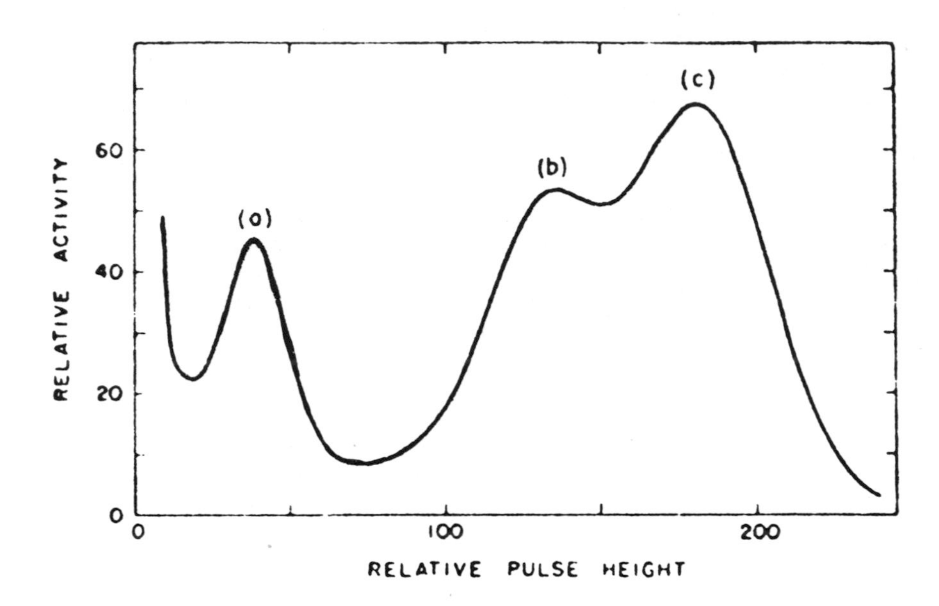

Figure 5 LS pulse height spectrum for ^{109}Cd. The peaks represent the 25.5 keV Ag K edge (a), and the 65.9 keV and 88 keV conversion electrons (b,c). The resolution (FWHM) at 25.5 keV is about 60 per cent (from Ref. 2, Figure 11).

standard of measurement expertise very high. The same applies to a different method first reported in 1966 (4) where pulses due to radionuclide decays are compared with light pulses whose intensity is modulated by an analogue device incorporating the shape of the relevant β spectrum as calculated from β decay theory. Here the accuracy achievable for activity measurements of ^{14}C and ^{35}S ($E_{\beta max}$ equals 159 keV and 167 keV respectively) is estimated to be close to 2.0% (c.i. 99) (12).

4. PROBLEMS WITH USING CERTIFIED STANDARDS

For the time being it cannot be expected that similar accuracies can be reached outside very well equipped standard laboratories. As a rule LS counting systems are calibrated with samples whose disintegration rate has been certified by a standard laboratory. To what extent the accuracy of the measurement of users approaches the certified accuracy of the activity of the standard depends principally on the awareness of users of the many possible sources of uncertainty (13). It would be rare that a known source of error would not be properly dealt with. What could be far more frequent is that sources of error which, given the present state of knowledge could have been known, are disregarded. If one looks at the expense and expertise brought into play by groups of specialists to achieve, say, an accuracy within ± 2% (c.i. 99) for a ^{14}C activity measurement one is a little surprised and perhaps more than a little sceptical when reading reports where similarly high accuracies are claimed with a very much more modest input in resources including relevant experience. The fact that independent verifications of results are rarely if ever practical outside standards laboratories explains at least in part why it is easy to make unrealistic estimates of accuracies.

5. COMPARATIVE PERFORMANCE OF SYSTEMS USED FOR ABSOLUTE COUNTING

Before making such comparisons it is necessary to define the area of interest. The principal concern continues to be with pure β emitters having end point energies below 200 keV. There are two groups of radionuclides in this category; those with properties which permit thin solid sources to be made and those where that cannot be done (6).

When thin solid sources can be made, e.g. for ^{35}S ($E_{\beta max}$ = 167 keV) and ^{63}Ni ($E_{\beta max}$ = 67 keV) 4π gas flow proportional counting yields more accurate absolute counting results than LS counting (1). To give a very brief explanation: proportional counting in gaseous detecting media yields a much more sharply defined response when counting low energy events, say $E < 3$ keV, than does LS counting. Reasons why LS counting faces difficulties in that energy region were summarised earlier. A few data about the performance of proportional counters will be given below.

Absolute activity determinations of pure β emitters have to be made by extrapolation procedures no matter which counting method is used. They are therefore subject to additional uncertainties which one seeks to minimise by extending the measurements to the highest counting efficiencies (see e.g. Figure 4) which include the largest fraction of low energy events. So far these procedures have been more successful for proportional than for LS counting and this is an important reason why results obtained in LS counters are invariably verified with reference to results obtained in 4π proportional counters, assuming this can be done and not the other way around (3, 8,12).

That nevertheless LS counting has its strongly established position is due largely to the many situations where 4π proportional counting of solid sources simply can not be used because it is completely impractical to make solid sources of an acceptable quality. The outstanding example is ^{3}H but there are many others. Here the LS counter has a virtual monopoly on measurements and, as was pointed out earlier, it has been used successfully also for absolute activity determinations (11,12).

So far the standards of activity for ^{3}H and ^{14}C are still those established by internal proportional gas counting. Difficulties with source making are avoided by introducing the radionuclides as components of the counting gas, e.g. as tritiated methane. The accuracies which have been attained are within 1.0% (c.i. 99) (14,15).

Internal gas counting is cumbersome for normal use and could never compete with LS counting whenever the latter method can be used. Absolute internal gas counting is, if anything, more costly in effort and equipment than are the absolute LS methods referred to earlier. However, one or two comments on the performance of the internal gas proportional counter may help to illustrate why conditions in gaseous media are more favourable for counting low energy events than conditions in liquid scintillants.

Assuming optimum conditions, i.e. skilled use of equipment incorporating PM tubes similar to the RCA type 8850, one can obtain threshold settings of around 0.7 keV and it is possible to resolve photon or electron energies down to about 2 keV (2). On the other hand the cut-off energy for internal gas proportional counting is within 0.05 keV, i.e. over ten times lower than what can be achieved in LS counters and the energy resolution is also proportionally higher (14).

It is apparent therefore that LS counting has not yet bridged the gap separating it from gas proportional counting as regards the response to low energy events. Nevertheless there are other aspects to the accuracy of counting apart from the response at the lowest energies and it is likely that, taking everything into consideration, the activities of ^{3}H and ^{14}C can now be measured by means of LS counting with about the same low uncertainty attained so far with internal proportional gas counters.

6. REFERENCES

1. R. Vaninbroukx and I. Stanef, "Present Status in the Field of Liquid Scintillation Counting", Proc. 1972 Int. Summer School on Radionuclide Metrology, Herceg Novi, Yugoslavia; Nucl. Instrum. Methods (to be published).

2. Houtermans, "Probability of Non-Detection in Liquid Scintillation Counting", Proc. 1972 Int. Summer School on Radionuclide Metrology, ibid; See also in same publication A. Williams, "After Pulses in Liquid Scintillation Counters".

3. R. Vaninbroukx and A. Spernol, Int. J. Appl. Rad. Isotopes, 16, 289 (1965).

4. J. Bryant, D.G. Jones and A. McNair, I.A.E.A. Symp., STI/PUB/139, SM 79-28, Vienna (1967).

5. K.F. Flynn, L.E. Glendinin and V. Prodi, Organic Scintillators and Liquid Scintillation Counting, p. 687, (D.L. Horrocks, Ed.) Academic Press (1971).

6. A.P. Baerg, Metrologia, 2, 23 (1966) and 3, 105 (1967)

7. A. Williams and I.W. Goodier, I.A.E.A. Symp., STI/PUB/139, SM 79-30, Vienna (1967).

8. J. Steyn and F.J.W. Hahne, Proc. Nat. Conference Nucl. Energy, p. 30, Pretoria, South Africa (1963).

9. G. Erdtmann and G. Herrmann, Int. J. Appl. Rad. Isotopes, 16, 301 (1965).

10. J. Steyn, I.A.E.A. Symp., STI/PUB/139, SM 79-16, Vienna (1967).

11. V. Kolarov, Y. LeGallic and R. Vatin, Int. J. Appl. Rad. Isotopes, 21, 443 (1970).

12. D.G. Jones and A. McNair, "Radioactivity Calibration Standards", NBS Special Publication No. 331, p. 37, National Bureau of Standards, Washington D.C. (1970).

13. G.C. Lowenthal, Atomic Energy in Australia (to be published).

14. A. Spernol and B. Denecke, Int. J. Appl. Rad. Isotopes 15, 241 (1964).

15. W.B. Mann, R.W. Medlock and O. Yura, Int. J. Appl. Rad. Isotopes, 15, 351 (1964).

MODERN TECHNIQUES AND APPLICATIONS IN CERENKOV COUNTING*

H. H. Ross
G. T. Rasmussen**

Analytical Chemistry Division
Oak Ridge National Laboratory
Oak Ridge, Tennessee, USA

ABSTRACT

Cerenkov counting, a relatively unknown technique in the laboratory five years ago, has recently become the method of choice for counting many beta-emitting nuclides on a routine basis. The interest in this method is due, in part, to the ease in employing conventional liquid scintillation counting equipment without modification. Unfortunately, this similarity in measurement routine and equipment has led some investigators to believe that procedures developed for organic scintillation counting are directly applicable to Cerenkov systems. Unfortunately, this is far from true. Therefore, the following presentation has a three-fold purpose. First, a general review of the Cerenkov physical process is presented. Particular attention is given to the mechanism of Cerenkov radiation production, energy input-output characteristics, and spacial phenomena. The unusual properties of Cerenkov systems and how they may be used to advantage in practical counting situations is stressed. Also, an attempt is made to delineate important points of departure between organic liquid scintillation and Cerenkov methods. A second objective is to describe special measurement techniques that have been applied to Cerenkov counting problems. Among those discussed are the use of waveshifters, solvent selection, quenching, and standardization procedures. Finally, we will present the results of an original study concerning a computer simulation of the Cerenkov process that we have developed to establish optimal counting conditions for dual isotope

---

*Research sponsored by the U. S. Atomic Energy Commission under contract with Union Carbide Corporation.

**Former student participant in Great Lakes College Association program at ORNL.

experiments. This technique is particularly interesting because it appears to be generally applicable to both Cerenkov and conventional liquid scintillation counting procedures. Use of a simple computer program allows one to quickly set-up his instrument for optimum isotope separation and also to evaluate the probable limits of error in the measurement.

Introduction

In the span of just a few years, Cerenkov counting techniques have become well established for the rapid assay of many beta and beta-gamma emitting radionuclides. Several authors have suggested that it is the method of choice for certain isotopes. Among the advantages of the techniques are extreme simplicity of sample preparation, the ability to count in aqueous systems without the use of organic fluors, and the high detection sensitivity that can be obtained. Another important consideration is that conventional liquid scintillation counting equipment can be used without modification. Many of the factors that influence the successful application of Cerenkov counting to practical problems have already been described (1,2). This paper will attempt to both review and extend the state of Cerenkov counting technology.

Fundamental Considerations

Cerenkov radiation is produced when a charged particle passes through a transparent medium (the Cerenkov generator) at a relative phase velocity greater than the speed of light in the same medium. Upon entering the generator, the particle experiences a strong slowing force and, during this period of rapid de-acceleration, electromagnetic radiation is generated. Cerenkov radiation has been described as the particle electromagnetic "shock wave" that is analogous to the boom created by hypersonic aircraft. Theoretically, the Cerenkov spectrum extends from microwave to X-ray frequencies; the portion of major concern to practical counting procedures lies in the ultraviolet and blue spectral regions. The most familiar display of Cerenkov radiation is the blue glow surrounding the core of swimming-pool reactors.

The basic physical condition that must be met for the formation of Cerenkov radiation is:

$$\beta n \geqslant 1 \qquad (1)$$

For relativistic electrons, β can be related to electron energy (E) and then combined with Equation 1 to yield:

$$\left[n^2 - \left(\frac{n}{\frac{E_{(KeV)}}{511} + 1} \right)^2 \right]^{1/2} \geqslant 1 \quad (2)$$

Where: β = velocity of particle (v)/speed of light (c)

n = refractive index of Cerenkov generator

E = energy of particle inducing Cerenkov radiation

When Equation 2 is solved for E, using the condition of equality and the refractive index of water, the result indicates that an electron must exceed 263 KeV in order to produce a Cerenkov response. Since many radionuclides emit beta particles significantly above the 263 KeV threshold, the legitimacy of using Cerenkov radiation for isotope assay in aqueous systems is obvious. It must be noted, however, that the photon yield from the Cerenkov process is quite small. Calculations have been made to describe the number of photons and the spectral distribution that results from a Cerenkov event (1,2) but perhaps one gets the best feeling for the magnitude of photon intensity by noting that a 1 MeV electron in water yields about the same number of photons as does an average tritium disintegration in an organic scintillator (3). Table I lists the detection efficiencies of several nuclides that have been determined experimentally (1,2,3,4,5,6,7,8,9).

Solvents

The selection of a solvent for organic liquid scintillation counting is often a formidable task, especially if particular sample solubility requirements are needed. For Cerenkov counting, no such problems exist; virtually any transparent liquid can be used as the Cerenkov generator. The primary solvent requirement is that it transmit radiant energy in the blue and near ultraviolet spectral regions with little attenuation. Even here, some compromises can be made if one is willing to work at reduced counting efficiency. The actual choice of a solvent is most often determined by what is convenient for the samples and the chemistry involved. Equation 1 shows that, as the refrac-

Table I.

Experimental Counting Efficiencies of Nuclides In Aqueous Systems

Isotope	Range of Counting Efficiency (%)	Isotope	Range of Counting Efficiency (%)
^{204}Tl	1.3 - 1.9	^{185}W	2.1
^{137}Cs	2.1	^{36}Cl	2.3 - 13
^{198}Au	5.4	^{170}Tm	5.6
^{210}Bi	6.8	^{47}Ca	7.5
^{40}K	14 - 25	^{24}Na	18
^{90}Sr-^{90}Y	22	^{86}Rb	23
^{32}P	20 - 48	^{90}Y	36 - 50
^{128}I	40 - 50	^{144}Ce-^{144}Pr	54 - 62
^{42}K	11 - 60	^{106}Ru-^{106}Rh	62 - 70

tive index of the generator is changed, the detection threshold also changes. It has also been demonstrated that the use of highly refractive liquids will increase the photon yield for a given particle energy (10). Occasionally, use is made of these characteristics to select a solvent to enhance the counting efficiency of a given nuclide or to change the detection energy threshold.

The most widely used solvent for Cerenkov counting is water. It is highly transparent to the Cerenkov emission and can act as an efficient solvent for a wide variety of inorganic and organic materials. Other peripheral considerations are its low cost, and lack of toxicity and flammability. However, for many biological materials, other solvents are preferred. Table II lists experimental relative counting efficiencies for ^{32}P in various solvents (8,9). The change in the relative counting efficiency in the above table is a direct reflection of the variability of the index of refraction combined with any optical absorption properties of the solvents. As the index of refraction increases, the energy threshold is lowered and thus, more of the beta spectrum is included in the measurement. The photon yield is enhanced also. In some cases, small inconsistancies can be seen; this is probably the result of experimental error. There are many practical applications for the use of these special solvents. For example, it has been noted (8) that chloroform is used to extract lipids (a severe quenching situation in organic scintillators) and formic acid is used as a solvent for defatted tissue. Acetone (5%) in chloroform has been used for counting organic radiobromides (11). Additionally, a solvent can be selected to count a nuclide that cannot even be detected in water. The low energy emitter ^{99}Tc has been counted with very low efficiency in a 95% glycerol solution (1) and ^{14}C has been measured in α-bromonaphthalene (12). These techniques are, however, rather limited in their scope of application.

The effect of solvent volume has been adequately discussed in the literature (1,2,3,9,13,14) and will not be repeated in detail here. Most investigators agree in general, if not in detail, that counting volume is a parameter that must be controlled for reproducible results. In all of the counting systems tested that use a "standard" size liquid scintillation vial, the volume-efficiency curve appears to go through a maximum at about 10 ml. The main

Table II.

Relative Counting Efficiencies Of ^{32}P In Various Solvents

Solvent	Volume Ratio (v/v)	Relative Counting Eff. (H_2O=1)
formic acid (98%)		0.96
formic acid-butyl acetate	1/1	1.00
methanol		1.08
chloroform		1.14
isopropanol-water	1/1	1.15
ethanol (95%)		1.16
ethanol-water	4/1	1.19
acetone		1.19
petroleum ether		1.21
chloroform-methanol	2/1	1.26
hexane		1.28
isopropanol		1.29
isopropanol-heptane	1/1	1.31
chloroform-isopropanol	2/1	1.32
heptane		1.38
benzene		1.59
toluene		1.66

disagreement among authors is the magnitude of the efficiency shift with volume and the energy of the nuclide being counted. We feel that the variability in the volume effect can be directly related to the optical geometry of a particular counting system. Perhaps the one general rule that can be formulated is that each investigator must establish the volume-efficiency function for his own instrument so that acceptable volume tolerance limits can be assigned. For a given liquid sample, one would like to use the maximum possible volume to extend the lower limit of detection. However, it is quite clear that the volume effect is much more pronounced in Cerenkov counting than the usual organic techniques.

Solid Cerenkov Generators

Thus far, the major consideration for Cerenkov generators has been liquids. There is no reason, however, that other transparent substances cannot be used. For large generating volumes, solids offer a number of advantages. Where Cerenkov detectors have been used to measure the radioactive products of an activation analysis, Lucite or Perspex generators have proven popular (15,16). One benefit of using these plastics is that they can be molded or machined into special shapes to suit the particular geometry required in the experiment. They also have excellent optical properties.

Other types of solids have also proven useful. One ^{32}P product was isolated on either paper or silica gel. For counting, the paper was suspended inside a counting vial and the response generated in the vial wall was counted. The silica gel was simply placed in the vial and the emission from the gel and the vial wall was measured (9). Results were good in both cases. In a similar experiment, a phosphorous compound was collected on a membrane filter and again suspended in the vial for counting (6). Other solid materials that have been used are polyethylene, sodium iodide, cesium iodide, quartz, ice, and crystals of many different types of organic compounds. With solids, as with liquids, the possible choices are almost limitless.

Geometry Effects

When Cerenkov radiation is generated, the emission takes place in a well defined spacial configuration. The practical effect of this phenomenon is covered in previously cited literature. However, two items are worth noting as being very useful techniques for routine assays. It has been found (4,9,13) that by substituting polyethylene counting vials for glass, a >25% increase in the counting efficiency can be obtained. Another investigator (5) shows that by adding Cab-O-Sil to the liquid generator, an increase of 5-10% can be realized. In both of these methods, the increased counting efficiency is probably due to the diffusing effect that is introduced. Since most of the commercial counting systems utilize some type of coincidence circuitry, a positive indication in two phototubes is needed for a count to be registered. The geometry of Cerenkov emission is such that one phototube will be favored for a given decay event. Normal reflections in the counting chamber will help to partially alleviate this effect. The additional diffusing ability of the polyethylene vial or the Cab-O-Sil provides an even more uniform distribution of the emitted photons.

Quenching

The worker in organic liquid scintillation counting is well aware of two forms of quenching, chemical and color. Chemical quenching (chemical interference with molecular transitions in the organic fluor) does not exist in Cerenkov spectroscopy. This is evidenced by the fact that strong acids such as sulphuric, hydrochloric, and perchloric acids and a wide variety of inorganic anions and cations show no significant reduction in the counting efficiency when compared with samples counted in pure water (3,5,13, 17). Unfortunately, color quenching (partial or total absorption of emitted photons) continues to be a significant quenching parameter. As might be expected, most of the standard quench correction techniques have been tested with Cerenkov systems; the published results give less than a clear picture of what can be accomplished with these methods.

Samples may color quench by either ultraviolet or visible absorption. For example, the very common NO_3^- and NO_2^- ions are strong ultraviolet absorbing species and cause moderate to severe quenching problems (13). This substantially limits the use of nitric acid solutions in counting.

Obviously, visibly colored solutions, especially reds and yellows, cause similar problems. One paper examines quench correction techniques in detail (18) and concludes that internal standard or de-colorization procedures are laborious, that the channels ratio method is too insensitive, and that the external standard is somewhat inaccurate but can possibly be used.

In spite of these findings, the external standard and the channels ratio method have both been used to obtain an assay correction of 98% $\pm$ 6% (19). Another investigator pointed out that if one considers the variability of the sample vial, the external standard can yield results with errors of about $\pm$ 2% (9). The channels ratio method has often been employed without difficulty. In one series of experiments designed to test various quench correction techniques for ^{42}K, the author found that the channels ratio was the best method (20).

Other less common techniques have been developed to deal with the color quenching problem. One worker has found that the degree of quenching can be related to the optical absorption of the sample as measured on a spectrophotometer (5). An interesting technique is the development of the "Cerenkov insert" (21,22). Here, the sample is placed in a thin-wall opaque holder which is then suspended in a transparent generating solution. Color quenching is completely avoided but, because of self absorption in the sample, high counting efficiency can only be obtained with energetic emitters.

It is clear from the above discussion that many different procedures can be used for color quench correction. No one technique (except perhaps the internal standard) appears to be universally applicable. It is suggested that each sample type be treated as an individual situation and that the best correction method be selected on the basis of experimental tests.

Waveshifters

All of the above discussion has been centered on the direct measurement of the Cerenkov emission. As noted previously, a significant portion of this emission occurs in the ultraviolet spectral region. These frequencies are easily absorbed in glass sample vials, in the face of glass photomultiplier tubes, or in the solvent itself. A second problem is the directional nature of the Cerenkov emission described under geometry effects. If the ultraviolet photons can be converted to the visible region and if these photons could be given an isotropic spacial distribution, the counting efficiency should be distinctly improved.

A possible answer to both of these problems rests in the use of a waveshifting compound in the Cerenkov generator similar to the use of POPOP in organic systems. Many investigators have already studied the characteristics and applications of waveshifting compounds and found them to be highly desirable (3,4,7,17,23,24,25,26). Table III lists some of the compounds that have been employed in both organic and aqueous media.

Although a significant increase in the counting efficiency can be obtained when using one of these compounds, it must be remembered that when they are employed, the system is no longer a purely Cerenkov process. Thus, the problem of chemical quenching becomes possible and large changes in the observed response can result. It is also true that visible quenchers can become more significant in a waveshifted system. The benefits of using a waveshifter must be carefully weighed against the possible deleterious side effects that may appear. Most authors agree, however, that waveshifters are particularly desirable when measuring low-energy emitters.

Applications

Cerenkov radiation has been employed in virtually every field of isotope assay problem. Samples such as plant and animal tissue, isotope preparations, environmental materials, reactor effluents, and biomedical fluids have been examined with excellent results. These applications are well documented in the open literature and will not be repeated here. However, some methods will be described that make use of the unusual properties of Cerenkov radiation and illustrate the unique characteristics that make the technique particularly valuable.

Since there is a distinct energy threshold below which no Cerenkov emission occurs, it is possible to determine the activity level of a small amount of an energetic nuclide in the presence of a much larger amount of a weaker emitter. This type of analysis can be performed with little or no interference from the predominant species. The simultaneous determination of ^{33}P and ^{32}P was accomplished by combining the techniques of Cerenkov and organic

Table III.

Waveshifting Compounds Used Successfully In Cerenkov Counting Systems

2-amino-6,8-naphthalene-disulfonic acid
2-naphthalene-3,6-disulfonic acid
2-naphthalamine-6,8-disulfonic acid
2-naphthol-3,6-disulfonic acid
2-naphthol-6,8-disulfonic acid
7-amino-1,3-naphthalene-disulfonic acid
4-methylumbelliferone
β-naphthol
β-naphthylamine
sodium salicylate

scintillation counting (27). First, the sample was assayed in an aqueous Cerenkov system to determine the ^{32}P concentration; ^{33}P was found to count with < 0.01% efficiency (probably zero). Then the sample was counted in an organic scintillator to determine total ^{33}P and ^{32}P. Subtracting the ^{32}P Cerenkov value from the total yielded the correct amount of each isotope. In a similar way, the ^{80g}Br and ^{80m}Br isotopes were counted successfully in the presence of ^{82}Br with only a small interference (11).

Both solid and liquid Cerenkov generators have been used for measuring the products from an activation analysis (7, 15,16,28,29,30). A particularly important aspect of this application is the very fast response of the detector; pulse pile-up and summation problems are greatly minimized. Some of the elements that have been determined are Be, Li, B, F, I, and O_2. The role of Cerenkov detectors in charged particle activation has been summarized (31). Some of the more unusual applications of Cerenkov detectors include the use of the ^{36}Cl emission to determine particle size (32), the absorption of Cerenkov photon emission to follow radiolysis in solution (33,34), the determination of the refractive index of liquid orthodeuterium by comparing its Cerenkov response with that of hydrogen (35), and the analysis of gaseous ^{85}Kr (36).

Dual Tracer Studies

In our laboratory, we have been concerned recently with the problem of determining two isotopes simultaneously in a Cerenkov system. The solution of this problem is somewhat more complex than that observed in organic scintillators because of the high degree of overlap in the Cerenkov spectra. For our study, we selected the isotopes ^{36}Cl and ^{32}P as being representative of a typical counting situation.

The effective determination of these two isotopes in a mixture requires that their Cerenkov spectra be sufficiently different to provide different relative counting efficiencies over two regions of the response function. If two single channel analyzers are set to count different portions of the response, it is possible by using a solution containing only chlorine-36, to determine counting efficiencies for this isotope in each channel. The efficiencies of phosphorous-32 in these same two channels can also be determined. Also, a count rate for a phosphorous-chlorine mixture can be determined in each of the two channels. Knowing the efficiency for each isotope in each channel and counting rates in each channel for a mixture, one can determine the disintegration rate of each isotope in the mixture using techniques developed previously for organic systems. The necessary equations are summarized in Table IV.

If the same stock solutions of phosphorus and chlorine are used to prepare samples containing a single isotope and also isotope mixtures, it is possible to determine the accuracy of a dual isotope measurement that is independent of the absolute accuracy of the stock solution assays. This quantity, hereafter referred to as the accuracy function, is the ratio of the experimentally determined disintegration rate of one isotope to the "known" disintegration rate for that isotope that is used in the calculations of the channel efficiencies. The accuracy function is a function of six variables which are the counting rates used to determine the efficiency of each isotope in each channel and the counting rates of the mixture in each channel. If the dual tracer method is effective, that is, if the equations describing the system are valid, the accuracy function for an isotope will equal 1 for the case where the amount of that isotope used is the same in both the sample used for the efficiency determination and the mixed sample. The accuracy function, appropriately, provides a convenient measure of the accuracy of the determination. It equals (or can be corrected to equal) one, when the system is "well-behaved," and deviations from unity indicate directly relative errors.

The equation for the accuracy function, because it is a function of six counting rates, can be used to estimate the effect of random statistical errors in specific gross counts on the precision of the determination. The variation in the accuracy function as a result of statistical variation in a given counting rate can be estimated as the product of the partial differential of the accuracy function with respect to that counting rate and the estimated error in the counting rate. The estimated error in the counting rate resulting from random statistical errors can be approximated as the product of the counting rate and the relative standard deviation of the gross count that was accumulated to determine that counting rate. A convenient estimate of the standard deviation of the gross count is its square root. The sum of the absolute values of the relative error in the accuracy function caused by random

Table IV.

Summary of Equations Used For Dual Isotope Analysis

$$\frac{G_{Cl,1}}{t_{Cl}D_{Cl}} = \frac{C_{Cl,1}}{D_{Cl}} = E_{Cl,1} \qquad \frac{G_{Cl,2}}{t_{Cl}D_{Cl}} = \frac{C_{Cl,2}}{D_{Cl}} = E_{Cl,2}$$

$$\frac{G_{P,1}}{t_P D_P} = \frac{C_{P,1}}{D_P} = E_{P,1} \qquad \frac{G_{P,2}}{t_P D_P} = \frac{C_{P,2}}{t_P D_P} = E_{P,2} \qquad \frac{G_{M,1}}{t_M} = C_{M,1}$$

$$\frac{G_{M,2}}{t_M} = C_{M,2} \qquad C_{M,1} = E_{Cl,1}D_{Cl} + E_{P,1}D_P$$

$$C_{M,2} = E_{Cl,2}D_{Cl} + E_{P,2}D_P$$

$$D^*_{Cl} = \frac{C_{M,1}E_{P,2} - C_{M,2}E_{P,1}}{E_{Cl,1}E_{P,2} - E_{Cl,2}E_{P,1}}$$

Substituting for efficiency terms and dividing by D_{Cl} yields the accuracy function:

$$A_{Cl} = \frac{D^*_{Cl}}{D_{Cl}} = \frac{C_{M,1}C_{P,2} - C_{M,2}C_{P,1}}{C_{Cl,1}C_{P,2} - C_{Cl,2}C_{P,1}} \quad ; \; \Delta A = \left[\frac{\partial A}{\partial C_{M,1}} \Delta C_{M,1}\right] +$$

$$\left[\frac{\partial A}{\partial C_{M,2}} \Delta C_{M,2}\right] + \left[\frac{\partial A}{\partial C_{Cl,1}} \Delta C_{Cl,1}\right] + \left[\frac{\partial A}{\partial C_{Cl,2}} \Delta C_{Cl,2}\right] + \left[\frac{\partial A}{\partial C_{P,1}} \Delta C_{P,1}\right] +$$

$$\left[\frac{\partial A}{\partial C_{P,2}} \Delta C_{P,2}\right] \quad ; \qquad \Delta C_{M,1} = C_{M,1}\frac{\sqrt{G_{M,1}}}{G_{M,1}} \text{, etc.}$$

Variables;		Subscripts;	
G:	gross counts	Cl:	chlorine
C:	counting rate	P:	phosphorous
t:	time of count	1:	channel 1
D:	disintegration rate "known"	2:	channel 2
D*:	experimentally determined disintegration rate	M:	mixture
A:	accuracy function		
E:	counting efficiency		

errors in the six counting rates provides an estimate of error based solely on the statistics of the system. Such an error estimate is quite useful in the selection of optimum channel combinations.

Chlorine-36 was obtained from the Isotopes Division at Oak Ridge National Laboratory, and the phosphorous-32 was prepared by irradiation of ammonium phosphate (dibasic) in the Oak Ridge Research Reactor. Conventional 4πβ counting techniques were used to determine absolute disintegration rates. The basic instrumentation used was a Packard Tri-Carb Liquid Scintillation Spectrometer (3000 series). All data were collected with the instrument operating in a coincident pulse height distribution mode. Counts were taken at preset times to yield between 10^5 and 10^6 gross counts in most counting situations. Background, which was determined to be approximately 10 counts per minute over the entire single channel analyzer range for the experimental conditions used, was considered negligible in all cases. Samples were prepared to give a total volume of about 8 ml of aqueous solution and were counted in standard glass counting vials at 1° C.

To accumulate a set of data, five different solutions were prepared, one of each of chlorine-36 and phosphorous-32 separately and three mixtures of the two in which the activity ratios were approximately one to one, one to ten, and ten to one. The solutions containing only one isotope were used to determine counting efficiencies in selected channels, and the mixtures were counted to determine the accuracy of the method.

The major variable parameters in the determination of two isotopes in a mixture are the gain and discriminator settings for the two single channel analyzers. The adjustment of these variables defines the two channels which are used to assay the mixture. While methods for the selection of optimum gain and channel combinations have been reported for dual tracer studies with liquid scintillation systems, it was not readily evident that these methods could be extended to Cerenkov counting systems. For example, a useful technique to optimize gain settings in dual tracer liquid scintillation systems utilizes an "efficiency isotope A vs. efficiency isotope B" plot for varying gain settings (37). However, when this technique, known as an Engberg plot, was applied to the phosphorous-32/chlorine-36 Cerenkov system, it provided relatively little guidance. (It should be noted that an Engberg plot for carbon-14 and tritium in liquid scintillation solutions prepared at the same time worked well for that combination of isotopes.) Experience indicated that a practical gain setting was one which allowed the entire spectrum of phosphorus-32, the stronger Cerenkov emitter, to be included completely in the range spanned by the single channel analyzer.

A more complex problem was that of selecting optimum channels for the simultaneous determination of the two isotopes. Any given channel may include a certain portion of each isotope's response spectrum, and thus it is possible to describe a channel in terms of the efficiencies that it defines for the two isotopes or in terms of the ratio of these efficiencies. In selecting channels, a variety of criteria may be applied. A channel may be set to maximize the chlorine/phosphorous efficiency ratio. Another channel may be set to minimize this ratio. Instead of counting efficiencies, ratios of other terms such as the figure of merit, which is the product of counting rate and efficiency, may be the basis for selecting channels. Another possibility is to select a channel which excludes entirely the lower energy β^- emitter, and counts only the stronger one. Selecting combinations of channels provides even more possibilities. Two channels may be selected so that they have a common boundary, no boundary, or boundary overlap.

To evaluate which combinations of which channels might be promising for application to the chlorine-36/phosphorous-32 Cerenkov system, an error analysis based on a number of different semi-hypothetical counting situations was initiated. The efficiencies of each isotope separately were determined over the whole range of the single channel analyzer in one-half volt increments. By summing the efficiencies of appropriate groups of increments, it was possible to predict efficiencies for both isotopes in hypothetical channels. Eight such channels were selected, and the efficiencies of phosphorous and chlorine associated with them were calculated. Each of these efficiencies was multiplied by 5×10^5. The resulting numbers were assumed to correspond to counts that would be tallied for a specific isotope in a given channel if it were exposed to a solution of that isotope that yielded five hundred thousand disintegrations over any arbitrarily selected time period. The numbers themselves provided estimates of counts that might actually be observed in efficiency determinations for the particular isotope in selected channels. Adding the phosphorous and chlorine numbers for a channel predicted the count for a mixture of the two with a one-to-one activity ratio. Counts were also predicted for one-to-ten and ten-to-one activity ratios, so that each hypothetical channel had five numbers associated with it--one for chlorine, one for phosphorous, and one for each of three mixtures. By selecting combinations of these eight hypothetiical channels it was possible to predict the errors in statistics for a particular combination of channels.

Of the twelve different combinations of these eight channels that were examined, the one for which the lowest errors on the basis of statistics were predicted, was a pair of channels with a common boundary in which the chlorine-36 efficiency was 3.0% in the lower channel and 0.53% in the upper channel, and the phosphorous-32 efficiency in the lower channel was 12.7% and in the upper channel 18.0%. An effort was made to duplicate these channels in the laboratory and to correlate the predicted and observed errors for the laboratory situation. The combination of channels found experimentally which closely approximated the hypothetical combination was a pair for which the phosphorous efficiencies were 13.0% in the lower channel and 17.7% in the upper channel. The chlorine efficiencies were 2.58% and .42% in the lower and upper channels respectively. Table V lists the predicted and observed errors for two combinations in laboratory trials. Another combination of hypothetical channels which was selected as being relatively good with respect to predicted error was also nearly duplicated in actual counting conditions, and these channels too showed roughly the same correlation between predicted and observed error. It must be remembered that predicted errors are maximum errors predicted on the basis of random errors in counting statistics while the observed errors reflect errors from all sources, such as errors in pipetting and variations in instrument response.

The data in Table V show that the largest error observed in any determination of phosphorus is <3%. The greatest observed error for chlorine (excluding the 10:1 P/Cl ratio) is <5%. Of course, predicted and observed errors can be reduced by extending the counting period.

In a preliminary way, we have applied this computer simulation technique to the more common situation of counting

Table V.
Comparison Of Experimental And Predicted Results For Two Isotopes

Condition # 1

Sample	Activity Ratio (P/Cl)	Phosphorous-32 % Of Error Observed	Phosphorous-32 % Of Error Predicted	Chlorine-36 % Of Error Observed	Chlorine-36 % Of Error Predicted
1	1:1	1.59	0.28	2.88	2.81
2	1:10	1.42	0.84	3.05	1.03
3	10:1	1.06	0.26	55.80	18.45
1	1:1	1.85	0.28	1.69	2.85
2	1:10	2.09	0.84	3.43	1.04
3	10:1	2.23	0.25	13.77	22.03

Condition # 2

Sample	Activity Ratio (P/Cl)	Phosphorous-32 % Of Error Observed	Phosphorous-32 % Of Error Predicted	Chlorine-36 % Of Error Observed	Chlorine-36 % Of Error Predicted
1	1:1	1.50	0.40	3.25	3.89
2		2.12	0.39	4.39	4.16
3		1.80	0.39	0.07	4.03
4		1.36	0.40	1.70	3.95
5		1.85	0.39	0.99	4.07
1	1:10	2.78	1.23	1.23	1.50
2		2.78	1.23	1.33	1.49
3		1.95	1.24	2.50	1.49
4		2.40	1.24	1.88	1.50
5		2.24	1.23	1.62	1.50
1	10:1	1.55	0.36	0.02	34.03
2		2.19	0.35	32.51	50.32
3		1.64	0.36	8.70	31.85
4		1.61	0.36	12.32	38.89
5		1.63	0.36	18.79	42.16

carbon-14 and tritium in organic scintillators. Over wide ranges of activity ratios and absolute activity levels, predicted and observed errors are in very close agreement. Thus, in our opinion, the computer simulation is generally applicable to the general problem of counting multiple isotopes that exhibit a continuous emission function. The great advantages of the method are the speed with which predictions can be obtained and the accuracy of those predictions.

References

1. H. H. Ross, Anal. Chem. 41, 1260 (1969).
2. R. P. Parker and R. H. Elrick in The Current Status of Liquid Scintillation Counting, P.110 (Edwin D. Bransome, Ed.). New York and London: Grune and Stratton (1970).
3. T. Iwakura, Radioisotopes (Tokyo) 18, 500 (1969).
4. A. Laüchli, Int. J. Appl. Radiat. and Isotopes 20, 265 (1969).
5. L. I. Wiebe, A. A. Noujaim, and C. Ediss, ibid. 22, 463 (1971).
6. S. Mizuno, K. Eguchi, and K. Yano, Radioisotopes (Tokyo) 18, 19 (1969).
7. F. L. Hoch, R. A. Kuras, and J. D. Jones, Anal. Biochem. 40, 86 (1971).
8. M. K. Johnson, ibid. 29, 348 (1969).
9. R. T. Haviland and L. L. Bieber, ibid. 33, 323 (1970).
10. I. M. Frank and I. G. Tamm, Dokl. Akad. Nauk. SSSR 14, 109 (1937).
11. G. P. Gennaro and K. E. Collins, J. Radioanal. Chem. 5, 387 (1970).
12. H. H. Ross in The Current Status of Liquid Scintillation Counting, p.123 (Edwin D. Bransome, Ed.). New York and London: Grune and Stratton (1970).
13. J. R. Robinson, Int. J. Appl. Radiat. and Isotopes 20, 531 (1969).
14. K. Asada, M. Takahashi, and M. Urano, Anal. Biochem. 48, 311 (1972).
15. M. Wiernik and S. Amiel, Trans. Am. Nuc. Soc. 12, 518 (1969).
16. H. R. Lukens, J. Radioanal. Chem. 1, 349 (1968).
17. V. K. Haberer, Atomwirtschaft 10, 36 (1965).
18. R. H. Elrick and R. P. Parker, Int. J. Appl. Radiat. and Isotopes 19, 263 (1968).

19. J. R. Brownell and A. Läuchli, ibid. 20, 797 (1969).
20. A. T. B. Moir, ibid. 22, 213 (1971).
21. P. E. Ballance and S. Johnson, Health Phys. 20, 447 (1971).
22. P. E. Ballance and S. Johnson, Planta 91, 364 (1970).
23. H. H. Ross in Organic Scintillators and Liquid Scintillation Counting, p.757 (D. L. Horrocks and C. T. Peng, Eds.). New York and London: Academic Press (1971).
24. E. Heiberg and J. Marshall, Rev. Sci. Instrum. 27, 618 (1956).
25. N. Porter, Nuovo Cim. 5, Series 10, 526 (1957).
26. G. Cosme, S. Jullian, and J. Lefrancois, Nucl. Instrum. and Methods 70, 20 (1969).
27. L. C. Brown, Anal. Chem. 43, 1326 (1971).
28. H. R. Lukens and J. E. Lasch, Int. J. Appl. Radiat. and Isotopes 15, 758 (1964).
29. J. P. Thomas and E. A. Schweikert, Radiochem. Radioanal. Lett. 9, 155 (1972).
30. M. Wiernik and S. Amiel, J. Radioanal. Chem. 5, 123 (1970).
31. F. Girardi, V. Camera, and E. Sabbioni, Radiochem. Radioanal. Lett. 2, 195 (1969).
32. J. Ashcroft, Int. J. Appl. Radiat. and Isotopes 20, 555 (1969).
33. Z. P. Zagorski and Z. Zimek, Nukleonika 15, 335 (1970).
34. Z. P. Zagorski and Z. Zimek, ibid. 16, 359 (1971).
35. D. S. Ayers, et. al., Physica 43, 105 (1969).
36. J. D. Buchanan, Radiochim. Acta 9, 218 (1968).
37. Y. Kobayashi and D. V. Maudsly in The Current Status of Liquid Scintillation and Counting, p.76 (Edwin D. Bransome, Ed.). New York and London: Grune and Stratton (1970).

BIOLUMINESCENCE MEASUREMENTS : FUNDAMENTAL ASPECTS, ANALYTICAL APPLICATIONS AND PROSPECTS

Eric SCHRAM

University of Brussels, Belgium

INTRODUCTION

Bioluminescence, more especially firefly bioluminescence, has been used for analytical purposes in several laboratories for about 20 years. During the initial period measurements were but seldom performed with quantum counters and the sensitivity of the method was therefore limited by that of the available photometers. After it had become clear that scintillation counters could adequately be used for this purpose (1) the method has been investigated further and its sensitivity for routine applications increased by several orders of magnitude. Modern scintillation counters, as used for the counting of soft beta emitters (carbon-14 and tritium), undoubtedly possess features that are useless for the mere measurement of bioluminescence : e.g. multiple counting channels, coincidence circuitry, external radioactivity standard etc. Their availability in most biochemical laboratories has nevertheless made their use quite popular for this purpose. On the other hand, it may be anticipated that simplified instruments, based on the same technology, will become more popular in the future. One should also realize that bioluminescence is only a special case of chemiluminescence and some of the conclusions mentioned here might therefore be extended to other fields than biochemistry. Such luminescence results from chemical reactions at room temperature, without absorption of light, whenever the free energy of reaction equals or exceeds that of electronic transition in one of the product molecules. However, even when this energy is sufficient to bring an electron in its lowest excited state, no luminescence will often be observed because of internal quenching or external quenching by the surrounding medium, more especially in polar media. This difficulty was circumvented by the living cell through the mediation of enzymes which show an extraordinary efficiency in the fulfillment of the most delicate and intricate chemical reactions. In firefly bioluminescence this efficiency is exceptionally high and the quantum yield even approaches unity, which means that all excited product molecules return to the ground state with the emission of light.

Chemiluminescence is a very widespread phenomenon encountered among all kinds of chemical substances : inorganic (phosphorus), organic (luminol, lucigenin) and biological (luciferins). Bioluminescence offers undoubtedly the most sophisticated aspects, be it only because of the existence of the enzymatic machinery. Its use as an analytical tool requires some familiarity with its reaction mechanisms and this problem will therefore be dealt with briefly in the first part of this paper in the case of the two best known systems : firefly and bacterial bioluminescence.

The literature on bioluminescence is quite abundant and has been reviewed by several authors : McElroy and Seliger (2), Cormier and Totter (3 and 4), Seliger and McElroy (5), Hastings (6 and 7), Strehler (8), Seliger and Morton (9), Chase (13). Further information is also to be found in the proceedings of symposia devoted to the same subject : McElroy and Glass (10), Johnson and Haneda (11), Hercules, Lee and Cormier (12) (only the publications that have appeared since 1961 are mentioned here). In the next paragraphs the accent will be laid on some important features of firefly and bacterial luminescence, more especially on the latest findings in this field.

Beside its biological significance bioluminescence offers two important aspects :

1° Biochemical aspects :

The following problems deserve special attention:

- Identification and structure of the excited molecule : some progress has been made in recent years but many difficulties were encountered because of the tiny quantities of material involved and the fact that the actual end-product may further react before it can be isolated. Sensitized fluorescence might also have to be taken into account.
- Reaction mechanism : bioluminescent reactions occur in

several steps which are not always easy to dissociate and more than one substrate may be involved; the enzyme plays a fundamental role not only in catalyzing the chemical reaction but also in creating the necessary micro-environment around the active site, in order to reduce quenching and stabilize the light-emitting species.

2° Physical aspects :

Bioluminescence is directly related to fluorescence phenomena and has therefore also retained the attention of physicists (see Seliger and Morton, 9).

From a thermodynamic point of view the emission of a photon requires about 50-60 Kcal per einstein, which is far more than the energy released in all known biochemical reactions, at least in a single step. Bioluminescent reactions must therefore differ from the other reactions usually encountered in the living cell.

In recent years bioluminescence has developed into an ultrasensitive tool for the measurement of biochemical substrates and enzymes. To understand this one should consider the sensitivity of modern quantum counters such as those used in most biochemical laboratories for the measurement of radioactivity by means of scintillating solutions. Taking into account the background of photomultipliers (10-100 pulses/sec), their efficiency for single photons (10-30%) and the quantum yield of bioluminescent reactions (e.g. 1%), it is theoretically possible to assay quantities of material down to 10^{-18} moles/sec (this evaluation is of course very rough). This sensitivity is lowered by a few orders of magnitude by the finite rate of luminescent reactions but on the whole the present method remains one of the most sensitive ever attained.

BACTERIAL LUMINESCENCE

A. Reaction mechanism

Luminous bacteria (Photobacterium fischeri or Achromobacter) were the first bioluminescent organisms to be used in biochemical analysis (Harvey, 14) and served to measure the oxygen production by plants under anaerobic conditions. The reaction requires a riboflavin compound (flavin mononucleotide or FMN) and an additional factor identified as a long-chain aldehyde which increases the light-yield by a factor ranging from 100 to 1000. The role of the latter as a substrate in the reaction has been proposed as early as 1955 by McElroy and Green (15) but not proved with some certainty until very recently. The role of reduced pyridine nucleotides was clarified around the same period when it was shown that they are merely used to reduce FMN and therefore useless in the presence of pre-reduced FMN.

The several steps involved in the mechanism of bacterial luminescence were described ten years ago by Hastings and Gibson (16). On the basis of former experimental evidence and of their own results they proposed the following scheme

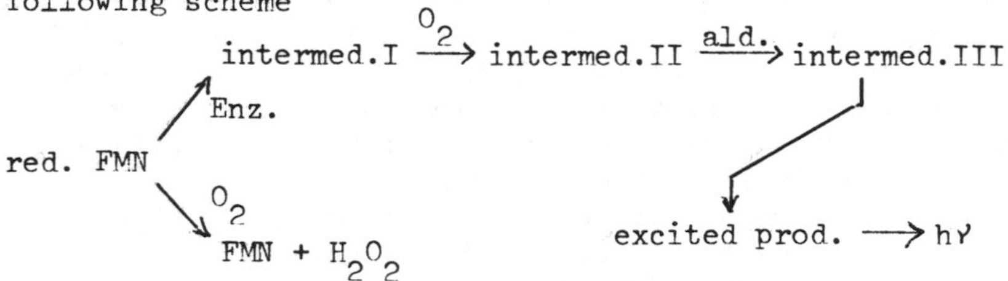

It is important to note that reduced FMN oxidizes autocatalytically in the presence of oxygen with a half-life of about 0.1 sec as shown by Gibson and Hastings (17). Taking into account that the half-life of the luminescence is of the order of 10 seconds and that the luminescence maximum occurs after about 2 seconds it must be concluded that the reduced FMN molecules are trapped very rapidly and efficiently by the enzyme and at least one long-lived intermediate may be postulated. The half-life of the luminescence is therefore independent of the enzyme concentrations over many orders of magnitude and the luminescence curve is determined solely by the reaction mechanism whereby the enzymatic complex gives rise to the excited species. The nature of the aldehyde, more especially its chain-length, is important in this respect as it will affect the decay-rate and quantum-yield of the luminescence reaction, whereas the brightness of the reaction is proportional to the FMN concentration. The absence of turn-over of the enzyme is only apparent since the addition of extra quantities of reduced FMN to an exhaused mixture gives rise again to light emission. On the other hand, the addition of fresh enzyme to a mixture after 3 seconds gives no additional light. The turn-over of bacterial luciferase was further studied by Erlanger, Isambert and Michelson (18) by fixing the enzyme to polyacrylic hydrazide. The stabilized enzyme showed all the main features of the native enzyme and could be used over and over without apparent limit. When kept at 4° C it remained stable in contrast with the native enzyme which tends to loose its activity rather rapidly.

One of the still debated questions is the ratio in which the several substrates react with each other. The consumption of aldehyde in particular has always been difficult to ascertain and the overall kinetic picture was often confused by the dominance of side-pathways. More recently massa spectrometry was used by Shimomura (19) to identify the end-products of aldehyde oxidation. Tetradecanoic and dodecanoic acid were obtained in roughly quantitative yields when using resp. tetradecanal and dodecanal for the light reaction. The observed quantum yield with respect to the aldehyde was found to be 0.17 ± 0.1 for chain lengths ranging from C_8 to C_{14}. No significant amount of carbon dioxide was produced. According to Hastings, Eberhard, Baldwin, Nicoli, Cline and Nealson (20) bacterial luciferase bears a single binding site for $FMNH_2$ and they propose therefore the following stoechiometry :

$$FMNH_2 + O_2 + RCHO \longrightarrow FMN + H_2O + RCOOH$$

Another approach was used by Lee (21) who deduced stoechiometric ratios from the oxygen consumption and from the relative substrate concentrations leading to maximum quantum yield. According to his scheme two reduced FMN molecules would react in a coordinated way.

B. Luminescent intermediate

Study and identification of luminescent intermediates has been based mostly on the comparison of the chemiluminescence spectrum with the fluorescence spectra of known substances and, of course, on the chemical identification of the end-product of the reaction. Spectral considerations may however be impaired by the fact that the spectrum of the emitting molecule is likely to be modified by its binding to the enzyme. Energetic considerations are also fundamental and should remind us of the fact that in the case of bacterial luminescence the oxidation of one $FMNH_2$ molecule by oxygen creates only 27 Kcal of energy which is insufficient to produce photons with a wavelength of 490 nm. Eley, Lee, Lhoste, Lee, Cormier and Hemmerich (22) observed that the luminescence spectrum matches more closely with that of flavin protonated at the N-1 position. By combining this observation with the need for an energy supply from the aldehyde Eberhard and Hastings (23) have recently proposed the following mechanism :

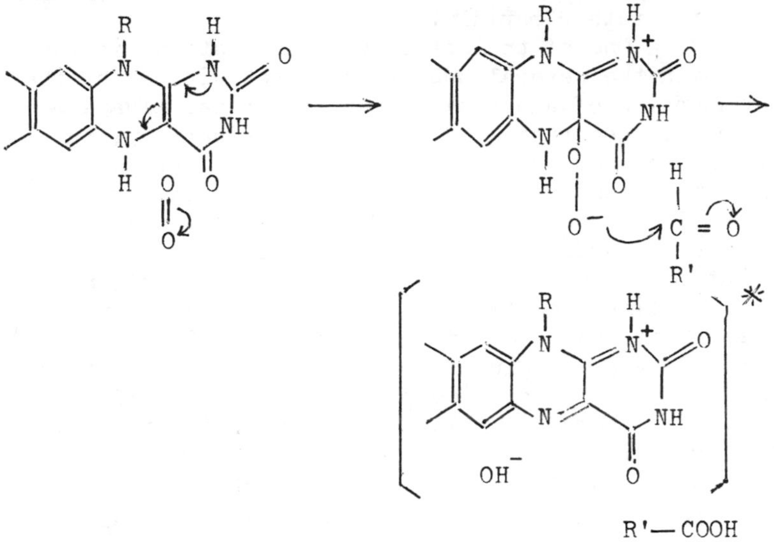

R'— COOH

C. Structure of the enzyme

The high quantum yields observed in bioluminescence may be ascribed to the specific role of the enzyme which protects the excited molecular species from quenching. The microenvironment surrounding the substrate in the active site appears therefore to be of paramount importance and gives additional intrest to the study of the enzyme structure. Bacterial luciferase was shown to consist of two sub-units (24,25) whereas kinetic experiments by Meighen and Hastings (26) showed the existence of a single flavin binding site per dimer.

Numerous mutants of bacterial luciferase have been isolated from which it could be deduced that both subunits are not functionally aequivalent (27). One of these mutants has been the subject of particular attention and was designated as MAV by Hastings, Weber, Friedland, Eberhard, Mitchell and Gunsalus (28) who isolated it some years ago. It differs slightly from Photobacterium fischeri with respect to the molecular weights and the composition of the subunits(25). It is further characterized by the longer life-time of the intermediate products, which makes it a "slow" luciferase. Optimum quantum yield is obtained with decanal as compared with tetradecanal for the Photobacterium fischeri strain.

FIREFLY LUMINESCENCE

A. Reaction mechanism

Firefly luminescence is characterized by the existence of a specific substrate, luciferin, which has become commercially available since a few years, and by the need for ATP. The system was developed into an assay method for this substance by Strehler and Totter (29).

The general reaction scheme of firefly bioluminescence may be summarized as follows (LH_2 = luciferin; L = dehydroluciferin)

$$E + LH_2 + ATP.Mg \rightleftharpoons E.LH_2AMP + MgPP \qquad (1)$$

$$E.LH_2.AMP + O_2 \longrightarrow Product + CO_2 + light \qquad (2)$$

$$E + L + ATP.Mg \rightleftharpoons E.L.AMP + MgPP \qquad (3)$$

Step (1) is reversible and shows a striking similarity with the activation of aminoacids by synthetases. Both the L and D isomers of LH_2 react with ATP although only the D-isomer is susceptible to give off a photon. ATP reacts as a complex with Mg++ which associates stoechiometrically with its β and γ phosphate groups, leaving the α phosphate group free to bind with the enzyme.

Oxidation of the activated substrate occurs in step (2). During many years it has been considered that L.AMP was the end-product of the reaction. It is indeed tightly bound to the enzyme ($K_{diss.}=5.10^{-10}$) and was thought therefore to cause end-product inhibition. Pyrophosphate, which inhibits step (1), would on the contrary reverse the inhibition caused by end-product accumulation (step 3) with the simultaneous release of ATP (8). Such cyclic mechanism was supposed to account for the long-lived tailing of the luminescence curve but in the light of more recent knowledge this mechanism has to be reconsidered.

The elimination of carbon dioxide was demonstrated to be roughly quantitative in the presence of luciferase (30) using C^{14}-labeled luciferin. AMP is probably ejected before the emission of light can occur (see also 31 and 32).

The activation of luciferin is a two substrate reaction and kinetic experiments performed by Denburg, Lee and McElroy (33) show the binding of LH_2 and Mg.ATP to occur in a random order. Luciferin enhances the binding of ATP slightly and vice-versa. The binding of ATP is influenced to a greater extent by the ionic strength of the medium (34) : The K_m for ATP.Mg increases with increasing ionic strength according to the Hofmeister series ($SCN^- > I^- \sim NO_3^- > Br^- > Cl^-$) but is not influenced by the nature of the cation (Na^+, K^+ or NH_4^+). The anion inhibition is instantaneously reversible and non competitive with respect to ATP.Mg; it may therefore be ascribed to a general ionic strength effect expressed as a decrease in attraction. The K_m of luciferin is not affected in this particular instance.

B. Luminescent intermediate

As stated above it has become evident that dehydroluciferin is not responsible for the observed luminescence in fireflies. Spectral studies have shown that at neutral or basic pH the quantum yield is 1 and the emitted light yellow-green (562 nm). In acid medium the quantum yield is depressed and the spectrum shifts to the red (maximum at 614 nm). This shift can be explained by the predominance, at low pH, of another distinct emitting species. The red emission observed in vitro as well as in vivo may be ascribed to the keto anion of decarboxylated luciferin (9 and 31); the yellow-green light observed in vivo would be emitted by the enol dianion (32).

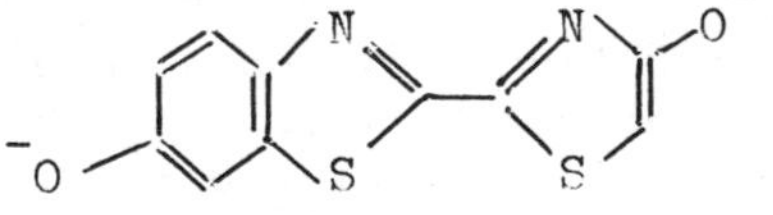

Enol dianion

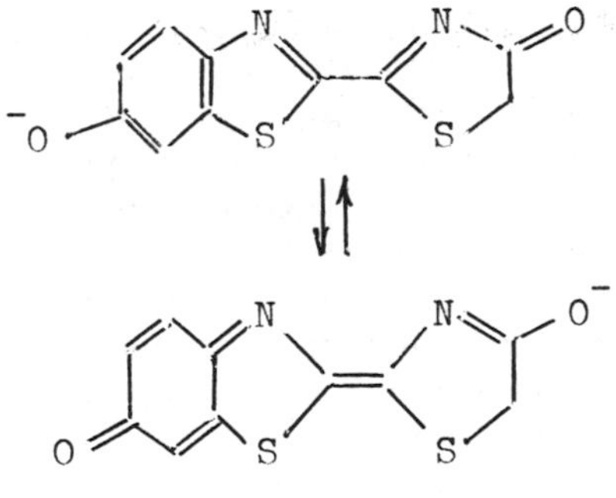

Keto anion

Similar and other considerationshave led to the conclusion that the most probable intermediate is a dioxetane ring, known to give highly efficient electronically excited fragments (35, 36, 37 and 38).

$$\begin{array}{ccccc}
\begin{array}{c} \mathrm{H} \quad \mathrm{AMP} \\ | \qquad | \\ \diagdown\mathrm{C}-\mathrm{C}=\mathrm{O} \\ | \\ \diagup\mathrm{CH_2} \end{array}
& \longrightarrow &
\begin{array}{c} \mathrm{O}-\mathrm{O} \\ | \qquad | \\ \diagdown\mathrm{C}-\mathrm{C}=\mathrm{O} \\ | \\ \diagup\mathrm{CH_2} \\ \\ +\ \mathrm{AMP} \end{array}
& &
\begin{array}{c} \left[\begin{array}{c} \diagdown\mathrm{C}{=}\mathrm{O} \\ | \\ \diagup\mathrm{CH_2} \end{array} \right]^{*} \\ \\ +\ \mathrm{CO_2} \end{array}
\end{array}$$

Some controversy still exists about the origin of the oxygen incorporated into the dioxetane ring, either molecular oxygen or water (39, 40).

C. Structure of the enzyme

Firefly luciferase has a molecular weight of 100,000 and consists of two subunits of molecular weight 50,000. At low concentration it is always dissociated. Using equilibrium dialysis and fluorimetric techniques it was possible to identify two sites for ATP (or AMP), two for LH_2, one for Mg.ATP and one for L.AMP per molecular weight of 100,000 (33, 34 and 41) whether the enzyme is dissociated or not. The subunits are similar but not identical. The ATP.Mg site is identical with one of the sites for ATP which acts as a competitive inhibitor with respect to ATP.Mg. The role of the second ATP site is not clear as yet and could possibly have a regulatory function. The enzyme binds only one AMP in the presence of ATP.Mg and none in the presence of an excess of ATP.

ANALYTICAL APPLICATIONS OF BIOLUMINESCENCE

A. Instrumentation and counting technique

One should first of all be aware of the fact that the analytical applications of bioluminescence are based on the measurement of reactions rates. On mixing of the reagents luminescence curves usually show a sharp rise followed by a more or less rapid decay. The decay-rate will depend on the specific reaction being measured and also on the experimental conditions. Many measuring instruments used up to now have been aimed at the measurement of the peak intensity of the luminescence curve, which used to be considered as the most representative of the substrate concentration. Analog instruments used for this purpose consist essentially of a photomultiplier with its associated electronics, connected to a fast recorder or oscilloscope. In a commercial version (Biometer, Dupont Instruments) the maximum is automatically recorded and displayed on a digital register.

Scintillation counters do not allow to record luminescence peaks of short duration (a few seconds or less). However, experiments performed with such counters have shown that one is not necessarily bound to the measurement of this peak. In the case of firefly luciferase and under the proper conditions, the luminescence decreases exponentially over a rather long period of time (half an hour or more), while remaining proportional to the ATP concentration (42, 43, 44, 45). This enables one to use the integral mode of counting which is, in practice, the only compatible with the use of scintillation counters, and at the same time makes flow-monitoring possible. A lin-log plot of the luminescence versus time gives straight lines which run perfectly parallel over a very wide concentration range and whose slopes are proportional to the enzyme concentration.

In the case of bacterial luminescence, when the reduced FMN is produced in situ by the oxidation of $NADH_2$ the concentration of the dehydrogenase associated with the luciferase can be adjusted in such a way that a flattened luminescence peak is obtained allowing for the necessary time to perform the countings (46, 47, 48). With some systems a steady state may be achieved whereby the reduced substrate is produced continuously in the presence of the luminescent system.

For the assay of ATP integration of the luminescence is currently started after a fixed time interval following mixing of the reagents. This principle has been automated in the "ATP Photometer" developed quite recently by JRB Inc., San Diego Calif. In our hands a time interval of a few minutes proved adequate, giving ample time to introduce the sample into the counter.

Although scintillation counters possess some features that are useless for bioluminescence measurements one can take advantage of their high performance as automated quantum counters

- they can be used for discrete as well as for flow measurements (when used in the "repeat" mode)
- they have a very high sensitivity
- the integral technique allows to obtain better statis-

tics; speed of mixing is also less critical
- they have a wide dynamic range
- discriminator, gain and high voltage settings permit an accurate choice of the best counting conditions
- advantage can be taken of the sample changer for automation purposes (49)
- digital results are obtained which can easily be used by data processing equipment (50).

Bioluminescence countings may often be impaired by the phosphorescence of the vials introduced into the counter. A fixed cell, remaining in the counter, was therefore developed together with a semi-automatic microtransferator for the introduction of small-sized samples. Transfer occurs by suction by means of a motor-driven microsyringe (51).

The choice of the high voltage and discriminator settings is a problem to which sufficient attention has not always been paid, when trying to achieve maximum sensitivity. The settings used for the scintillation counting of tritium are not adequate, contrary to what has sometimes been advocated. Overlapping of the photomultiplier noise with the luminescence spectra is an important factor since one operates with the coincidence disabled and at room temperature where the noise is evidently higher. Higher voltages are not necessarily better as discussed in one of our previous papers (51) and may give rise to memory effects at high counting frequencies. High counting rates may further be at the origin of erratic results due to double photon pulses (43).

Bioluminescence measurements are preferably performed around room temperature. This temperature may be stabilized in a very simple way by using small sample tubes inserted into regular scintillation vials filled with water (43, 48).

Partial automation of bioluminescence assays was achieved by means of a conventional auto analyzer equipment (52) or a Hamilton precision liquid dispenser (49), in both cases in conjunction with a liquid scintillation spectrometer.

B. General methodology

1. Firefly luminescence : The assay of ATP has been dealt with by many authors since this method was introduced by Strehler and Totter in 1952 (29). Whether one measures the initial luminescence flash,or the light intensity after a fixed delay, in both cases results could be obtained which were shown to be proportional to the concentration of ATP which acts as the limiting component in the system. It is of course essential that this concentration does not exceed half-saturation of the enzyme, but this is not likely to be the case. Indeed, the half-saturation concentration of firefly luciferase is 35 μg/ml (10^{-7} mole/ml), i.e. several orders of magnitude more than the quantities actually assayed.

The sensitivity of the method is determined by that of the measuring equipment but is also affected by the residual luminescence of commercial enzyme preparations. Unfortunately purification of the enzyme by crystallization (54) is neither practical nor economical on a small scale. In common practice the blank value is usually lowered by ageing the enzymatic preparation before use. Methods have also been suggested, based on gel filtration (55, 56, 57). Luciferin, dehydroluciferin and ATP are easily separated from luciferase on low porosity gels, and sensitivities of 10^{-14} - 10^{-15} mole ATP can be attained with the purified preparations. Higher porosity gels may also be used to free the luciferase from other interfering enzymes which affect the specificity of the reaction. Nielsen and Rasmussen (55) were able to lower the adenylate kinase (myokinase) content 500 times but the NDP kinase (transphosphorylase) is removed less efficiently. These enzymes are responsible for the production of ATP in the presence of resp. ADP and other nucleoside triphosphates. Since the ATP is formed but gradually after the addition of the luciferase, interference by these enzymes is indicated by a modification of the shape of the luminescence curve. However, in most routine cases the results will not be impaired.

Several methods have been tested for the extraction of ATP from biological material : n-butanol (57), trichloracetic acid (58), perchloric acid (59), ethanol (58, 60, 61), tris-buffer, etc. Perchloric and trichloracetic acid interfere with the luminescence and should therefore be avoided and tris-buffer is usually to be preferred.

Interfering authocyanins and phenolic compounds extracted with the ATP from plants could be eliminated by means of polyvinylpyrrolidone (58).

When assaying ATP in living material it is important to ascertain that its content has not been modified during sampling, storage or extraction (59, 61, 62). The best results are obtained by injecting the sample rapidly into a greater volume of boiling water or buffer taking care to keep the temperature as close as possible to 100° C. Thanks to the extremely high sensitivity of the ATP assay the dilution effect will in general not be harmful.

2. Bacterial luminescence

Bacterial luciferase is specific for FMN but its intrest lies also in the fact that it occurs associated with a dehydrogenase that makes the system suitable for the measurement of reduced NAD in the presence of an excess of FMN (44, 46, 47, 48, 63, 64, 65). The same system reacts also with NADP but to a lower extent.

Reduced FMN being rapidly oxidized in the presence of oxygen its assay must be performed after reduction "in situ" by means of $NaBH_4$ (63) or in the presence of an excess of $NADH_2$ (FMN being in this case the limiting component).

The sensitivity of the present method reaches far beyond that of fluorimetric techniques (10^{-13} - 10^{-14} mole, 47, 64). Moreover, many substances likely to interfere in fluorimetric assays will not affect the bioluminescence. The proportionality between substrate concentration and luminescence holds over many orders of magnitude as for ATP luminescence.

A method has been proposed recently by Brolin et al. (65) for the assay of pyridine nucleotides in mixtures. Selective measurements of either NADH or NADPH is performed after oxidation of the unwanted nucleotide by means of a highly specific enzymatic system. An alternative method is to reconstitute the reduced nucleotide to be measured, resulting in a continuous light emission.

Both bacterial and firefly luciferase are but poorly stable and their efficiency for lightproduction should therefore be tested with each batch of samples. The effect of interfering substances will also preferably be taken into account by the addition of an internal standard.

C. Analytical applications

Many of the actual but also potential applications of firefly bioluminescence have already been reviewed by Strehler and Totter as early as 1954 (53) and more recently by Strehler (8). They can be subdivided according to the substance being assayed:

1. assay of the luminescence substrate (ATP)
2. assay of related substrates which can be converted directly into ATP (ADP, AMP) or may yield ATP in the presence of ADP (phosphoenolpyruvate, creatinephosphate, nucleosidetriphosphate, etc.). The method has been extended recently to c-AMP (66). In some cases the substrate concentration may be deduced from the consumption of ATP (e.g. glucose)
3. assay of enzymes : e.g. hexokinase, pyruvate kinase, creatine kinase, etc.

Recent applications are concerned with the measurement of phosphodiesterase (67) and ATP-sulphurylase (68, 69).

Although bacterial bioluminescence was introduced more recently as an analytical method its potential field of application is at least as broad as that of firefly luciferase. Many compounds which are either convertible in dehydrogenase reactions or can lead to a dehydrogenase step can indeed be subjected to analysis.

NADH is currently produced in separate reactions preceding the light yielding reaction. The following substances were for instance assayed by Brolin et al. (64).

- glucose : NADPH was formed in a coupled hexokinasedehydrogenase reaction. Sensitivity : 0.5-6.0 pmole in aliquot of 1.1 µl.
- malate : the unfavourable equilibrium for the NADH formation was overcome by using an NAD analog and hydrazine as a trapping agent.
- NAD : reduced with ethanol and alcoholdehydrogenase in the presence of semicarbazide as a trapper.

In our laboratory (Gerlo, unpublished results) the oxidation of glucose-6-P was performed in presence of the luminescent system, giving rise to a rather steady light yield.

During recent years bioluminescence has been involved in the determination of biomass and its application to soil and water ecology. With this in mind systematic studies were undertaken on the ATP level and its variations in bacteria (57, 70) and other microorganisms, chiefly marine algae (62). Biomass determination on the basis of ATP content is a simple and rapid technique compared with slower and more tedious classical methods as plate-counting and microscopic examination. It is based on the assumption that 1°/ no ATP is associated with non-living particulate material, 2°/ that the ratio of ATP to other parameters is fairly constant. As far as this last assumption is concerned it was shown that in marine bacteria the ATP content ranges from 0.3 to 1.1 % of the cell carbon, the average being 0.4 % (70). Similar studies on marine algae (62, 71, 72) showed a considerable uniformity in ATP concentration in diverse algae ranging in size from less than 1 µg C/cell to 215,000 µg C/cell. The average value of 0.35 % relative to C is very close to that reported for bacteria. Biomass profiles of sea-water could in this way be obtained on 1-2 liter samples down to depths of 3500 m. (73, 74). The samples were filtered on 47 mm membranes of 0.45 µ pore-size and immediately immersed in boiling tris-buffer to extract ATP and inactivate the enzymes (61).

ATP determinations in lake sediments were performed by Lee et al. (75, 76).

CONCLUSIONS

Bioluminescence has been used up to now for the assay of various substances and many more are now awaiting to be assayed by the same method. The conditions for each specific substance need of course to be standardized as for any other enzymatic assay, but it may be anticipated that the array of applications will be extending steadily, together with the further development of suitable instrumentation.

It should be observed that ATP, FMN and NAD(P) are involved more or less directly in all biochemical reactions and that a great number of substrates and enzymes are therefore likely to be assayed in some way by the techniques of bioluminescence. Because of their high sensitivity only minute samples are needed and they might therefore appear very suitable in the future in the field of clinical chemistry.

REFERENCES

1. E. Tal, S. Dikstein & F.G. Sulman, Experientia 20, 652 (1964).
2. W.D. McElroy & H.H. Seliger in Advances in Enzymology (F.F. Nord, ed.) Interscience, 25, 119 (1963).
3. M.J. Cormier & J.R. Totter, Ann. Rev. Biochem. 33, 431 (1964).
4. Id., in Photophysiology (A.C. Giese, ed.) Acad. Press, 4, 315 (1968).
5. B.L. Seliger & W.D. McElroy, "Light, Physical and Biological Action", Acad. Press, (1965).
6. J.W. Hastings, in Current Topics in Bioenergetics (D.R. Sanadi, ed.) Acad. Press 1, 113 (1966).
7. Id. Ann. Rev. Biochem. 37, 597 (1968).
8. B.L. Strehler, in Methods in Biochemical Analysis (D. Glick, ed.) Interscience, 16, 99 (1968).
9. H.H. Seliger & R.A. Morton, in Photophysiology (A.C. Giese, ed.) Acad. Press, 4, 253 (1968).
10. W.D. McElroy & H.B. Glass (Eds.), Symposium on light and life, Johns Hopkins Univ. Press (1961).
11. F.H. Johnson & Y. Haneda (Eds.), Bioluminescence in Progress, Princeton Univ. Press (1966).
12. D. Hercules, J. Lee & M.J. Cormier (Eds.) Chemiluminescence and Bioluminescence, Plenum Press (in press). **
13. A.M. Chase, in Photophysiology (A.C. Giese, ed.) Acad. Press, 2, 389 (1964).
14. E.N. Harvey, Plant Physiol. 3, 85 (1928).
15. W.D. McElroy & A.A. Green, Arch. Biochem. Biophys. 56, 240 (1955).
16. J.W. Hastings & Q.H. Gibson, J. Biol. Chem. 238, 2537 (1963).
17. Q.H. Gibson & J.W. Hastings, Biochem. J. 83, 368 (1962).
18. B.F. Erlanger, M.F. Isambert & A.M. Michelson, Biochem. Biophys. Res. Commun. 40, 70 (1970).
19. O. Shimomura, F.H. Johnson & Y. Kohama, Proc. Nat. Acad. Sci. U.S.A. 69, 2086 (1972).
20. J.W. Hastings, A. Eberhard, T.O. Baldwin, M.Z. Nicoli, T.W. Cline & K.H. Nealson, in Chemiluminescence and Bioluminescence (D. Hercules, J. Lee & M.J. Cormier, eds.) Plenum Press, in press. **
21. J. Lee, Biochemistry 11, 3350 (1972).
22. M. Eley, J. Lee, J.M. Lhoste, C.Y. Lee, M.J. Cormier & P. Hemmerich, Biochemistry 9, 2902 (1970).

23. A. Eberhard & J.W. Hastings, Biochem. Biophys. Res. Commun. 47, 348 (1972).
24. J.M. Friedland & J.W. Hastings, Proc. Nat. Acad. Sci. U.S.A. 58, 2336 (1967).
25. A. Gunsalus-Miguel, E.A. Meighen, M.Z. Nicoli, K.H. Nealson & J.W. Hastings, J. Biol. Chem. 247, 398 (1972).
26. E.A. Meighen & J.W. Hastings, J. Biol. Chem. 246, 7666 (1971).
27. T.W. Cline & J.W. Hastings, Biochemistry 11, 3359 (1972).
28. J.W. Hastings, K. Weber, J. Friedland, A. Eberhard, G.W. Mitchell & A. Gunsalus, Biochemistry 8, 4681 (1969).
29. B.L. Strehler & J.R. Totter, Arch. Biochem. Biophys. 40, 28 (1952).
30. P.J. Plant, E.H. White & W.D. McElroy, Biochem. Biophys. Res. Commun. 31, 98 (1968).
31. T.A. Hopkins, H.H. Seliger, E.M. White & H.W. Cass, J. Am. Chem. Soc. 89, 7148 (1967).
32. E.H. White, E. Rappaport, T.A. Hopkins & H.H. Seliger, J. Amer. Chem. Soc. 91, 2178 (1969).
33. J.L. Denburg, R.T. Lee & W.D. McElroy, Arch. Biochem. Biophys. 134, 381 (1969).
34. J.L. Denburg & W.D. McElroy, Arch. Biochem. Biophys. 141, 668 (1970).
35. W. Adam, J.C. Liu, G. Simpson & H.C. Steinmetzer in Chemiluminescence and Bioluminescence (D. Hercules, J. Lee & M.J. Cormier, eds.) Plenum Press, in press.**
36. W.D. McElroy & M. De Luca, ibid. **
37. F. McCapra, ibid. **
38. Th. Wilson, ibid. **
39. M. De Luca & M.E. Dempsey, Biochem. Biophys. Res. Commun. 40, 117 (1970).
40. Id., in Chemiluminescence and Bioluminescence (D. Hercules, J. Lee & M.J. Cormier, eds.) Plenum Press, in press. **
41. R.T. Lee, J.L. Denburg & W.D. McElroy, Arch. Biochem. Biophys. 141, 38 (1970).
42. E. Schram, Arch. Int. Physiol. Biochim. 75, 894 (1967).
43. Id., in The Current State of Liquid Scintillation Counting (E.D. Bransome, ed.), Grune and Stratton, p. 129 (1970).

44. E. Schram, R. Cortenbosch, E. Gerlo & H. Roosens, in Organic Scintillators and Liquid Scintillation Counting (D.L. Horrocks & C.T. Peng, eds.) Acad. Press, p. 125 (1971).
45. P.E. Stanley & S.G. Williams, Anal. Biochem. 29, 381 (1969).
46. E. Gerlo & E. Schram, Arch. Int. Biochim. Biophys. 79, 200 (1971).
47. P.E. Stanley, Anal. Biochem. 39, 441 (1971).
48. Id., in Organic Scintillators and Liquid Scintillation Counting (D.L. Horrocks & C.T. Peng, eds.) Acad. Press, p. 607 (1971).
49. R.H. Hammerstedt, Anal. Biochem. 52, 449 (1973).
50. G.J.E. Balharry & D.J.D. Nicholas, ibid. 40, 1 (1971).
51. E. Schram & H. Roosens, in Liquid Scintillation Counting, Heyden & Son, 2, 115 (1972).
52. L. Dufresne & H.J. Gitelman, Anal. Biochem. 37, 402 (1970).
53. B.L. Strehler & J.R. Totter, in Methods of Biochemical Analysis (D. Glick, ed.) Interscience 1, 431 (1954).
54. A.A. Green & W.D. McElroy, Biochim. Biophys. Acta 20, 170 (1965).
55. R. Nielsen & H. Rasmussen, Acta Chem. Scand. 22, 1757 (1968).
56. H. Rasmussen & R. Nielsen, ibid. 22, 1745 (1968).
57. E.W. Chappelle & G.V. Levin, Biochem. Med. 2, 41 (1968).
58. G. Guinn & M.P. Eidenbock, Anal. Biochem. 50, 89 (1972).
59. H.A. Cole, J.W.T. Wimpenny & D.E. Hughes, Biochim. Biophys. Acta 143, 445 (1967).
60. H. Holmsen, E. Storm & H.J. Day, Anal. Biochem. 46, 489 (1972).
61. O. Holm-Hansen & Ch. R. Booth, Limnol. Oceanogr. 11, 510 (1966).
62. O. Holm-Hansen, Plant and Cell Physiol. 11, 689 (1970).
63. E.W. Chappelle, G.L. Picciolo & R.H. Altland, Biochem. Med. 1, 252 (1967).
64. S.E. Brolin, E. Borglund, L. Tegner & G. Wettermark, Anal. Biochem. 42, 124 (1971).
65. S.E. Brolin, C. Berne & U. Isacsson, Anal. Biochem. 50, 50 (1972).
66. R.A. Johnson, J.G. Hardman, A.E. Broadus & E.W. Sutherland, Anal. Biochem. 35, 91 (1970).

67. B. Weiss, R. Lehne & S. Strada, Anal. Biochem. 45, 222 (1972).
68. G.J.E. Balharry & D.J.D. Nicholas, Anal. Biochem. 40, 1 (1971).
69. P.E. Stanley, in Chemiluminescence and Bioluminescence (D. Hercules, J. Lee and M.J. Cormier, eds.) Plenum Press, New York (in press). **
70. R.D. Hamilton & O. Holm-Hansen, Limnol. Oceanogr. 12, 319 (1967).
71. R. Daumas & M. Fiala, Marine Biol. 3, 243 (1969).
72. B.R. Berland, D.J. Bonin, P.L. Laborde & S.Y. Maestrini, Marine Biol. 13, 338 (1972).
73. R.D. Hamilton, O. Holm-Hansen & J.D.H. Strickland, Deep-Sea Research 15, 651 (1968).
74. O. Holm-Hansen, Limnol. Oceanogr. 14, 740 (1969).
75. C.C. Lee, R.F. Harris, J.D.H. Williams, D.E. Armstrong & J.K. Syers, Soil Sci. Soc. Amer. Proc. 35, 82 (1971).
76. C.C. Lee, R.F. Harris, J.D.H. Williams, J.K. Syers & D.E. Armstrong, Soil Sci. Soc. Amer. Proc. 35, 86 (1971).

** Editorial Note. This volume has now been published. Chemiluminescence and Bioluminescence (M.J. Cormier, D.M. Hercules & J. Lee, eds.) Plenum Press, New York (1973).

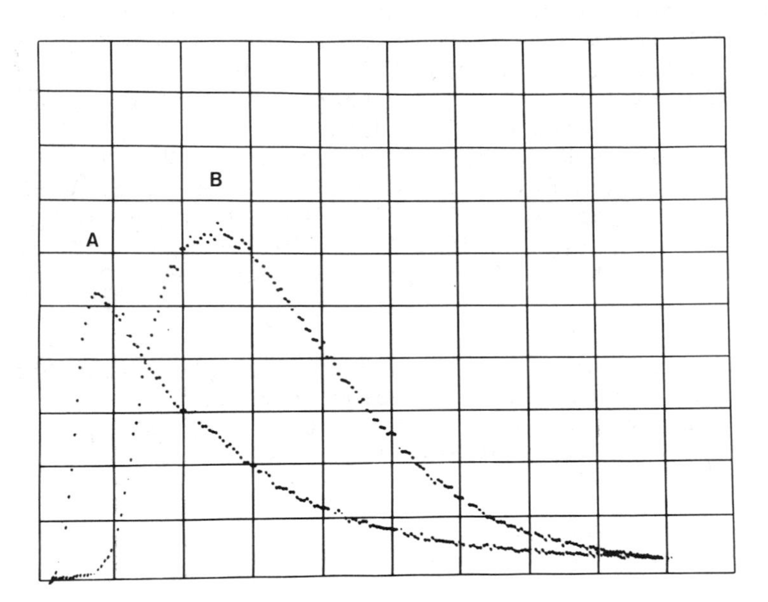

Comparison of photomultiplier background (A) and bioluminescence (B) spectra. Curves obtained with a SA41 multichannel analyzer connected to a SL20 scintillation counter (Intertechnique)

BACTERIAL BIOLUMINESCENCE AND ITS APPLICATION TO ANALYTICAL PROCEDURES

John Lee, Charles L. Murphy, George J. Faini and Terry L. Baucom

Department of Biochemistry, University of Georgia
Athens, Georgia 30602, U. S. A.

INTRODUCTION

Bioluminescence techniques using scintillation counters for detection are finding increasing applications in biological trace analysis. One of the most sensitive chemical assay procedures known is the determination of adenosine triphosphate (ATP) by measuring the light emitted when it is added to purified extracts of the firefly (*1-5*). With suitable precautions less than 10^{-15} moles of ATP can be measured (*2*).

Assay procedures based on the bioluminescence system from certain marine bacteria have also been developed (*5-9*). Changes in the bioluminescence from whole cells have been used to monitor the effects of ionizing radiation (*10,11*), air pollutants (*12*) and anaesthetics (*13*), not to mention the classical work on the effect of temperature, pH, pressure and narcotics on whole cells (*14*).

The bacterial bioluminescence reaction is quite different from that of the firefly. It involves a reaction between reduced flavin mononucleotide ($FMNH_2$), a long-chain aliphatic aldehyde such as dodecanal (RCHO), oxygen and the enzyme (E) bacterial luciferase. By measuring the quantum yields of each component, that is the number of photons emitted per molecule involved, it has been established that the complete reaction is (*15,16*):

$$2\ FMNH_2 + RCHO + 2O_2 \rightarrow 2\ FMN + H_2O_2 + H_2O + RCOOH$$

The $FMNH_2$'s add with O_2 in two sequential steps to make some type of oxidized flavoprotein, which subsequently reacts with the aldehyde to give luminescence (*17*). The light emission appears to come from a protonated flavin molecule ($FMNH^+$), since the spectral emission distribution precisely corresponds to the fluorescence from this species (*18,19*). This implies that the FMN is bound to a site on the luciferase having a high acidity, a finding of some consequence for the theory of enzyme catalysis.

With partially purified luciferases procedures have been devised for the rapid and accurate assay of FMN, flavin adenine dinucleotide (FAD) and nicotine adenine dinucleotide (NAD) (*6,8,9*). The sensitivity is not as good as for ATP in the firefly assay, being only about 10^{-14} moles for NADH (*8*). Still, the method is remarkably sensitive and specific, and the results of some of our recent work on the mechanism show that it possesses the capacity of being improved.

In this paper we show how the choice of type of luciferase and degree of purification can be quite important in developing assay procedures. Using a highly purified, high specific activity luciferase, together with another protein fraction, which is not yet characterized but is separated in the purification of luciferase, we demonstrate that an improved sensitivity to NADH, below 10^{-15} moles, may be achieved.

EXPERIMENTAL METHODS

Types of Bacteria. All the bioluminescent bacteria are of marine origin and may be found either free-living or symbiotic with certain fish. The type we have used for many of our studies is *Photobacterium fischeri* (PF) which was obtained from M. J. Cormier (University of Georgia) in 1965, which was derived from type number 7744 of the American Type Culture Collection (ATCC), Washington, D.C. The type *Achromobacter fischeri* (AF) we obtained from F. H. Johnson (Princeton University) and the type 7744 was originally derived from this. The type MAV was obtained from J. W. Hastings (Harvard University) but he is unsure of its origin. The type A13 is a symbiotic bacteria from the "silver macrourid" fish, isolated by J. Paxton (Australian

Museum) and J. Fitzgerald (Monash University).

Preparation of Luciferase. Bacterial strains were maintained on solid agar medium (*18*). For enzyme extraction the cells were inoculated into liquid medium (250 cc), grown for 24 hours and used as an inoculum for 400 liters (Fermacell, New Brunswick Scientific Co., New Brunswick, N.J.). After 24 hours growth the cells were harvested on a Sharples refrigerated centrifuge. The cells were then disrupted in a continuous-flow French press, the debris removed by centrifugation and the proteins salted out with ammonium sulfate (80% saturation). The luciferase was purified by column chromatography on Sephadex G-75, DEAE-cellulose (DEAE-32) and DEAE-Sephadex (A-50). The detailed procedures are described elsewhere (*19*). The luciferase gave a single band on disc electrophoresis, two subunit bands on SDS gels and very close to one equilibrium binding stoichiometry with pure $FMNH_2$ (*20*).

Chemicals. FMN (87%) was obtained from Fluka, A. G., Buchs, Switzerland and was further purified on a DEAE column (*21*). Four minor bands separated from the bulk of the material (FMN) and two had slight activities in the light reaction at about the level expected for riboflavin and FAD. The $FMNH_2$ was prepared by photoreduction of FMN in the presence of ethylenediamine tetra-acetic acid (EDTA, 20 mM, pH 7) taking care to exclude oxygen. The aldehydes were from Chemsamp Co., Columbus, Ohio and were repurified by vacuum distillation. All other chemicals were the best available commercial grades and were used without further purification.

Light Reactions and Measurement. The bioluminescence was initiated by rapid addition of $FMNH_2$ to a cuvette containing luciferase and aldehyde in aerated buffer (1 cc). The light was detected by a photomultiplier in an optical setup the same as or similar to that previously described (*15,22*). For intensity measurements the photomultiplier output signal was amplified by a picoammeter (Keithley 414 A) which drove a strip chart recorder (Esterline Angus Speedservo) and total light was measured by passing the output of the photomultiplier to an operational integrating amplifier circuit. Both systems were calibrated for absolute light intensity (photons sec^{-1} μA^{-1} or photons $volt^{-1}$)

by use of the luminol chemiluminescence reaction as a light standard (*22,23*).

A typical chart recording of a light reaction is shown in Figure 1. This was obtained with the luciferase type A13. The insert shows the light intensity (I) decays approximately logarithmically with time only over the first decade.

For the NADH stimulated reaction, NADH was added to the cuvette containing luciferase, aldehyde, NADH dehydrogenase and a yellow protein fraction to be described later. A light flash was obtained which was similar to Fig. 1 but on about a ten times longer time scale.

Emission spectra were measured on an absolutely calibrated fluorescence monochromator and the details of the measurements will be described elsewhere (*24*).

RESULTS

Quantum Yields. The oxidation of $FMNH_2$ by molecular oxygen is both fast and complex (*25*). When the $FMNH_2$ is added to the luciferase (E) to initiate the light ($h\nu$) reaction, there will be competition for it between O_2 and E:

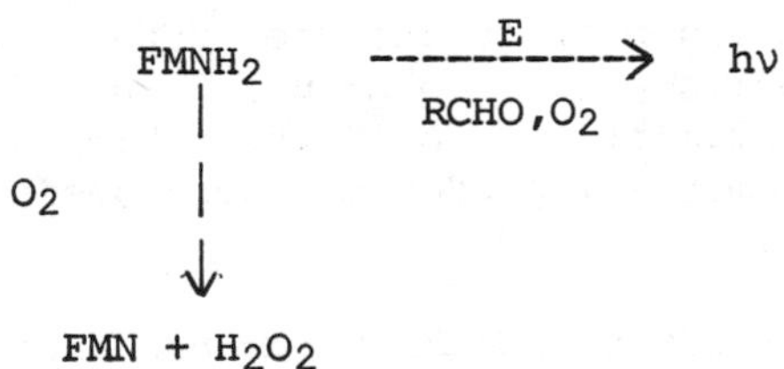

The dashed lines are used to indicate that this is composed of a series of chemical steps.

At a sufficiently high concentration of E, the light path can be made to outcompete the autooxidation route and the luminescence utilization of $FMNH_2$ will then be optimal. It has been shown that this simple scheme represents reality since as the concentration of E is increased, the quantum yield of bioluminescence with respect to $FMNH_2$, $Q_B(FMNH_2)$, increases to a maximum saturating value (*15,*

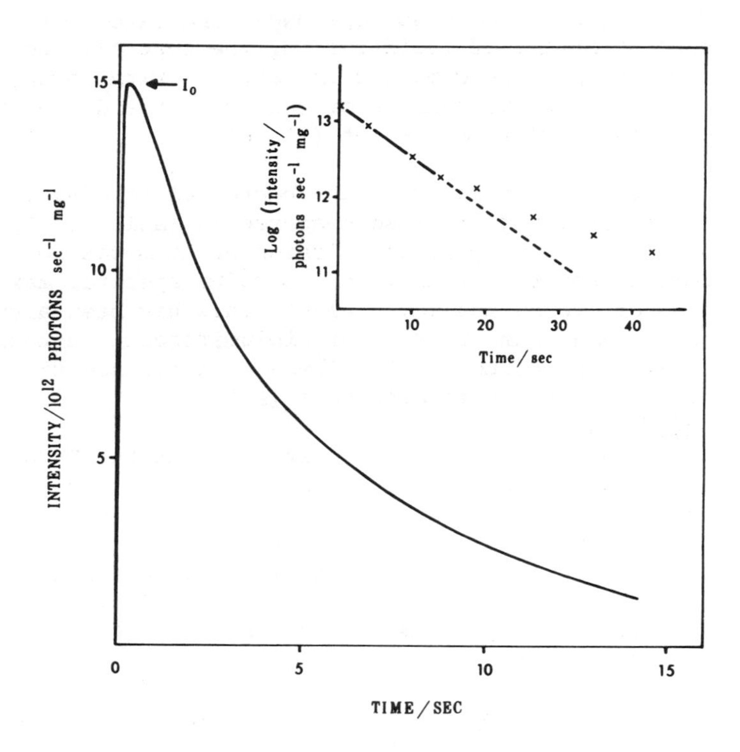

Figure 1. Time dependence of light intensity from reaction of $FMNH_2$, RCHO and O_2 with luciferase of the type A-13 (23°C, pH 7). I_o is the initial light intensity.

16).

Figure 2 shows how $Q_B(FMNH_2)$ increases with increasing concentrations of luciferase of the type A13 and AF. The maximum Q_B's are compared in Table I, with those previously reported for luciferases of the type PF and MAV and are all seen to be quite different (*15,16*).

Other properties of the bacterial *in vivo* and *in vitro* light reactions are also compared in Table I. Although A13 has an *in vivo* spectral emission maximum considerably different from the others, the *in vitro* spectral maxima all cluster around the same value. This has been also noted for a large number of other bioluminescent bacteria by Seliger and Morton (*26*). These emission spectra will be presented in detail elsewhere (*24*).

The flash height I_o observed under the $FMNH_2$ assay conditions is a measure of the specific activity of the luciferase. The specific activity of AF luciferase is thirty times that of MAV, yet the Q_B's differ only by a factor of two. Thus Q_B is a minor factor in determining the flash height, and the nature and velocity of the rate-determining step are more important and differ between the different types.

TABLE I. Characteristics of Bioluminescence from the Different Bacteria.

Bacteria Type	Light per cell (*in vivo*) $h\nu\ sec^{-1}$	Spectral Emission Maximum (cm^{-1}) *In vivo*	*In vitro*	I_o $10^{12}\ h\nu\ sec^{-1}mg^{-1}$	Q_B ($FMNH_2$)
PF	60	20400	20400	90	0.05
MAV	50	20400	20200	5	0.027
AF	300	20300	20000	140	0.057
A13	2000	21000	20000	15	0.021

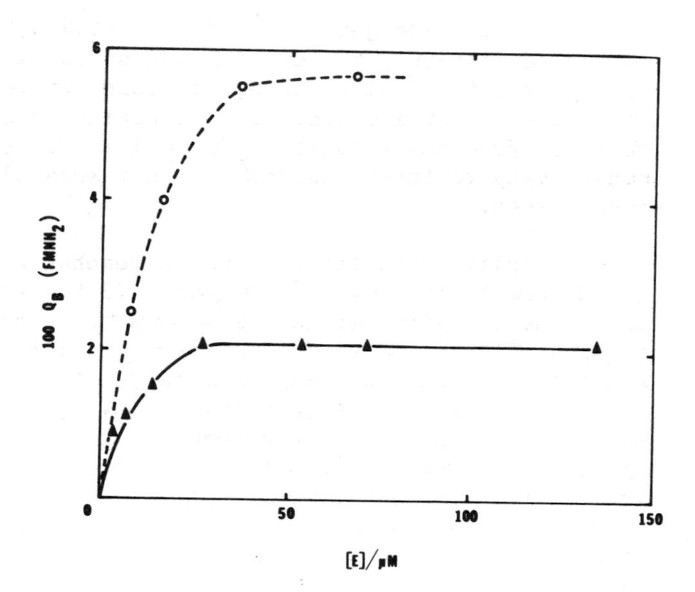

<u>Figure 2</u>. Change of quantum yield with respect to $FMNH_2$, $Q_B(FMNH_2)$ with luciferase (E) concentration for type A-13 (▲) and AF(o).

If the luciferase reactions differ in the nature of the rate determining step, the stationary state concentrations of certain intermediates in the chemical pathway will not be the same. Thus the light reactions may differ in susceptibility to the influence of external agents, such as temperature, pH or concentrations of radical scavengers.

Radical scavengers such as butylated hydroxy-toluene (BHT; 2,6-di(t-butyl) p-cresol) and sodium formate are seen in Figure 3 to provide strong quenching of the light reaction with A13 luciferase. The quenching constant K is obtained from the relation $I_o/I_o^c = 1 + KC$. Such a strong sensitivity to these quenchers is not seen with the other luciferases.

In culture the light emission reaches a maximum at the end of the logarithmic growth phase (*27*). At this point the light emission per cell also reaches a maximum and these are tabulated as photons sec^{-1} $cell^{-1}$ in Table I. Although the cells are approximately all the same size the remarkable fact is that one cell of A13 produces 2000 $h\nu$ sec^{-1} whereas its nearest competitor is 300. Yet the I_o's are in the reverse order which raises the question of whether the substrates might not be different between the *in vivo* and *in vitro* situations. We shall present some preliminary data in the following section to show that this may be so.

Spectral Properties of Luciferases. Previous workers studying the purification of bacterial luciferases have established that the enzyme contains no bound metals or other co-factors (*28-30*). Nevertheless the absorption spectra of luciferase reveals the presence of minor amounts of pigments and some of these have been separated and partially characterized (*31,32*). The absorption spectrum of purified PF luciferase (before the final column step) is shown in Figure 4. Basically the same features are shown here as reported by others (*28-30*) except that here the spectrum is run against a high concentration of bovine serum albumin (BSA) in the reference cuvette to compensate for the very high degree of light scattering exhibited by luciferase. This technique provides an absorption spectrum much less distorted by scattering than reported by others.

In Figure 4 the absorption of the oxidized and

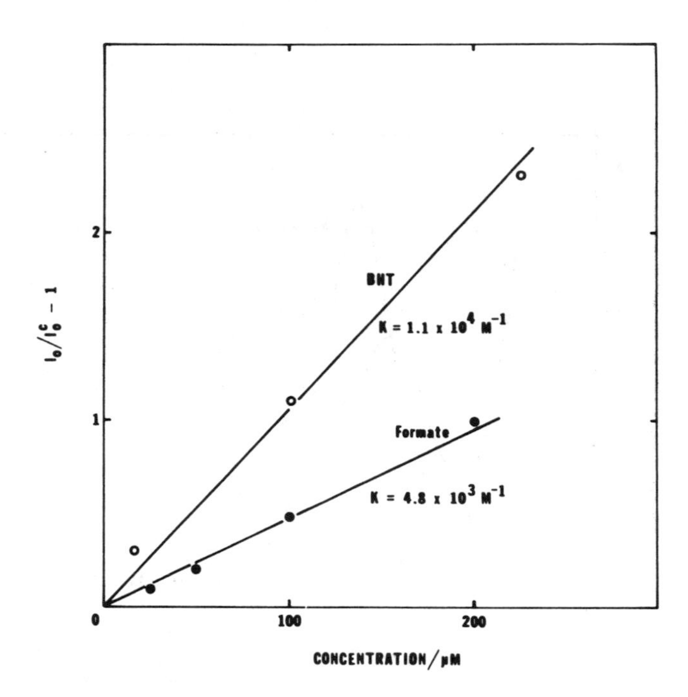

Figure 3. Quenching of initial light intensity I_o with A-13 luciferase by radical scavengers. I_o^c is the I_o at scavenger concentration C.

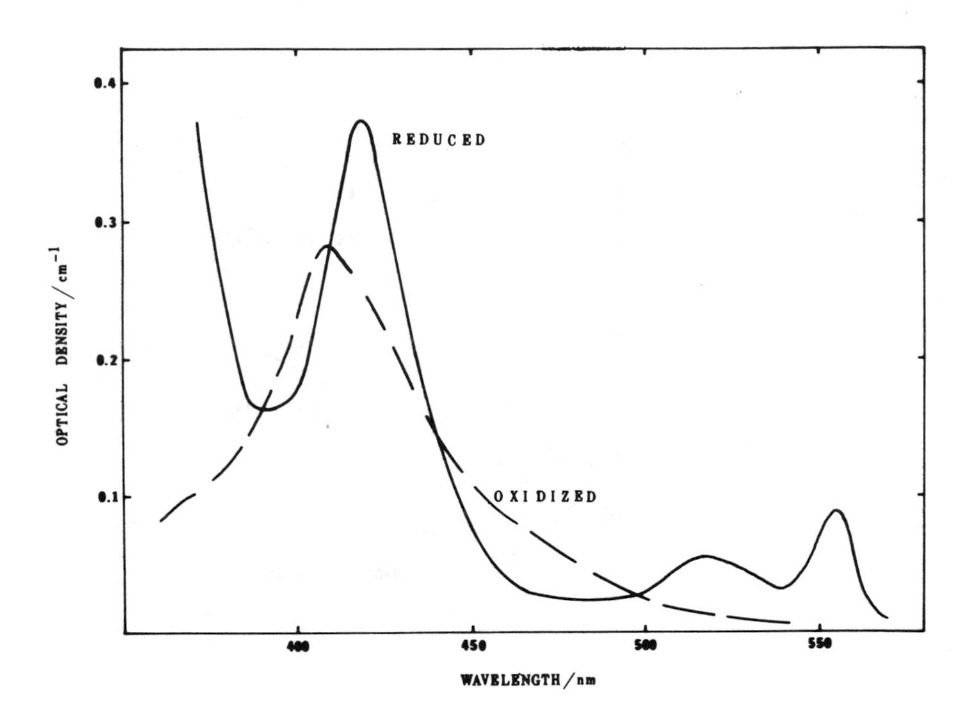

Figure 4. Absorption spectrum of purified bacterial luciferase (10 mg/ml). The reference cuvette contains BSA (10 mg/ml) to remove the distortion due to scattering.

dithionite reduced luciferase preparation are observed to be different. First there is a loss of absorbance at 450 nm. There is also a weak fluorescence in the region of 520 nm (450 nm excitation) for the oxidized but not reduced material so that we can attribute this contribution to a flavoprotein, present at the level of about one mole percent of the luciferase. This impurity is not detectable in preparations of the MAV luciferase which would place it at a level less than 0.1 mole percent.

A second feature is the appearance of a narrow peak around 419 nm and minor bumps at 518 and 555 nm in the dithionite reduced spectrum. These are characteristic of cytochrome and indeed a cytochrome band does follow the luciferase down the G75 column and no doubt contaminates it a little. The absorption spectrum of this dithionite reduced cytochrome (Fig. 5) is characteristic of cytochrome c_1, but estimates of its molecular weight place it at 30,000-50,000, rather different from values of 300,000 measured for other c_1 (*33*).

Comparison of Light Reactions. It is generally believed that $FMNH_2$ and aldehyde act as substrates *in vivo* as they have been shown to *in vitro*, although their presence in these bacteria has not been firmly established. Indeed the presence of free FMN/$FMNH_2$ in the cytoplasm would not be expected in principle, although, as we have shown, quantities of flavins bound as flavoproteins are certainly there.

An NADH dehydrogenase linked to FMN can be extracted and purified (*34*) and contaminates the luciferase preparations. Thus the $FMNH_2$ may be generated in the cell by

$$NADH + H^{+} + FMN \longrightarrow NAD + FMNH_2$$

This reaction may also be used for the *in vitro* assay, i.e. the addition of NADH and FMN to luciferase and RCHO, generates a persistent low level luminescence. Under the normally used reaction conditions the $Q_B(FMNH_2)$ in this reaction is less than 10^{-4}, since most of the $FMNH_2$ is oxidized by O_2 before it reacts with the luciferase.

The cell could overcome the auto-oxidation problem in

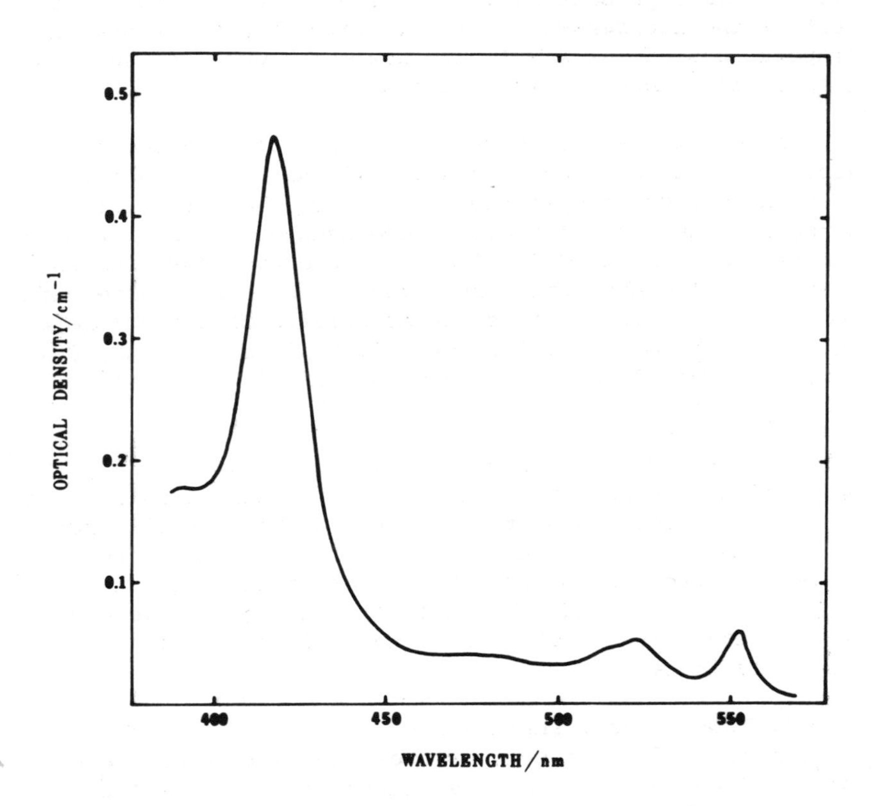

Figure 5. Absorption spectrum of a soluble cytochrome isolated from *Photobacterium fischeri* reduced with dithionite.

two ways. The luciferase comprises more than 1% of the total soluble protein in the cell. This places its concentration in the cytoplasm at values adequate to outcompete the O_2 for $FMNH_2$ and so achieve the maximum $Q_B(FMNH_2)$ as shown in Fig. 2 (*15*). Alternatively, perhaps the cell does not use FMN at all and this possibility prompted us to look for a flavoprotein in the extracts which might show light activity with luciferase.

A yellow band elutes from DEAE at a 0.15 M salt concentration in the purification scheme of PF luciferase. It has not been further purified but the absorption and fluorescence of this fraction show the presence of flavoproteins.

This fraction substitutes for FMN in the light reaction with NADH, dehydrogenase and luciferase. It is not reduced under the conditions of EDTA/light showing that its activity is not due to release of free FMN.

In Table II the activities of luciferases are compared. The advantage of using a fresh preparation either partially or fully purified is readily seen to give rise to many times greater activity in both $FMNH_2$ and NADH assays.

TABLE II. Comparison of Luciferase Activities.

Luciferase	Specific Activity/10^{12} hν sec^{-1} OD(280)$^{-1}$		
	$FMNH_2$ reaction	NADH reaction with	
		FMN	yellow fraction[b]
PF	90	0.3	1.2
Sigma[a]	0.6	.005	
Worthington[a]	3	.014	

[a]Commercial preparations.

[b]Isolated during luciferase purification.

The yellow fraction greatly stimulates the NADH activity and part of this comes from an increased level of dehydrogenase in this fraction. The details of this reaction are still being investigated.

A number of other FMN flavoproteins (flavodoxin, dihydro-orotic dehydrogenase and glycollate oxidase) were substituted for FMN in the NADH assay and were found not to show activity except for what was attributable to free FMN from denatured material.

Bioluminescence Assay for FMN and NADH. Having established conditions for optimizing Q_B's, we thought these techniques could be adapted to improve the bioluminescence assay for FMN and NADH. In Figure 6 is shown the I_o obtained as a function of added $FMNH_2$ or NADH under the following conditions. The FMN is determined in a 1 cc mixture of FMN, luciferase (PF, 0.5 mg/ml), dodecanal (10λ, methanol saturated) by the addition of dithionite or $NaBH_4$ in optimal amounts (*9*). The sensitivity is limited by the presence of an enzyme blank even in the purest preparations, equivalent to about 10^{-12} moles of FMN.

The NADH is added to 1 cc of luciferase (PF, 0.5 mg/ml), dodecanal (10λ, methanol saturated), yellow fraction (0.5 cc), FMN (5 μM). There is no enzyme blank and the method appears only limited at this point by photomultiplier noise, which is equivalent to 10^{-16} moles NADH. This can probably be reduced another two orders of magnitude.

The sensitivity for NADH using this assay is about 100 times better than reported previously (*5,7,8*) and is even better than for ATP using the firefly assay (*2*).

DISCUSSION

In our studies of the mechanism of bacterial bioluminescence we have developed techniques of optimizing the light output. First of all the luciferase must be sufficiently purified and present in sufficient concentration to outcompete the very rapid auto-oxidation of $FMNH_2$. Second, luciferases from different sources differ greatly in initial light output, so that if the light flash (I_o) method is to be used for assay it is very important to choose an enzyme that has a maximum activity. Total light however

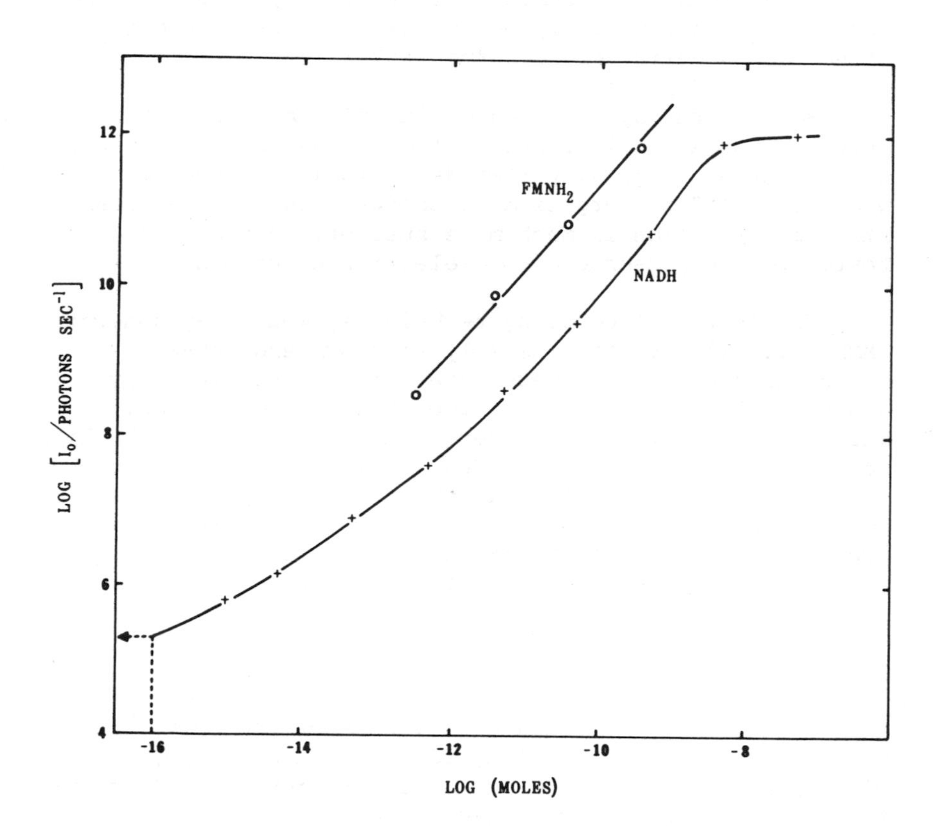

Figure 6. Initial light intensity I_o on addition of $FMNH_2$ (o) or NADH(+) to PF luciferase under the appropriate assay conditions. The arrow indicates the photomultiplier noise level and corresponds to 10^{-16} moles NADH.

does not vary as much from one type to another.

Practical considerations also would make the susceptibility to external quenchers a less desirable feature. The types MAV and A13 are strongly inhibited by excess aldehyde for instance and this may be the same radical scavenging effect as is definitely observed with A13.

However the application of luminous bacteria to the detection of air pollutants (*12*) and other substances which have a strong propensity towards radical type reactions, would make A13 the organism of choice. *In vitro* at least, its light reaction is much more susceptible than the others tested to the presence of one-electron acceptors.

In the use of coupling techniques, which may generate $FMNH_2$ directly but more usually via NADH and NADH-dehydrogenase, past workers have not been sufficiently aware of the competition for the $FMNH_2$ between O_2 and luciferase. Again this competition may be overcome by high concentrations of luciferase but we have proposed here that there may be a substrate that the bacteria itself uses, possibly a flavoprotein which is less readily reoxidized from the reduced state by oxygen, and consequently provides a more efficient coupling between NADH dehydrogenase and luciferase.

ACKNOWLEDGEMENTS

We thank R. A. Makula and J. Linn of the University of Georgia Fermentation Facility for their assistance, R. W. Miller, M. L. Salin and G. R. Bell for providing flavoproteins tested and J. Fitzgerald and J. Paxton, J. W. Hastings, F. Johnson and M. J. Cormier for bacterial cell types.

REFERENCES

1. B. L. Strehler *in* "Methods of Enzymatic Analysis," p. 559 (H. U. Bergmeyer, Ed.) Academic Press, New York 2nd Ed. (1965).
2. E. W. Chappelle and G. V. Levin, *Biochem. Med. 2*, 41 (1968).
3. P. E. Stanley and S. G. Williams, *Anal. Biochem. 29*, 381 (1969).

4. E. Schram, R. Cortenbosch, E. Gerlo and H. Roosens, *in* "Organic Scintillators and Liquid Scintillation Counting," p. 125 (D. L. Horrocks and C-T. Peng, Eds.) Academic Press, New York (1971).
5. P. E. Stanley, *ibid.*, p. 607.
6. P. E. Stanley, *in* "Chemiluminescence and Bioluminescence," p. 494 (M. J. Cormier, D. M. Hercules and J. Lee, Eds.) Plenum Press, New York (1973).
7. E. W. Chappelle, G. L. Picciolo and R. H. Altland, *Biochem. Med. 1*, 252 (1967).
8. P. E. Stanley, *Anal. Biochem. 39*, 441 (1971).
9. E. W. Chappelle and G. L. Picciolo, *in* "Methods in Enzymology," Vol. XVIII, Part B, p. 381 (D. B. McCormick and L. D. Wright, Eds.), Academic Press, New York (1971).
10. A. P. Jacobson and K. A. McDermott, *J. Lumin. 3*, 419 (1971).
11. I. A. Lerch, *Rad. Res. 45*, 63 (1971).
12. W. F. Serat, J. Kyono and P. K. Mueller, *Atmos. Environ. 3*, 303 (1969).
13. D. C. White, B. Wardley-Smith and G. Adey, *Life Sci. 12*, 453 (1973).
14. F. H. Johnson, H. Eyring and M. J. Polissar, "The Kinetic Basis of Molecular Biology," John Wiley, New York (1954).
15. J. Lee, *Biochemistry 11*, 3350 (1972).
16. J. Lee and C. L. Murphy, *in* "Chemiluminescence and Bioluminescence," p. 381 (M. J. Cormier, D. M. Hercules and J. Lee, Eds.) Plenum Press, New York (1973).
17. J. Lee and C. L. Murphy, *Biochem. Biophys. Res. Commun. 53*, 157 (1973).
18. M. Eley, J. Lee, J.-M. Lhoste, C. Y. Lee, M. J. Cormier and P. Hemmerich, *Biochemistry 9*, 2902 (1970).
19. C. L. Murphy, M. S. Thesis, University of Georgia (1973).
20. J. Lee and C. L. Murphy, *Biophys. J. 13*, 274a (1973).
21. V. Massey and B. E. P. Swoboda, *Biochem. Z. 338*, 474 (1963).
22. J. Lee and H. H. Seliger, *Photochem. Photobiol. 4*, 1015 (1965).
23. J. Lee, A. S. Wesley, J. F. Ferguson and H. H. Seliger, *in* "Bioluminescence in Progress," p. 35 (F. H. Johnson and Y. Haneda, Eds.) Princeton University Press, Princeton, N.J. (1966).

24. G. J. Faini, C. L. Murphy and J. Lee, manuscript in preparation (1973).
25. V. Massey, G. Palmer and D. Ballou, *in* "Flavins and Flavoproteins," p. 349 (H. Kamin, Ed.), University Park Press, Baltimore (1971).
26. H. H. Seliger and R. A. Morton, *in* "Photophysiology" Vol. 4, p. 253 (A. C. Giese, Ed.) Academic Press, New York (1968).
27. E. N. Harvey, "Bioluminescence," Academic Press, New York (1952).
28. S. Kuwabara, M. J. Cormier, L. S. Dure, P. Kreiss and P. Pfuderer, *Proc. Natl. Acad. Sci. U.S. 53*, 822 (1965).
29. J. W. Hastings, W. H. Riley and J. Massa, *J. Biol. Chem. 240*, 1473 (1965).
30. M. J. Cormier and S. Kuwabara, *Photochem. Photobiol. 4*, 1217 (1965).
31. G. W. Mitchell and J. W. Hastings, *Biochemistry 9*, 2699 (1970).
32. K. Matsuda and T. Nakamura, *J. Biochem. 72*, 951 (1972).
33. A. Lehninger, "Biochemistry," Worth Publishers, New York (1970).
34. K. Puget and A. M. Michelson, *Biochemie 54*, 1197 (1972).

USE OF BIOLUMINESCENCE PROCEDURES AND LIQUID SCINTILLATION SPECTROMETERS FOR MEASURING VERY SMALL AMOUNTS OF ENZYMES AND METABOLITES

Philip E. Stanley

Department of Agricultural Biochemistry,
Waite Agricultural Research Institute,
The University of Adelaide,
Glen Osmond, South Australia.

ABSTRACT

The design and operation of a specially constructed analytical bioluminescence cell is described. The unit, which will fit into the detector chamber of a liquid scintillation spectrometer, is used to measure flash heights as well as to follow the production of light in bioluminescence assays.

The value of assaying metabolites and enzymes by a coupling to a bioluminescence reaction is illustrated by the measurement of picomole amounts of pyrophosphate and the enzyme ATP-sulphurylase using the firefly luciferin-luciferase system for determining adenosine triphosphate (ATP).

A new assay is described for determining adenosine 3'-phosphate 5'-phosphate (PAP) by the bioluminescence system of the sea pansy (Renilla reniformis). The sensitivity of the assay is around 1 picomole (10^{-12} mole). The procedure may also be used for measuring adenosine 3'-phosphate 5'-sulphatophosphate (PAPS, or "active sulphate") since this compound is readily hydrolysed to PAP.

INTRODUCTION

There is now a wide interest in the use of the liquid scintillation spectrometer as a quantum counter for measuring the single photons produced in analytical bioluminescence assays. The availability of photomultipliers with low-noise photocathodes enables the instrument to be used at room temperature, which is suitable for enzyme reactions. Thus, bioluminescence techniques are not only extremely sensitive but also very specific and easy to perform with equipment which is readily available.

In the field of analytical bioluminescence, two basic reactions have been developed: i) ATP, using the firefly luciferin-luciferase system (1,2,3,4) and ii) FMN* and NADH*, using the dehydrogenase-luciferase complex of the marine bacterium Photobacterium fischeri (2,3,5). They have been used mainly for measuring static levels of these compounds, but more recently the use of enzyme-coupled reactions in which ATP or NADH is either produced or utilized has enabled dynamic measurements to be made, e.g. ammonia (6) and APS (7). In this paper a new coupled system for measuring pyrophosphate and ATP-sulphurylase (ATP-sulphate adenylyl transferase E.C. 2.7.7.4) will be described.

A study of the metabolism of the sulphur nucleotides (APS* and PAPS) in biological systems has now been facilitated by bioluminescence procedures. Thus the one for APS has been described (7) and in the present paper a bioluminescence system from the sea pansy Renilla reniformis (8,9 10) is used for detecting PAP.

When assaying very small amounts of ATP or NADH (less than a picomole) it is better to measure "flash height" since light production occurs only for a few seconds (11). A unit specially designed to fit the detector chamber of a spectrometer is described which allows for the monitoring of samples from the time of mixing reactants.

MATERIALS AND METHODS

Analytical Bioluminescence Cell. The unit has been designed to fit a Model 3375 Packard Tri-Carb Liquid Scintillation Spectrometer but a simple modification will eanble the device to fit other units. The cell, which is illustrated in Fig. 1, consists of an opaque 5 ml polypropylene reaction/mixing vessel G (3.6 x 1.65 cm) fixed

---

*Abbreviations: APS, adenosine 5'-sulphatophosphate; FMN, flavin mononucleotide; NADH, reduced nicotinamide adenine dinucleotide.

at one end of a metal tube E (7.8 x 2.85 cm), which can be manually inserted into the detector chamber of the spectrometer so that G is positioned directly between the photocathodes of the photomultipliers. Extraneous light is prevented from reaching the photomultiplier tubes by three O-rings D and a large flange H (5.0 x 1.0 cm) which facilitates the removal of the unit from the detector chamber. A screw at the collar F (1.6 x 2.85 cm) attaches the reaction vessel G to the tube E in a gas-tight closure. The insert tubes A, made of 16 gauge stainless-steel tubing, conduct and mix the reactants, which are introduced into the tubes with syringes via the taps. Tube C, which extends to the bottom of the reaction vessel G, is used to extract the spent mixture, and B enables G to be kept at atmospheric pressure. The tube may also be used to introduce other gases.

The cell is introduced into the detector after switching off the high-voltage supply to the photomultipliers. The procedure is similar to that for fitting a flow-cell, as described by the manufacturer of the scintillation unit. Extraneous light reaching the photomultipliers will be evident as a high background in a tritium out-of-coincidence setting and this effect must be rectified immediately to avoid permanent damage to the photomultipliers.

The unit is used as follows: 1 ml of reactant is placed in each of two 2 ml glass syringes (fitted with Luer locks) together with an air bubble (0.25 ml). One syringe contains the purified luciferase and necessary cofactors and the other the energy source. The syringes are then attached at A, locked, and the pistons depressed simultaneously so that the reactants are mixed quickly. The air bubble in the syringes ensures that the tubes are emptied and that the reactants are thoroughly mixed. A small electrical switch is fitted to one syringe so that on depressing the piston, the switch closes, thus initiating a synchronising pulse to trigger multiscaling with the multichannel analyzer set at 100 msec per channel. The analyzer was connected to the ratemeter output of the RED channel set at 70-300, 100% amplification and out of coincidence. When multiscaling is complete, the counts in each channel are read and the flash height, occurring about 700-1000 msec after multiscaling had commenced, is measured.

Instruments. The bioluminescence assays reported here

were made in Packard 3375 and 3390-544 and Nuclear Chicago Mark II and Unilux III liquid scintillation spectrometers, all operating at 20° ± 0.5°. The Nuclear Chicago instruments were operated in the print-selected display mode, thus enabling counting to continue even during print-out. This is especially convenient during short counting periods, since very little information is lost when the counting rates are changing quickly.

A 200-channel Packard Spectrazoom Multichannel Analyzer, operated in multiscale, was used to follow bioluminescence reactions. It was operated at 100 or 1000 msec per channel. Alternatively, reactions were followed with a Packard ratemeter and chart recorder using a range setting of 3×10^5 or 10^6, a time constant of 1 second and a speed of 5 cm per minute.

Chemicals. Firefly lanterns and luciferin were obtained from Sigma Chemical Co., St. Louis, U.S.A. Preparation of crude firefly luciferin-luciferase, purified firefly luciferase and APS have been described elsewhere (1,12,13). The bioluminescence enzyme system from Renilla reniformis was prepared by extracting 10 g of an acetone powder of Renilla with 40 ml buffer (100 mM phosphate, 1 mM EDTA and 1 mM β-mercaptoethanol, pH 7.5) for 2 hr. The supernatant fraction remaining after centrifuging at 20,000g for 1 hr was passed through a Sephadex G-25 column to remove endogenous PAP and used without further purification (14).

Assay of Pyrophosphate. All solutions were filtered through a 0.22μ Millipore filter prior to use. Glassware was cleaned in hot detergent, washed four times in double-distilled water, dried at 100° and kept in an oven at 100° until required. Vials were not exposed to direct sunlight or fluorescent lighting, both of which result in phosphorescence of the glass. The room housing the spectrometers was lit only with tungsten lamps. The instrument was switched out of coincidence and a window set in the lower third of an unquenched tritium setting.

A scintillation vial containing 1 ml 50 mM Tris-HCl, 1 ml 15 mM sodium arsenate and 1 ml 5 mM $MgCl_2$ (pH 8.5) was equilibrated for 10 min at 20°. The vial was then counted for 2 min (6 x 0.1 min on the Packard units, 50 x 0.04 min on the Nuclear Chicago models). The appropriate amounts of ATP-sulphurylase and firefly extract and 2 nmoles APS were

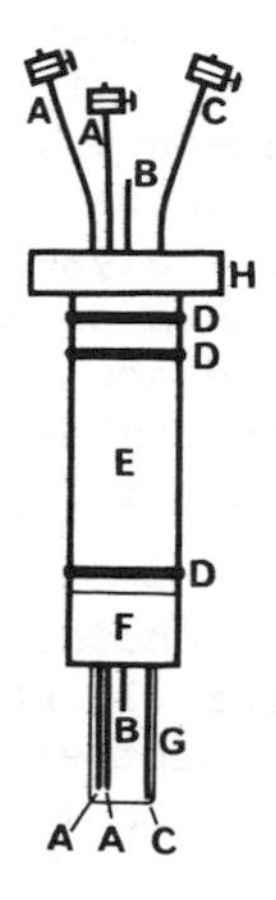

Fig. 1. A diagramatic representation of the analytical bioluminescence cell. See text for details.

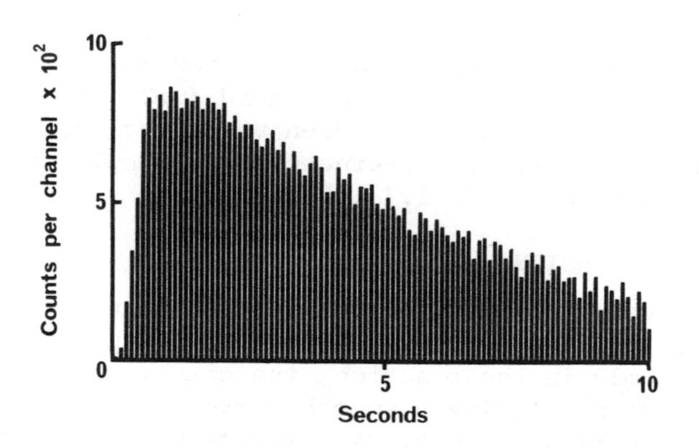

Fig. 2. Use of the analytical cell. A time course for the ATP-firefly luciferase-luciferin (5×10^{-14} moles ATP). Multichannel analyzer used in multiscale (100 msec per channel).

then added and a 2-min counting period initiated. Little difference in light output was observed during this time. The addition of pyrophosphate (1-100 pmoles) caused an immediate increase in light output which was monitored for a further 2 min. As well as a digital read-out, a continuous analogue display was available using the ratemeter and recorder or alternatively the multichannel analyzer operated in multiscale (1 sec/channel). The amounts of ATP-sulphurylase and firefly extract were varied according to the activity of the preparations. ATP-sulphurylase was readily assayed using a reaction mixture containing 2 nmoles APS and 20 nmoles pyrophosphate.

Assay of PAP (adenosine 3'-phosphate 5'-phosphate. A scintillation vial containing 2 ml of buffer (10 mM potassium phosphate, 1 mM Na-EDTA and 1 mM β-mercaptoethanol, adjusted to pH 7.5) was equilibrated at 20° for 10 min and then counted for 2 min (see pyrophosphate assay for spectrometer setting). The reaction was started by quickly adding PAP (1-100 pmoles) in a small vol (< 50 µl), 1 µl 0.3 mM luciferyl sulphate (in 0.01 M potassium phosphate pH 7.5 : ethanol, 1:1) and 50 µl Renilla enzyme system. The sample was then counted for a further 2 min, during which the light output increased linearly with time. The assay was calibrated internally by adding a standard of 10^{-10} moles PAP.

PAPS can be measured by the above procedure since it is readily hydrolysed to PAP in dilute acid (0.2 N HCl) at 37° for 30 min. This procedure has been used successfully for measuring the activity of APS-kinase (ATP:adenylyl-sulphate 3'-phosphotransferase E.C. 2.7.1.25).

RESULTS AND DISCUSSION

Some data used for measuring flash height in the analytical cell are presented in Fig. 2. Since the photons are produced randomly in the reaction, the counts recorded in each channel do not follow a smooth curve. To obtain an estimate of the flash height, the average was taken of the counts in the twenty channels immediately following the channel containing the maximum counts. For each batch of purified luciferase and luciferin it is necessary to make a calibration curve of ATP *vs* flash height over the desired range. The reproducibility of the assay depends on the level of ATP and the activity and blank of the enzyme.

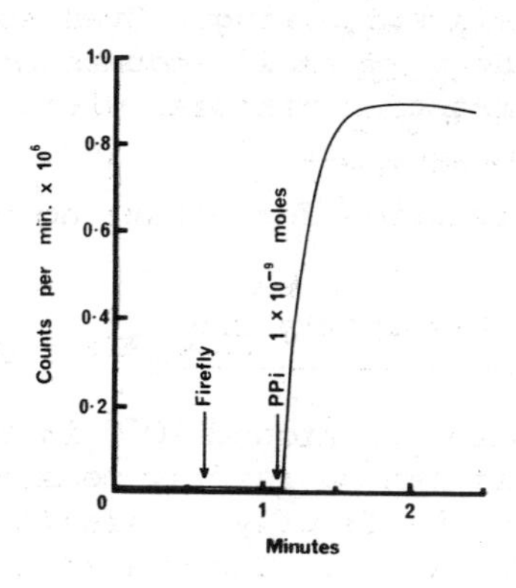

Fig. 3. Reaction sequence for measuring pyrophosphate

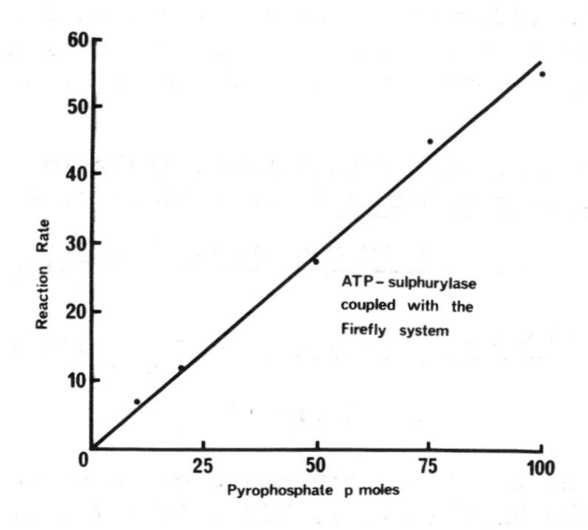

Fig. 4. A calibration curve for pyrophosphate (see text for details)

Thus, for a picomole (10^{-12}) the standard error from 25 measurements was 6.2%, while at a femtomole (10^{-15} mole) it was 14.8%. These errors might be reduced by using a mechanical or pneumatic means of depressing the piston, thus achieving a more consistent mixing. Even so the unit is valuable for measuring very small amounts of ATP (and NADH) or for working with reaction mixtures where considerable luciferase inhibition is apparent.

The basis of the reaction for measuring pyrophosphate is as follows:-

$$\text{ATP + sulphate} \underset{}{\overset{\text{ATP-sulphurylase}}{\rightleftharpoons}} \text{APS + pyrophosphate}$$

The equilibrium constant is around 10^{-8} in favour of ATP production, so that it can be readily measured in a dynamic system by coupling to the firefly luciferin-luciferase bioluminescence system (1,7). The reaction can be followed in an analogue form using the multichannel analyzer in multiscale mode or the ratemeter and chart recorder, as shown in Fig. 3. A calibration curve for pyrophosphate against reaction rate or the rate of photon production is presented in Fig. 4. Thus, this procedure is several orders of magnitude more sensitive than the standard colorimetric methods. Because this system involves an enzyme reaction, it is affected by various inhibitory compounds. Internal standards are employed to assess this and also to correct for any coloured material which might absorb photons.

The reaction sequence resulting in light emission in the sea pansy Renilla reniformis is as follows (8,9):-

$$\text{Luciferyl sulphate} \xrightarrow{\text{L sulphokinase}} \text{L + sulphate}$$

$$\text{L} + O_2 \xrightarrow{\text{Luciferase}} \text{products} + h\nu \; (\lambda_{max} \; 485 \text{ nm})$$

L = luciferin

The structure of the luciferin has been reported recently (10). Since the K_m for PAP is 7.3×10^{-8} M (8), the system provides a very sensitive assay for not only this compound but also for PAPS, as illustrated in Fig. 5 where 1 to 100 picomoles of PAP are measured with the reaction sequence presented in Fig. 6. This assay has been used recently to

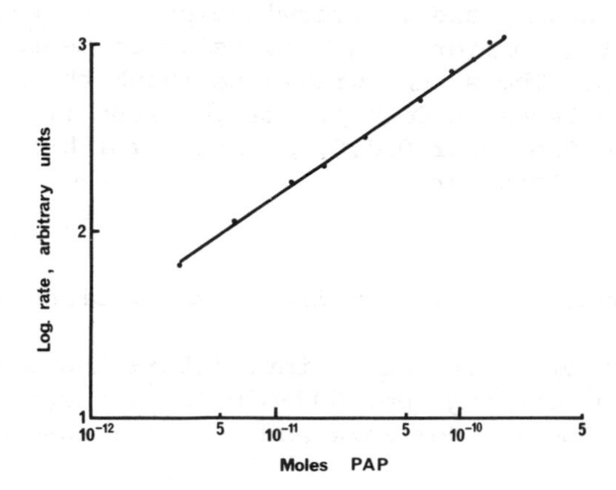

Fig. 5. Reaction sequence for measuring adenosine 3'-phosphate 5'-phosphate (PAP).

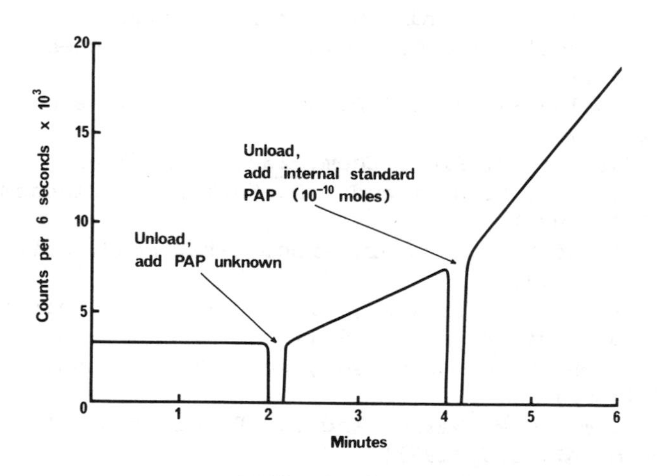

Fig. 6. A calibration curve for adenosine 3'-phosphate 5'-phosphate (PAP) using the bioluminescence system of _Renilla reniformis_.

show that higher plants synthesise PAPS from sulphate (15).

Acknowledgments. It is a pleasure to acknowledge the gift of an acetone powder and luciferyl sulphate of Renilla from Professor Milton Cormier, The University of Georgia, Athens, Georgia, U.S.A. The author wishes to thank the Australian Wheat Industry Research Council for generous financial assistance and Professor D.J.D. Nicholas for his encouragement throughout this work.

REFERENCES

1. P.E. Stanley and S.G. Williams, Anal. Biochem. 29, 381 (1969)
2. P.E. Stanley in Organic Scintillators and Liquid Scintillation Counting, pp. 607-620 (D.L. Horrocks and C.-T. Peng, Eds.). New York and London : Academic Press (1971)
3. E. Schram, R. Cortenbosch, E. Gerlo and H. Roosens in Organic Scintillators and Liquid Scintillation Counting, pp. 125-135 (D.L. Horrocks and C.-T. Peng, Eds.). New York and London : Academic Press (1971)
4. E. Schram in The Current Status of Liquid Scintillation Counting, pp. 129-133 (Edwin D. Bransome Jr., Ed.). New York and London : Grune and Stratton (1970)
5. P.E. Stanley, Anal. Biochem. 39, 441 (1971).
6. D.J.D. Nicholas and G.R. Clarke, Anal. Biochem. 42, 560 (1971)
7. G.J.E. Balharry and D.J.D. Nicholas, Anal. Biochem. 40, 1 (1970)
8. M.J. Cormier, J. Biol. Chem. 237, 2032 (1962)
9. M.J. Cormier, K. Hori and Y.D. Karkhanis, Biochemistry 9, 1184 (1970)
10. K. Hori and M.J. Cormier, Proc. Natl. Acad. Sci. U.S. 70, 120 (1973)
11. S.E. Brolin, E. Borglund, L. Tegner and G. Wettermark, Anal. Biochem. 42, 124 (1971)
12. R. Nielson and H. Rasmussen, Acta Chem. Scand. 22, 1257 (1968)
13. C.A. Adams, G.M. Warnes and D.J.D. Nicholas, Anal. Biochem. 42, 207 (1971)
14. M.J. Cormier, Personal Communication
15. P.E. Stanley, B.C. Kelley and D.J.D. Nicholas, Manuscript submitted to Biochem. J.

CHOICE OF COUNTING VIAL FOR LIQUID SCINTILLATION: A REVIEW

Kent Painter
Amersham/Searle Corporation
Arlington Heights, Illinois, USA

ABSTRACT

Various types of counting containers have been used in liquid scintillation counters. The most commonly used types are standard 20 ml capacity glass, polyethylene and nylon vials. A comparison is made of the various counting parameters, such as background, efficiency, E^2/B, permeability, external standard ratio, wall adsorption, durability and inertness to solvents. A number of new miniature counting vials are described and warnings are given regarding external standard ratio quench correction, volume dependency, static electricity and geometry. The advantages and disadvantages of various types of liquid scintillation caps are discussed. Special emphasis is given to recognizing and avoiding errors produced by poor choice of vial, with specific examples from the literature.

INTRODUCTION

In many ways liquid scintillation counting has become so routine that few papers in bioanalytical research discuss or reference counting procedures or materials. Even papers dealing specifically with liquid scintillation often fail to mention the counting container. Unfortunately, we have encountered many instances where the choice of counting vial was based solely upon habit or economy with little thought given to possible adverse consequences. It has been frequently noted that the counting bottle can be a major factor in liquid scintillation counting accuracy and reproducibility (1-6).

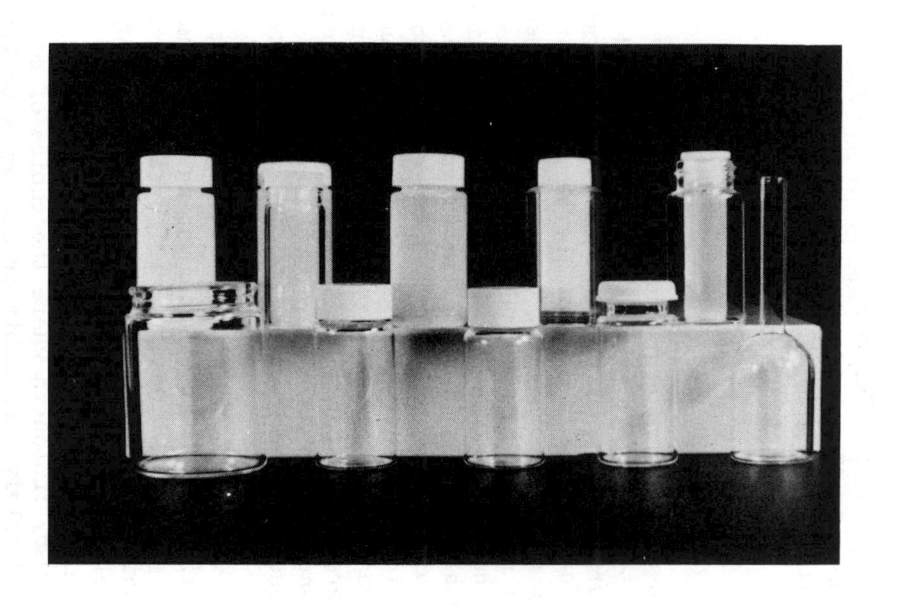

Fig. 1. <u>Liquid Scintillation Counting Vials</u>
Top row left to right: (a) Polyethylene (b) 10 ml Glass Insert for Polyethylene Holder (c) Nylon (d) Minivial™ (e) Polyethylene Insert for Glass Holder. Bottom row left to right: (f) 60 ml Glass (g) Borosilicate Glass (h) General Purpose Glass (i) Quartz (j) Flame-seal Borosilicate.

It is the purpose of this paper to review the history and development of the various counting containers now in use and to compare the relative merits of currently used systems.

HISTORY

Due to limitations of cocktail capacity for aqueous samples, the first commercial liquid scintillation counters introduced in the early 1950's were designed to accommodate rather large 60-85 ml weighing bottles (7-12) (Fig. 1). In some systems vials were optically coupled to the photomultiplier tube by the use of a silicone fluid (7,10,13,14), but by the mid-1950's reflective counting chambers of polished aluminum proved convenient and superior in performance to fluid coupled systems.

It soon became apparent that the use of smaller counting containers would allow the photomultiplier tubes to be positioned closer together, thereby reducing light path distance losses and resulting in improved counting efficiencies. Backgrounds due to natural radioactivity in glass were also reduced due to the smaller mass of the counting container. The first successful liquid scintillation spectrometer manufacturer introduced in 1953 an ordinary 5 dram (20 ml) medicine bottle which became the accepted standard in the industry (15). Kimble Opticlear medicine vials were used initially (12), followed soon after by vials from Wheaton (10,12,16).

The standard 5 dram medicine bottle has been modified over the years to accommodate various instrument designs. A flat or slightly concave bottom was required and various height and bottom dimensions were necessary for sample sensing switches, group select indicators, and sample-changer elevator mechanisms. A drawing of a vial recently proposed as a standard by the International Electrotechnical Commission (17) is shown in Fig. 2. It should be noted that at present this standard has not been accepted or approved. Use of a counting bottle of improper dimension can result in a costly repair bill (18).

GLASS COUNTING VIALS

Glass counting bottles are of two general types: borosilicate glass and general purpose soda-lime or flint glass. The major differences between the two types are the higher cost and lower background count rate of borosilicate

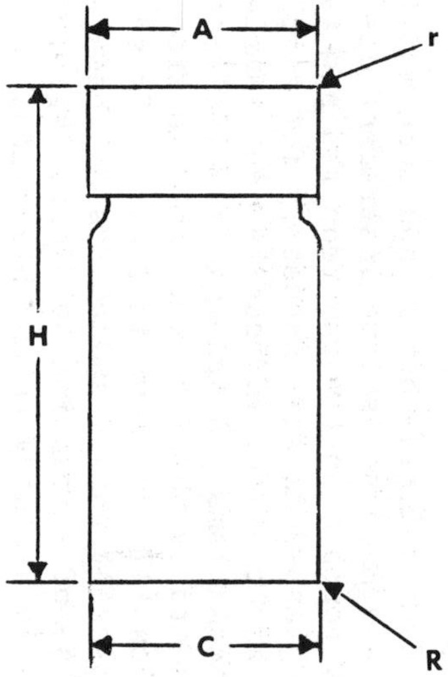

45(Secretariat)166
March 1973

	Dimension values min mm	max mm
A	25.0	(1)
C	26.0	30
H	58.0	63.0
r	0.4	3.6
R		1.5

(1) Maximum not to exceed bottle diameter C.

Fig. 2. Recommendation Concerning Standard Dimensions of Test Bottles for Liquid Scintillation Counting-International Electrotechnical Commission.

glass (19,20). With general purpose glass vials, tritium backgrounds of 50 cpm are common, but good borosilicate glass bottles are generally in the range of 18-22 cpm. It should be emphasized that background count rates of glass vials differ considerably from manufacturer to manufacturer (21) and, more importantly, from lot to lot (4,20,22 23). Samples of each new shipment of vials should be checked and lot numbers recorded. We recommend purchasing a large quantity from a single lot.

Variations in mass and wall thickness can cause serious reproducibility problems, especially with the external standard ratio method of quench correction (3,24). Laney (29)has reported that background count rate of various containers is a function of mass rather than composition. A value of 1.3 cpm/g was nearly identical for several types of plastic and glass vials studied. Paix (23) has shown that the major variation between different sources of glass is due to natural radium-226 and its decay products. This was demonstrated by placing a crushed glass bottle in a low-level gamma spectrometer. Potassium-40 appears to be of lesser importance in background contribution of glass bottles.

Glass counting vials offer the advantages of inertness to strongly acidic or basic solvents, such as NCS* and Hyamine hydroxide†, resistance to permeability and deformation, low susceptibility to photoluminescence and compatibility to the external standard ratio method of automatic quench correction. Glass vials allow the analyst to be able to visually inspect samples (25). It should be noted that figures of merit for samples containing large volumes of water are not vastly different for general purpose and borosilicate glass vials when counting energetic beta emitters. This is due to the fact that the higher background of the general purpose vial is quenched in the same manner as the sample (26-28) and stresses the importance of using an identical sample minus activity for accurate background determination. We find very often that analysts use an unquenched background standard supplied with instruments, rather than an identical blank sample for determining background count rates. This usually results in a high estimate of background count

*Trademark Amersham/Searle
†Trademark Rohm and Haas

rate, but in some samples the opposite effect has been noted(29,4).

There are three disadvantages to the use of glass counting vials, only one of which can be regarded as serious. The first, which is really a nuisance, is breakage. Many of you, I am sure, have lost a sample from time to time by dropping a bottle. At one time a rather large vial manufacturer had difficulty with a borosilicate vial which would snap off at the neck when the slightest pressure was used to tighten the cap. Fortunately, the problem only occurred in about 5% of the bottles, but the problem did exist for several years.

The second disadvantage of glass bottles is cost. However, one must balance the value of glass versus the disadvantages of other types of containers, particularly with regard to accurate quench correction, and with the accuracy required for the experiment. We feel that glass vials are the vial of choice for critical counting applications where accuracy and reproducibility are required. For most counting applications standard glass vials can be used rather than the more expensive borosilicate glass vials.

The third and only serious disadvantage of glass is vial wall adsorption.

Recognizing vial wall adsorption can be a problem in itself. Sometimes it would be manifest in a decreasing count rate, but chemiluminescence, phase separation or precipitation can also be the cause. Chemiluminescence can be identified by comparing single phototube versus coincidence counts. Companies now offer instrument options which automatically compare such data and reject a sample which is chemiluminescent, but not one which is adsorbing or precipitating.

Usually an adsorbing sample can be distinguished from a precipitating sample by comparing glass to nylon or polyethylene counting vials which do not have polar surfaces (30). Another method is to shake vigorously or vortex the vial to see if the count rate increases by redissolving the sample. It should be emphasized that the mass of labelled sample is usually quite small, and one cannot rely on a visual check to ensure that precipitation is not occurring.

Phase separation can be a serious problem for the newer detergent cocktails which are quite temperature/phase dependent, particularly in cases where samples are prepared under ambient conditions and counted in refrigerated systems (31). We feel much more comfortable with glass bottles where we can "see" the sample, even though visual evidence is not the only criterion one must use for determining the homogeneity of a sample.

Another simple check for adsorption is to pour the scintillator solution from the original vial to a second vial, fill the original vial with fresh scintillant and recount. If adsorption is occurring, the original vial will retain a good percentage of the original counts.

The use of a double ratio method of quench correction (32,25) is the preferred screening method to check for precipitated, phased, adsorbed, or any non-homogeneous sample. If a sample is adsorbing, the external standard ratio will usually indicate a higher counting efficiency than the sample channels ratio. D. S. Glass (34) has shown that modern spectrometers with on-line computers can simply and routinely perform this check for the analyst.

Certain types of compounds, such as lipids, cations, multivalent anions, proteins and basic amino acids have been frequently reported to adsorb (30,35-44).

Several methods have been used to overcome vial-wall adsorption problems. Use of polyethylene or nylon containers usually prevents adsorption. Adding an excess of the corresponding non-radioactive carrier compound (30,35,37) results in a competition for the binding sites on the vial wall (45). If sufficient carrier material is present to saturate the vial walls, adsorption is minimized. For inorganic radioactivity standards, The Radiochemical Centre has found that 100 μg of carrier material per ml is needed to prevent appreciable adsorption.

Another method to overcome the problem of adsorption is to siliconize the vial walls (30,37,46,47). It should be noted that if a sample adsorbs to the counting vial, it has probably adsorbed to walls of flasks, pipettes and all other glass surfaces with which it has been in contact. All glassware should be siliconized or carrier added at the earliest possible stage in the assay.

Use of complexing agents (NCS*, Di(2-ethylhexyl)phosphate, EDTA) can prevent vial wall adsorption, particularly of the inorganic ions, such as $^{22}Na^{+}$ and $^{32}PO_4^{-3}$.

*Trademark Amersham/Searle

Although photoluminescence has been reported as a problem with glass counting bottles (10,48), we have not experienced any difficulties, as it is standard practice for samples to remain in the counter for four hours before counting. In our experience vials prone to photoluminescence are:

nylon>>>>polyethylene>glass

We always like to use a glass vial, but one must be aware of the problem of vial wall adsorption.

Two additional glass vials should be mentioned. Quartz counting vials were recognized many years ago because of the improved properties of light transfer or resultant higher counting efficiency and also lower backgrounds (14, 20). Although quartz vials can still be obtained from Packard, their cost ($12.00 each) is prohibitive for most applications.

Etched or sandblasted glass vials have been used to improve pulse heights and counting efficiency (49-51). A 3-5% increase in counting efficiency using frosted vials has been reported (52). Although frosted vials were commercially available from Beckman, presently there is no known commercial source. The $0.25 cost of the vials apparently was not worth the extra 3-5% in performance.

POLYETHYLENE/POLYPROPYLENE VIALS

Bell (53) noted in 1957 that chlorotrifluoropolyethylene bottles counted with the same efficiency as glass, but lowered the background 3- to 4-fold. Horrocks and Studier (54) and Rapkin and Gibbs (50,55,56) evaluated the application of polyethylene counting containers, pointing out several advantages and a number of disadvantages.

Polyethylene has the advantage of highest counting efficiency and lowest background of any commercial 20 ml bottle (55,57,58). Other advantages include low cost, combustibility for disposal purposes, non-breakability, non-polar surfaces which resist adsorption, and low susceptibility to photoactivation by strip lighting or sunlight.

The most serious disadvantage of polyethylene counting vials is permeability of the vial walls to many chemicals, but in particular, toluene and xylene. Rapkin and Gibbs (55) indicated that the leakage rate for toluene was approximately .7%/day. Lieberman and Moghissi (58) found a somewhat similar weight loss rate for xylene and toluene (but little loss of water and dioxane), and further pointed out that there was considerable variation from manufacturer to manufacturer. One polypropylene counting bottle tested in our laboratories showed a 47% weight loss in six days at 22°C.

Leakage of toluene and xylene poses a serious environmental problem for workers continually exposed to vapors. It is recommended that polyethylene bottles be discarded immediately after use. A further hazard is fire, particularly when quantities of filled bottles are stored in poorly ventilated areas, such as storerooms or closets (22).

A radiological hazard also exists, particularly when penetrating substances such as toluene and n-hexadecane are being used as in internal standards.

The permeation of other radioactive substances through polyethylene containers has been reported to us many times. As an example, certain enzymatic assays are carried out directly in liquid scintillation bottles. A carboxyl-^{14}C labelled substrate and a decarboxylase enzyme are used to measure a reaction rate by the liberation of $^{14}CO_2$. When polyethylene bottles are used, $^{14}CO_2$ tends to diffuse into the vial wall while the blank reaction is being carried out. When scintillator solution is added, $^{14}CO_2$ diffuses back into solution, resulting in high blanks and non-reproducible results (59).

Hansen (38) reported that when polyethylene vials, some with cocktail but no activity and some with cocktail plus Toluene-^3H, were placed in a liquid scintillation counter, the "background" count rate in the vials containing no activity rapidly increased with time.

Many polyethylene bottles swell appreciably as toluene permeates the vial wall. Serious instrument repair bills can result if such a vial jams the sample-changer mechanism. Heating polyethylene bottles usually softens the vial to the point where it is not useable, and polyethylene bottles are not recommended for use with strongly basic solubilizers, like Hyamine hydroxide* and NCS†. Opacity of most polyethylene bottles prevents direct inspection of the sample -- a serious disadvantage.

A more important effect of permeability noted by Bush(25)

*Trademark Rohm and Haas
†Trademark Amersham/Searle

and further studied by Laney (4) and others (60,61) is that once toluene or xylene has begun to permeate the vial walls, changes occur in the external standard ratio. The vial wall becomes more "efficient" when it is saturated with primary solvent. Although one study indicates that this variation stabilizes after 5 hrs. (61), another recommends two days (4). It should be noted that the time required for the external standard ratio to stabilize with polyethylene vials varies considerably from manufacturer to manufacturer and from batch to batch. If polyethylene vials must be used, the sample channels ratio method (33) of quench correction should be used.

Polyethylene and polypropylene vials differ considerably in wall thickness and the density used in manufacture. The physical characteristics vary from nearly transparent to opaque, and from soft and pliable to rigid.

In general, we have not found that the rate of permeation through vials from various suppliers is very different. It takes longer for the solvents to permeate the thick-walled, high-density bottle initially, but once the molecules have permeated the outer surface, the rate of solvent loss is not very different from manufacturer to manufacturer.

Curtis (61) has studied relative transpiration rates for a number of solutions stored in polyethylene, polypropylene, high-density polyethylene, teflon and glass containers.

We have not seen any real advantage of polypropylene as opposed to polyethylene, although one report (61) prefers polypropylene. In our experience the significant difference in plastic vial performance is the source of supply and not the plastic employed.

Spurious counts due to photoluminescence do not appear to be a serious problem with polyethylene bottles, providing they are placed in a counter several hours prior to counting (18,57). Static electricity has been reported with polyethylene bottles (4,57,63)but we have never found this to be a serious problem except in very dry climates. Treating vials with an anti-static agent, avoiding rubbing vials with cloth and preventing vials from continuously circulating through the counter can prevent photoluminescence due to static electricity.

In Cerenkov counting polyethylene vials are preferred to glass because of significant background reduction (64-67); nylon vials are unsuitable for the aqueous samples normally used.

NYLON VIALS

Nylon counting vials offer all the advantages of polyethylene vials -- unbreakable, disposable, extremely low background (68), adsorption resistance -- without the disadvantage of toluene or xylene permeability.

One major disadvantage of a nylon vial is that it softens and deforms when used with strongly polar solvents, such as cocktails containing large amounts of methanol, Triton X-100, water and some of the pre-mixes (58). Any commercial cocktail that is designed to hold large amounts of water probably will not work well with nylon vials.

A second disadvantage of nylon is its high susceptibility to photoluminescence. Nylon vials when exposed to the same light source result in 2-3 orders of magnitude higher luminescence than polyethylene or glass and the decay time is considerably longer.

Nylon vials are useful in applications where the cocktail contains minimum amounts of polar solvents, providing precautions are taken to protect the bottles from exposure to sunlight and fluorescent lighting.

TEFLON VIALS

Calf (69,70) has used teflon counting vials to achieve higher E^2/B than polyethylene for low-level counting, and teflon is much more resistant to leakage than polyethylene (62).

There is no known commercial source of teflon vials, but one would expect them to be very expensive, perhaps more so than borosilicate glass.

MINIATURE VIALS

Recent improvement in sample preparation, which allows the analyst to count up to 40 per cent aqueous sample, has generated considerable interest in the use of smaller counting vials. The first of the commercial versions, the Mini-vial™*, consists of a holder the size of a standard 20 ml counting vial which has been bored down the center to accommodate a 7 ml polyethylene vial.

The Mini-vial™ has a history of difficulties. The first holder design was simply a Lucite tube with an open

*Trademark Nuclear Associates

bottom. The shape of the holder caused serious jamming problems in some counters, and the holder was so short it passed beneath many sample sensing devices.

A second version was produced which featured an enclosed bottom, but the radius of curvature at the bottom edge of the holder would not trip some sample sensing switches. Further difficulties with spurious luminescence due to static electricity have been reported to us many times.

A third version said to overcome the static electricity and geometry difficulties is now being produced. Although we have not tested the newer version, we are not aware of any difficulties with the latter.

A further difficulty with the Mini-vial™ is destruction of the rather expensive holder by toluene permeating the polyethylene vial. If a vial is left in holder more than a few hours, it fuses to the holder, and the holder frequently deforms. The polyethylene insert is subject to all the disadvantages described for standard 20 ml polyethylene bottles.

Two novel miniature vial combinations are shown in Fig. 1. The first used a standard 3 dram (10 ml) glass vial and a 24 mm polyethylene vial as a holder. This combination is very inexpensive and offers all the advantages of a glass system. The glass vial fits snugly in the polyethylene holder when the cap is screwed on.

Another miniature vial combination employs a standard 22 mm glass vial as a holder, for which Sterilin designed a 5 ml polyethylene insert vial. Although this combination is economical, it possesses all the inherent disadvantages of a polyethylene counting vial and its use is not recommended.

COUNTING BOTTLE CAPS

Several types of bottle caps are available and the choice can seriously affect results. It was discovered quite early that aluminum foil-lined caps would increase counting efficiency 1-2% by reflecting photons escaping from the solution surface back down towards the photomultiplier tubes. It should be noted, however, that foil and cork-lined caps are incompatible with strongly basic cocktails

*Trademark Nuclear Associates

containing NCS* or Hyamine†. Teflon or polyethylene lined caps are preferred in these applications.

It was noted in 1965 that liquid scintillation counting vials with white screw caps resulted in higher, more erratic background count rates than liquid scintillation counting vials containing black caps (71). Further studies have shown that the higher background count rates are due to photoactivation of the caps by sunlight and laboratory fluorescent lighting (48).

As a typical example, when an empty 20 ml glass counting vial with white cap was allowed to drop into the counting chamber while the lid of the counter remained open, 1500 counts were recorded in the tritium channel during the first 0.01 minute, even with the coincidence mode of the spectrometer in operation. Under the same conditions, black-capped vials gave only 140 counts.

The phosphorescence of white caps appears to decay to normal levels very quickly, usually within two minutes (72). However, with the increasing popularity of multi-user counters, it is possible that while one individual's samples are counting, a second user may be opening the lid to unload or load samples. It is recommended that the counting cycle be stopped before opening the counter, and that the counter be reset 2-3 minutes after the lid has been closed.

The use of black caps on liquid scintillation vials, however, will minimize the problem of spurious counts due to photoluminescence (73).

Snap-on or push-in polyethylene caps are available for certain vials, but these almost inevitably cause fluid to be squeezed out of the vial when inserted. These caps can only be conveniently used one time and should not be used with the internal standard method of quench correction. For reasons of protection from radiation contamination, we do not allow snap-on or push-in plastic caps in our laboratories.

Complete sealing of commercial screw-cap counting vials is nearly an impossible task (71,74) mainly due to an irregular top edge. All vials leak primary solvent vapors, polyethylene, as previously discussed, being the worst offender. We find taping the space between the lower edge

*Trademark Rohm and Haas
†Trademark Amersham/Searle

of the cap and the shoulder of the bottle with teflon convenient for samples we may wish to re-count at a later date or for samples of long counting times. Care should be taken not to use thick tape which may cause jamming of the sample changer mechanism if the diameter becomes too large. Alternatively, flame-sealable low-background liquid scintillation glass bottles are commercially available (Fig. 1) but the laborious task of sealing the bottles precludes routine use.

EFFECT OF VIAL GEOMETRY IN COUNTING CHAMBER

Garfinkel (71), Stanley (5) and Laney (4) have described significant error introduced in the assay by the differential positioning of a counting vial in the counting chamber. When using mini-vials, it is imperative that holders are of reproducible dimensions both in terms of concentricity and height of sample vial.

VIAL FILLING DEVICES

Filling large numbers of counting vials can be a tedious, and sometimes hazardous, task. Several devices are available which reduce exposure to solvent vapors and speed bottle filling.

The least expensive filling device is the tilting dispenser offered by many suppliers in adjustable and fixed volumes. Cost is generally about $15.00. A more accurate and rapid filling mechanism is the Repipet*, which sells for about $75.00. This device screws on a standard 4 l bottle and can fill a counting bottle accurately in a few seconds.

A much more elaborate (and expensive) device is the Multi-Jet 100†, which fills ten vials simultaneously by one stroke of a lever, or an entire case of vials in less than one minute. The cost, however, is $595.00 per unit.

EFFECT OF VIAL ON PULSE HEIGHT SPECTRUM

While it is commonly recognized that the shape of a beta energy spectrum changes with sample composition, it should be noted that energy spectrum shifts when using vials of different composition can cause an increase or decrease in counting efficiency, changes in ratio and changes in spill-

*Trademark Labindustries
†Trademark Isolab

over factors (50,60,66,75). Optimal gain and discriminator settings should be established with the type of vial to be used in the assay.

EFFECT OF VOLUME ON COUNTING EFFICIENCY

Several investigators have shown that optimum counting efficiency in a 20 ml counting vial plateaus between 10-15 ml where the efficiency is very little affected by volume (4,60,66,67,76-78). Rummerfeld and Goldman (3) have shown efficiency of mini-vials to be much more sensitive to small volume changes. In the above studies if height of sample in vial had been plotted against counting efficiency, nearly identical curves would have been obtained.

SPECIAL TYPES OF VIALS

Schram (79) has reviewed counting vials which can be used as flow cells in conventional liquid scintillation spectrometers. Ashcroft (80) constructed a glass vial to count gamma emitting nuclides by an external sample method, and these vials are now commercially available. Schram (81) has suggested a modified counting bottle for bioluminescence assays.

WASHING AND REUSING COUNTING VIALS

Washing and reusing counting bottles is risky business, unless each vial is filled with fresh scintillant and counted prior to the addition of the radioactive sample. We have investigated reusing scintillation bottles several times and have found that if one considers the cost of technician time, counter time and the chance of a contaminated bottle affecting the precision of data, reusing standard counting bottles is false economy. In no case should polyethylene, polypropylene or nylon vials be reused, and all caps should be discarded after a single use.

Rummerfeld and Goldman (76) have discussed the economics of vial washing versus the accuracy required for the experiment. Harris and Friedman (82) have described a simple rinsing and cleaning apparatus for use with glass counting bottles. Drosdowsky and Egoroff (83) designed a simple decontamination unit which can clean up to 55 vials per hour. Several vial washing units are available commercially, among them the Refluxowasher (Buchler Inst.) which has been evaluated by Kushinsky and Paul (84).

It is interesting to note that in the studies mentioned

above radioactive steroids were employed. Since steroids are known to be fairly soluble in non-polar aromatic solvents and tend not to adsorb to glass (30,85), those contemplating reuse of counting vials should not presume that similar results will be obtained with other types of samples.

SUMMARY

The choice of counting container can drastically affect not only the optimum conditions of the assay, but also the accuracy and reproducibility of counting data.

For the majority of applications we recommend the use of a glass counting vial, and the sample channels ratio method of quench correction. The exception is extreme low level counting where polyethylene or teflon must be used to achieve high figures of merit.

In the future I believe the ability to count large volumes of aqueous samples and the high specific activity labelled compounds now available will spur manufacturers to develop, with only minor modifications to counting chamber geometry and sample changer, counters which will be designed to accommodate 5-10 ml bottles.

ACKNOWLEDGEMENTS:

I should like to acknowledge the invaluable assistance of Bart Laney and Dave Hansen (Nuclear-Chicago); Jim Gibbs (Packard); Murray Vogt (Isolab); and Mike Gezing, Dorothy Gabbei and JoAnn Jacobsen (Amersham/Searle).

BIBLIOGRAPHY

1. Birks, J.B., and Poullis, G.C., in Liquid Scintillation Counting (Crook, M.A., Johnson, P., and Scales, B., Eds.) Heyden, London, 1972, p. 1.
2. Kalbhen, D.A., in Liquid Scintillation Counting, (Dyer, A., Ed.) Heyden, London, 1971, p. 127.
3. Cavanaugh, R., in The Current Status of Liquid Scintillation Counting (Bransome, E.D., Ed.) Grune and Stratton, New York, 1970, p. 293.
4. Horrocks, D.L., in Tritium (Moghissi, A.A., and Carter, M.W., Eds.) Messenger Graphics, Phoenix, 1973, p. 30.
5. Stanley, P.E., in Liquid Scintillation Counting, (Crook, M.A., Johnson, P., and Scales, B., Eds.) Heyden, London, 1972, p. 285.
6. Spratt, J.L., in Liquid Scintillation Counting, (Crook, M.A., Johnson, P., and Scales, B., Eds.) Heyden, London, 1972, p. 245.
7. Rapkin, E., in The Current Status of Liquid Scintillation Counting, (Bransome, E.D., Ed.) Grune and Stratton, New York, 1970, p. 45.
8. Utting, G.R., in Liquid Scintillation Counting, (Bell, C.G., and Hayes, F.N., Eds.) Pergamon, New York, 1958, p. 67.
9. Guinn, V.P., in Liquid Scintillation Counting, (Bell, C.G., and Hayes, F.N., Eds.) Pergamon, New York, 1958, p. 166.
10. Davidson, J.D., and Feigelson, P., Int. J. Appl. Radiat. Isotopes, 2, p. 1 (1957).
11. Packard, L.E., in Liquid Scintillation Counting, (Bell, C.G., and Hayes, F.N., Eds.) Pergamon, New York, 1958, p. 50.
12. Davidson, J.D., in Liquid Scintillation Counting, (Bell, C.G., and Hayes, F.N., Eds.) Pergamon, New York, 1958, p. 88.
13. Hodgson, T.S., Gordon, B.E., and Ackerman, M.E., Nucleonics, 16, p. 89 (1958).
14. Bernstein, W., Bjerknes, C., and Steele, R., in Liquid Scintillation Counting (Bell, C.G., and Hayes, F.N., Eds.) Pergamon, New York, 1958, p. 74.
15. Rapkin, E., in Liquid Scintillation Counting, (Crook, M.A., Johnson, P., and Scales, B., Eds.) Heyden, London, 1972, p. 61.

16. Radin, N.S., in Liquid Scintillation Counting, (Bell, C.G., and Hayes, F.N., Eds.) Pergamon, New York, 1958, p. 108.
17. International Electrotechnical Commission, Technical Committee No. 45 (Nuclear Instrumentation), Draft 45 (Secretariat) 166, March 1973.
18. Horrocks, D.L., in Survey of Progress in Chemistry, Vol. 5, (Scott, A.F., Ed.) Academic Press, New York, 1969, p. 185.
19. Agranoff, B.W., in Liquid Scintillation Counting, (Bell, C.G., and Hayes, F.N., Eds.) Pergamon, New York, 1958, p. 220.
20. Agranoff, B. W., Nucleonics, 15, P. 106 (1957).
21. Kasida, Y., and Iwakura, T., Radioisotopes (Japan), 11, p. 257 (1962).
22. Kobayashi, Y., and Maudsley, D.V., in Methods in Biochemical Analysis (Glick, D., Ed.), 17, p. 31 (1969).
23. Paix, D., Int. J. Appl. Radiat. Isotopes, 19, p. 162 (1968).
24. Hagashimura, T., Int. J. Appl. Radiat. Isotopes, 13, p. 308 (1962).
25. Mueller, E, in The Current Status of Liquid Scintillation Counting, (Bransome, E.D., Ed.) Grune and Stratton, New York, 1970, p. 181.
26. Schram, E., Organic Scintillation Detectors, Elsevier, New York, 1963.
27. Scales, B., Anal. Biochem., 5, p. 489 (1963).
28. Wang, C. H., in The Current Status of Liquid Scintillation Counting (Bransome, E.D., Ed.) Grune and Stratton, New York, 1970, p. 305.
29. Laney, B. H., in Tritium, (Moghissi, A.A., and Carter, M.W., Eds.) Messenger Graphics, Phoenix, 1973, p. 156.
30. Litt, G. J., and Carter, H., in The Current Status of Liquid Scintillation Counting, (Bransome, E.D., Ed.) Grune and Stratton, New York, 1970, p. 156.
31. Turner, J.C., Int. J. Appl. Radiat. Isotopes, 19, p. 557 (1968).
32. Bush, E.T., Int. J. Appl. Radiat. Isotopes, 19, p. 447 (1968).
33. Bush, E.T., Anal. Chem., 35, p. 1024 (1963).
34. Glass, D.S., in Organic Scintillators and Liquid Scintillation Counting, (Horrocks, D.L., and Peng, C.T., Eds.) Academic Press, New York, 1971, p. 803.

35. Davison, P.F., and Andersson, L.P., Anal. Biochem., 47, p. 253 (1972).
36. Allison, J.M., Monro, A.M., and Offerman, J.L., Anal. Biochem., 47, p. 73 (1972).
37. Robel, E., Anal. Biochem., 51, p. 137 (1973).
38. Hansen, D.L., Nuclear-Chicago Corporation, DesPlaines, Illinois, U.S.A., personal communication.
39. Johnson, E.P., and Lowenthal, G.C., Int. J. Appl. Radiat. Isotopes, 23, p. 196 (1972).
40. Brandreth, Dale A., and Johnson, R.E., Science, 169, p. 864 (1970).
41. Hayes, F.N., U.S.A.E.C. Document LA-1639 (1963).
42. Blanchard, F.A., and Takahashi, I.T., Anal. Chem., 33, p. 975 (1961).
43. Dobbs, H.E., in Organic Scintillators and Liquid Scintillation Counting, (Horrocks, D.L., and Peng, C.T., Eds.) Academic Press, New York, 1971, p. 669.
44. Snyder, F., and Stephens, N., Anal. Biochem., 4, p. 128 (1962).
45. Tuck, A.G., and McNair, A., Int. J. Appl. Radiat. Isotopes, 23, p. 395 (1972).
46. Petroff, C.P., Nair, P.P., and Turner, D.A., Int. J. Appl. Radiat. Isotopes, 15, p. 491 (1964).
47. Parkinson, P.I., and Medley, E., in Liquid Scintillation Counting, (Crook, M.A., Johnson, P., and Scales, B., Eds.), Heyden, London, 1972, p. 109.
48. Scales, B., in Liquid Scintillation Counting, (Crook, M.A., Johnson, P., and Scales, B., Eds.) Heyden, London, 1972, p. 101.
49. Gordon, B.E., and Curtis, R.M., Anal. Chem., 40, p. 1486 (1968).
50. Rapkin, E., and Packard, L.E., Proc. Univ. New Mexico Conf. on Organic Scintillation Detectors, U.S. Govt. Printing Office, Washington, 1961, p. 216.
51. Schwerdtel, E., Int. J. Appl. Radiat. Isotopes, 17, p. 479 (1966).
52. "Frosted Vials", Beckman Instrument Company, Fullerton, Calif., S70462-1069-150A.
53. Bell, C.G., in Liquid Scintillation Counting, (Bell, C.G., and Hayes, F.N., Eds.) Pergamon, New York, 1958, p. 157.
54. Horrocks, D.L., and Studier, M.H., Anal. Chem., 33, p. 615 (1961).
55. Rapkin, E., and Gibbs, J.A., Int. J. Appl. Radiat. Isotopes, 14, p. 71 (1963).

56. Rapkin, E., and Gibbs, J.A., Packard Instrument Company Bulletin, Number 9, 1965.
57. Moghissi, A.A., Kelley, H.L., Regnier, J.E., and Carter, M.W., Int. J. Appl. Radiat. Isotopes, 20, p. 145 (1969).
58. Lieberman, R., and Moghissi, A.A., Int. J. Appl. Radiat. Isotopes, 21, p. 319 (1970).
59. Berry, J.A., Carnegie Institution of Washington, Stanford, California, U.S.A., personal communication.
60. Haviland, R.T., and Bieber, L.L., Anal. Biochem., 33, p. 323 (1970).
61. Johanson, K.J., and Lundqvist, H., Anal. Biochem., 50, p. 47 (1972).
62. Curtis, G.J., Rein, J.E., and Yamamura, S.S., Anal. Chem., 45, p. 996 (1973).
63. Moghissi, A., in The Current Status of Liquid Scintillation Counting, (Bransome, E.D., Ed.) Grune and Stratton, New York, p. 86 (1970).
64. Lauchli, A., Int. J. Appl. Radiat. Isotopes, 20, p. 265 (1969).
65. Parker, R.P., and Elrick, R.H., Int. J. Appl. Radiat. Isotopes, 17, p. 361 (1966).
66. Gould, J.M., Cather, R., and Winget, G.D., Anal. Biochem., 50, p. 540 (1972).
67. Parker, R.P., and Elrick, R.H., in The Current Status of Liquid Scintillation Counting, (Bransome, E.D., Ed.) Grune and Stratton, New York, p. 110 (1970).
68. Kobayashi, Y., and Maudsley, D.V., in The Current Status of Liquid Scintillation Counting, (Bransome, E.D., Ed.) Grune and Stratton, New York, 1970, p. 76.
69. Calf, G.E., Int. J. Appl. Radiat. Isotopes, 20, p. 611 (1969).
70. Calf, G.E., in Organic Scintillators and Liquid Scintillation Counting, (Horrocks, D.L., and Peng, C.T., Eds.) Academic Press, New York, p. 719, 1971.
71. Garfinkel, S.B., Mann, W.B., Medlock, R.W., and Yura, O., Int. J. Appl. Radiat. Isotopes, 16, p. 27 (1965).
72. Laney, B., Nuclear-Chicago Corporation, DesPlaines, Illinois, personal communication.
73. Painter, K., and Gezing, M. (in press).
74. Burleigh, R., in Liquid Scintillation Counting, (Crook, M.A., Johnson, P., and Scales, B., Eds.) Heyden, London, p. 139, 1972.
75. Butterfield, D., and McDonald, R.J., Int. J. Appl. Radiat. Isotopes, 23, p. 249 (1972).

76. Rummerfield, P.S., and Goldman, I.H., Int. J. Appl. Radiat. Isotopes, 23, p. 353 (1972).
77. Price, L.W., Laboratory Practice, April, 1973, p. 277.
78. Clausen, T., Anal. Biochem., 22, p. 70 (1968).
79. Schram, E., in The Current Status of Liquid Scintillation Counting, (Bransome, E.D., Ed.) Grune and Stratton, New York, 1970, p. 129.
80. Ashcroft, J., Anal. Biochem., 37, p. 268 (1970).
81. Schram, E., in The Current Status of Liquid Scintillation Counting, (Bransome, E.D., Ed.) Grune and Stratton, New York, 1970, p. 129.
82. Harris, J.E., and Friedman, L., Anal. Biochem., 30, p. 199 (1969).
83. Drosdowsky, M., and Egoroff, N., Anal. Biochemistry, 17, p. 365 (1966).
84. Kushinsky, S., and Paul, W., Anal. Biochem., 30, p. 465 (1969).
85. "Super Hot Tritiated Steroids", Amersham/Searle Corp., Arlington Heights, Illinois, U.S.A., (1971).

WORKSHOP SESSION

Chairman   Dr. J. Coghlan

SUBJECT: *ABSOLUTE ACTIVITY MEASUREMENTS USING LIQUID SCINTILLATION COUNTING*

Dr. Ross   I would like to describe and get some ideas from the audience concerning the determination of absolute activity in liquid scintillation counting. The absolute activity of a β-γ emitter can be measured using the coincidence procedure first described in 1940 (Rev. Sci. Inst. 11, 167, 1940), where the source is placed between a β- and a γ-detector and counts from each are registered on individual scalers and a third scaler monitors counts coincident in the other two. Then the absolute β-activity equals the β-count rate times the β-detector efficiency and the γ-count rate times the γ-detector efficiency equals the absolute γ-activity. The counts in the coincidence channel are equal to the absolute activity times the two efficiencies. Thus,

$$C_\beta = A_o \cdot eff_\beta$$

$$C_\gamma = A_o \cdot eff_\gamma$$

$$C_c = A_o \cdot eff_\beta \cdot eff_\gamma$$

$$\text{Then } A_o = \frac{C_\beta \cdot C_\gamma}{C_c}$$

At the Northwestern Conference in 1957 Dr. V.P. Guinn saw the analogy between the β-γ system and the liquid scintillation counter, which had two phototubes each looking at the singles count, and coincidences were also available. He gave data for ^{14}C, ^{32}P and ^{3}H and using the analogous equations got results 1-2% high and said it could be used with colour or chemical quenching. So apparently here was a technique not requiring channels ratio or external standardization. As far as I know the technique was not reported again until five years later when a group at the National Laboratory - Richland, Washington - confirmed Guinn's results. In 1970 there was a report from a group at Saclay that also appears to confirm the work. If three groups say the procedure works, why are we using quench correction techniques?

Dr. Lowenthal Unless the geometry is 4π, the coincidence method does not work as well as other methods. The procedure used by the Saclay group is a good technique but is complex and expensive.

Dr. Ross Does their method not involve the coincidence method?

Dr. Lowenthal Not really.

Dr. Ross Then I am wrong to include the work at Saclay, but Guinn and the Richland group say it works. My question is, does it? Most results published have a correction around 1-2% and for many applications this is satisfactory, especially if we con't have to do any quench correction.

Dr. Lowenthal At an international conference last year when these methods were discussed at length, the liquid scintillation coincidence procedure was not singled out. However, the gas proportional system is better.

Mr. Polach I am sure Dr. Lowenthal doesn't imply gas counters are 100% efficient because they are not since the end effect must be considered. In my opinion there is only a difference in degree of efficiency between gas and liquid scintillation counting. Both should be equally suitable for absolute measurements.

Dr. Birks Let's get things clarified. The β-γ method is a perfectly good established method and you will find it in my book of 1964.

Dr. Ross I'm sorry I went to the original reference, John.

Dr. Birks The β-γ procedure depends on a coincidence in time between the β and the γ from the same atom. The formula then applies. However, in liquid scintillation what we are observing is the coincidence between two parts of the same scintillation and then it does not apply.

Dr. Ross   Would you, Dr. Birks, say that in singles the activity is equal to the count rate time the efficiency?

Dr. Birks   But they are both counting the same event, i.e. the β-particle.

Mr. Hartley   The equation holds for liquid scintillation for only one condition. That is for the condition when you emit exactly 2 photons per event. When the number gets to 10-15, it falls down. That is why it sort of works at the bottom end and it's not too far out.

Mr. Laney   I agree with Mr. Hartley that the third equation is true for the 2-photon case. However, for larger numbers of photons, multiplying probabilities is invalid since the number of photons arriving at one detector is dependent upon those arriving at the other.

Dr. Ross   I asked this as a loaded problem. When Guinn proposed it, we tried it too and found it did not work, especially with quenching. Then when a sister laboratory got it to work, we tried again but still couldn't get it to work. John Birks is of course quite right - the equations are not valid in this case but it is important to indicate why. In the β-γ procedure, the individual events are uncorrelated and thus the coincident detection probability can be correctly represented as the product of the individual efficiency terms. In the liquid scintillation counter, coincident events occurring in two tubes are most definitely correlated and the simple combination of the efficiency terms is not correct. We are now convinced that the technique does not work and I have a lot of data to back this up.

SUBJECT: *LESSER PULSE HEIGHT ANALYSIS*

Mr. Laney   The "lesser" is an improved method of pulse height analysis for efficiency determination by pulse height shift. In 1960, Baille (Int. J. Appl. Rad Isotopes 8: 1, 1960) presented a method of efficiency determination based upon pulse height shift which has become the accepted method of automatic efficiency determination. The method correlates pulse height shift with counting efficiency in what has become known as a quench correction curve. Pulse height shift is determined indirectly by taking a ratio of the counts in one portion of the spectrum with the counts in another part of the spectrum.

Some of the investigators who have shown the ratio method is in error for samples containing colour quenchers include: Baille; Bush (Anal. Chem. 35: 1024, 1963); Neary and Budd (Liquid Scintillation Counting, ed. E.D. Bransome, Grune and Stratton, New York, 1970, p. 273); Noujaim, Ediss and Wiebe (Organic Scintillators and Liquid Scintillation Counting, ed. D. Horrocks and C.-T. Peng, Academic Press, 1971, p. 705). Neary and Budd and Noujaim *et al.* showed that the summed pulse height spectrum has greater pulse amplitude for colour quenched than chemically quenched samples of the same efficiency. They showed that efficiency determination based upon pulse height shift using pulse summation tends to over-estimate the coincidence detection efficiency of colour quenched samples.

In session 2, Dr. Noujaim presented a paper comparing summed and lesser methods of pulse height analysis. Errors in efficiency determination of colour quenched samples were dramatically reduced using the lesser because spectra of chemical and colour quenched samples are more nearly identical with the lesser. Even external standard lesser spectra are nearly identical.

The "lesser" and how it restores the colour quench spectra will now be described.

The block diagram of a single channel beta coincidence spectrometer is shown in Figure 1. Pulse height signals X and Y from multiplier phototubes PM1 and PM2 respectively are applied to coincidence detector, COIN, and functional block, f. Coincidence detector, COIN, produces a digital enabling signal to the output logic if the X and Y signals are coincident. Functional block, f, combines the X and Y signals for pulse height analysis by discriminators LL and UL. If the amplitude of pulses generated by functional block f are greater than the LL and less than the UL and are coincident, a pulse will be generated to increment the scaler. Counting efficiency in the scaler channel is

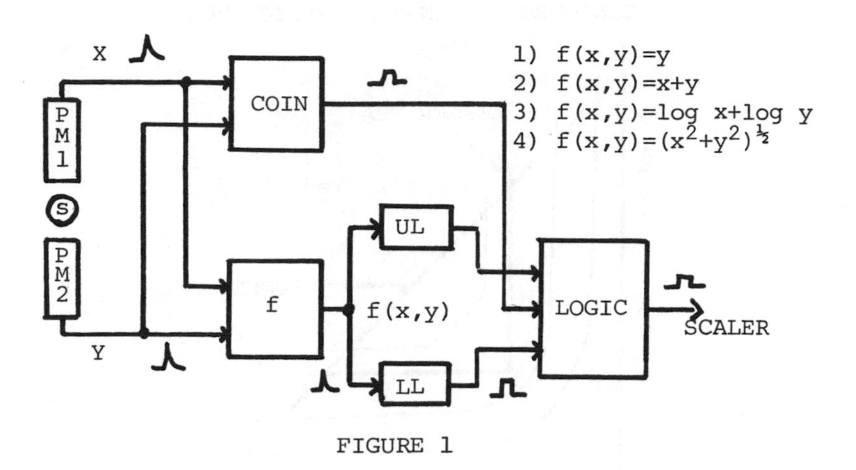

FIGURE 1

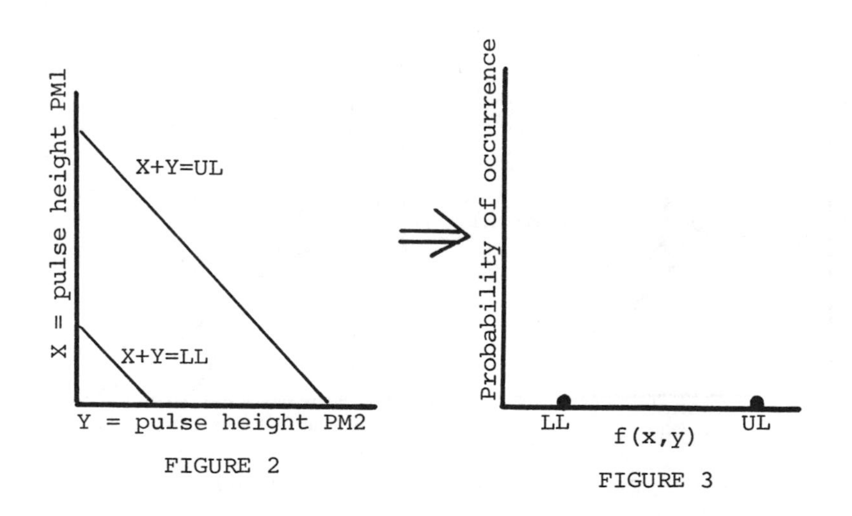

FIGURE 2

FIGURE 3

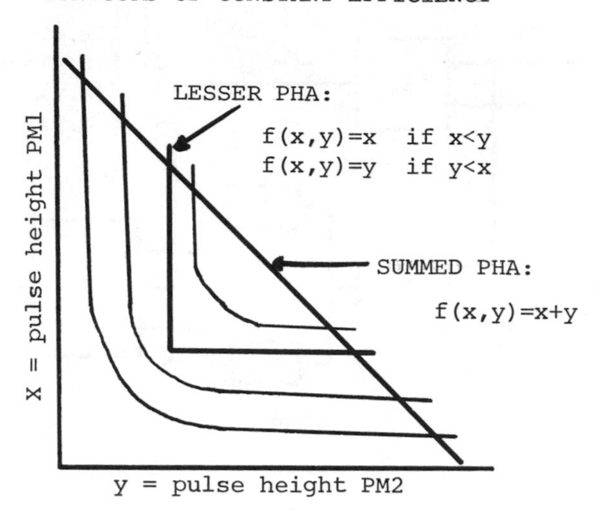
CONTOURS OF CONSTANT EFFICIENCY
LESSER PHA:
f(x,y)=x if x<y
f(x,y)=y if y<x
SUMMED PHA:
f(x,y)=x+y
X = pulse height PM1
y = pulse height PM2

FIGURE 4

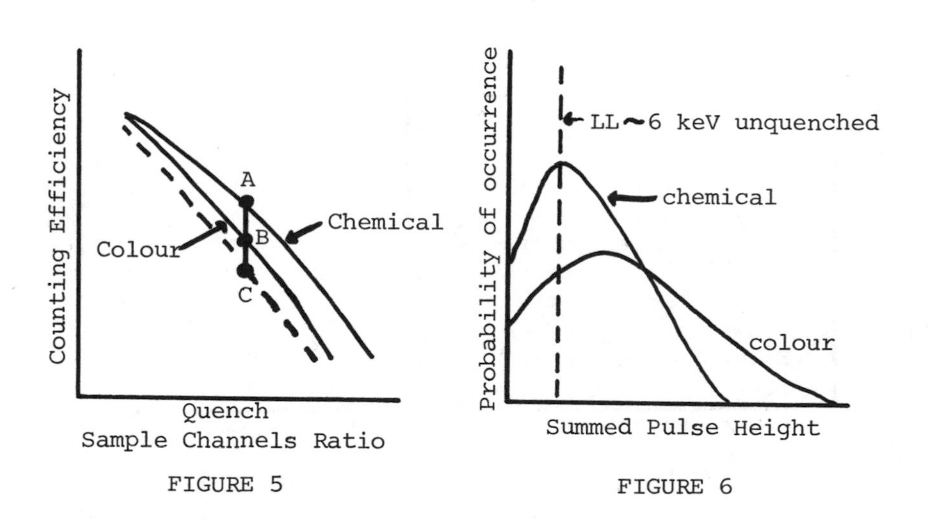
Counting Efficiency
A
B
C
Colour
Chemical
Quench
Sample Channels Ratio
Probability of occurrence
LL~6 keV unquenched
chemical
colour
Summed Pulse Height

FIGURE 5

FIGURE 6

therefore dependent upon f, LL, UL and COIN, whereas detection efficiency is determined by COIN only.

Many functions can be generated by functional block f. Early counters utilized only one input (equation 1) called the data signal. Signals from the other phototube were used only for the coincidence gate. In the mid-1960's pulse summation became prevalent (equation 2). One commercial instrument employing logarithmetic phototubes did pulse height analysis by equation 3. Other functions, for example the circular transform (equation 4), can be generated. Functional block f is thus used to convert the two phototube signals into a single variable for discrimination.

Figures 2 and 3 show how a two parameter coordinate system is transformed into a single parameter coordinate system. Figure 2 is a two parameter pulse height coordinate system. Each coincident event is located in the XY plane by its X and Y pulse height. Probability of occurrence (counts) increases in the Z direction. Curves of X + Y = LL and X + Y = UL shown in Figure 2 become points on the abscissa of the single parameter coordinate system in Figure 3 for f (x,y) = X + Y.

Contours of constant efficiency are shown in two parameters in Figure 4. One can see from Figure 4 that the lesser transform more nearly follows the contours of constant efficiency than does the sum. Therefore each point on the abscissa of the single parameter coordinate system more nearly represents constant efficiency with the lesser transform. In other words, the single parameter pulse height spectrum using the lesser transform is more independent of the spectral distribution in the two paramater plane. Coincident pulses from the two phototubes tend to be unequal for colour quench samples. Therefore the spectra tend to spread along the axis of the two parameter plane. If one follòws the line of constant X + Y in Figure 4 away from X = Y, efficiency declines while the summed pulse height remains the same. This is why the summed spectra from colour quenched samples have a higher pulse height for the same efficiency than from a chemically quenched sample.

Mr. Kreveld I would like to ask if you have experimented with different types of photomultipliers, optical chambers and external standards. I understand your explanation and am also indebted to Dr. Kobayashi for a clear explanation. However, it seems to me that in practice it may not be so in other instruments. (At this point Mr. Kreveld showed a set of graphs of external standard ratio *vs* efficiency for nitromethane, carbon tetrachloride and yellow, red and blue dyes taken from data obtained on a Packard 3375; see Figure 7.) The colour and chemical quenched curves are coincident over a considerable external standard range. This suggests that photomultipliers, optical chambers and the type of external standard are important.

<u>Dr. Wiebe</u> We reported in 1970 that there were significant differences when you applied chemical quenched correction curves to colour quenched samples. The instrument we were using was a Picker Liquimat which has a caesium-137 external standard and we found errors of considerable magnitude. Since then we have used the Searle Analytic Mark II in both the lesser and summed methods. As we showed in the first session we found significant improvements in quench correction with the lesser. There is one other paper of which I am aware and I believe it was by Lang at the San Francisco Symposium in 1970 (Organic Scintillators and Liquid Scintillation Counting, ed. D. Horrocks and C.-T. Peng, Academic Press, 1971) who I believe used a Packard unit fitted with a radium source. I think that what is important is the selection of the channel windows, especially for the external standard, may play the same role in improving the summed method. At this stage I'm not convinced that it is a function of the external standard.

<u>Mr. Laney</u> It is possible to achieve colour correction with summation under certain conditions by careful selection of the lower level discriminator. Summed sample channels ratio quench curves of carbon-14 counted without a lower discriminator are shown in Figure 5. Quenched samples have lower efficiency and higher ratios than less quenched samples. Since colour quenched samples have larger summed pulse height than chemically quenched samples of equal efficiency (Figure 6), colour quenched samples fall on a separate quench curve which is lower in ratio. One can see from Figure 6 that by raising the lower level discriminator the counting efficiency of the chemically quenched sample decreases more rapidly than for the colour quenched sample. A chemically quenched sample

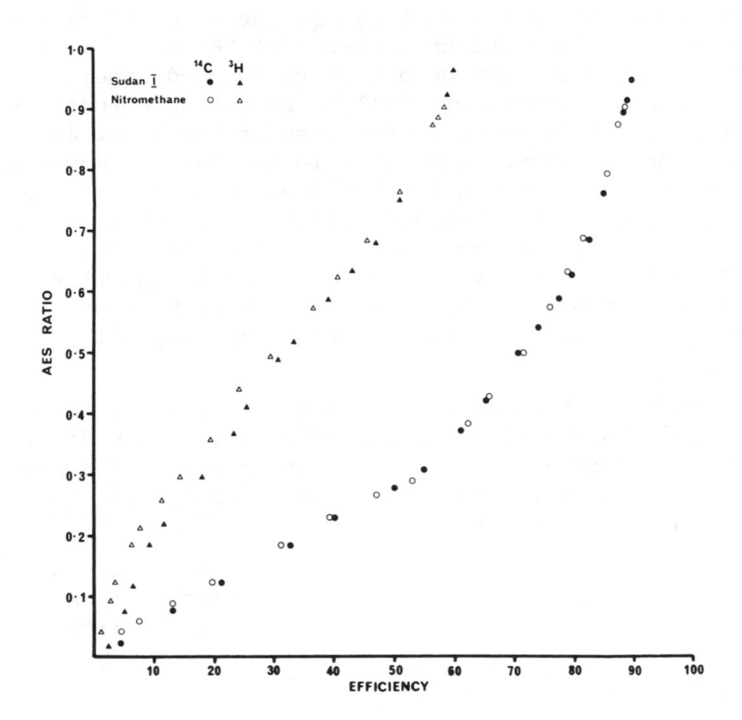

FIGURE 7. External standardization quench correction curves for ^{3}H and ^{14}C on Packard "Tri-Carb" Model 3375 Liquid Scintillation Spectrometer. Nitromethane is a chemical quencher. Carbon tetrachloride (also a chemical quencher) quench correction curves lie on top of the nitromethane curves. The quench curves for dyes (Sudan III, Brilliant Oil Blue and Dimethylaminoazobenzene) other than Sudan I (yellow) follow the Sudan I curve very closely to an external standard ratio of 0.3 (1.00 represents unquenched) but are not shown here to preserve clarity of presentation.

at efficiency "A" in Figure 5 can be moved to efficiency "C" while the colour quenched sample at "B" moves to position "C". The resultant quench curve (dashed line), though lower in efficiency, will be almost independent of quencher. The difference in the loss at the lower level is thus made to balance the error in the ratio. However, there is only one setting for the lower discriminator which will achieve this balance between the ratio and the difference in the spectral shape. On the other hand, efficiency determined by the lesser pulse height transform is more independent of the choice of window because the lesser spectrum is more independent of the type of quencher.

Dr. Kobayashi I would like to ask Mr. Laney if the lesser pulse height analysis procedure diminishes discrimination in the double-label case. The feature of conventional pulse summation is that it sharpens spectra and better separation results.

Mr. Laney One would certainly expect the lesser to have poorer energy resolution. We were surprised to find only 1 or 2% difference. I do not know why it is so little.

Dr. Kobayashi With lesser pulse height analysis you throw away half of the signal so you are back to where you were when one tube was used for data and the other was for coincidence as in the original Packard design.

Mr. Laney There is quite a difference between using the lesser and using the original data-gate system. With the lesser system, for each event a decision is made based upon the pulse height from each tube. The smallest pulse is used because it gives the more valid data for efficiency determination. With the data-gate system, pulse height information from the gate tube is never used.

Dr. Kobayashi But you are still analyzing one signal.

Dr. Noujaim I would like to comment on Mr. Laney's proposal that if you select a lower discriminator for the external standard you could reduce the error between the colour and chemical correction curves. This discriminator applies to only one isotope and one fluor solution. With the lesser pulse height analysis you can use the external standard settings for tritium and ^{14}C and for ^{45}Ca and ^{35}S although we haven't tested the last two. This is very convenient and is thus valuable for multi-user operation.

Dr. Birks I want to talk about the mathematical description that we were given. I got the impression that the efficiency curves or contours you drew in the X-Y plane were in fact the same curves you got from your lesser pulse height analysis. Is there any mathematical reason for this - I mean do they both have to be square in that same kind of way? Do all instruments have the same kind of efficiency curves? Is this independent of the photomultiplier, of the fluor, of the system, of the degree of quenching? If this is so, do you have any proof of this?

Mr. Laney I could not go into how the contours of coincidence detection shown in Figure 4 are derived in the time permitted. However, if one examines the probability of detection at each phototube based upon photon input and combines them to determine the probability of detection for a coincidence system, one obtains the contours shown in Figure 4. The reason the lesser transfer function is "squared up" is because it is easier to generate. We found little improvement when we rounded the corners to approximate the actual contours. We believe this is because there is too much statistical deviation at low pulse amplitudes where the rounding occurs to distinguish between the two cases.

Dr. Birks In this Symposium, Malcolm and Stanley presented a computer-model for the liquid scintillation process. I have now seen the computer-generated data and it would be interesting to compare your data with data generated by the model. They have included colour quenching and they know the number of photons going to each phototube and they can pick out the lesser pulse height. It strikes me that this approach would be quite interesting.

Mr. Malcolm It will not be at all difficult to include lesser pulse height analysis in the model; it will take only about 2 man-hours. What we will need, however, is experimental data to validate our model.

Mr. Kreveld I think a very objective measurement is the pulse height shift rather than the external standard ratio. I think it is a valid point to make that, depending on the electronic circuitry in the counter, these ratios will look and perhaps even correlate differently with efficiency. So it's possible for someone's data to look very good and someone else's to look very poor. Glazunov and Fleischman, who were early workers in the field of external standardization, in fact correlated their efficiency data with pulse height shift only. I would also like to point out that on the external standard-efficiency data which I have presented, the ratio of 0.3 (on the Packard 3375) represents a pulse height shift of about 10:1, or rather to 0.1 of the original spectrum.

SUBJECT: *STANDARDS FOR BIOLUMINESCENCE ASSAYS*

Professor Schram The problem I want to raise is that of having standards for single photon counting just as you have them for regular liquid scintillation counting. With photometers, standards consisting of liquid scintillation sources can be used since this procedure involves integration of photons and a small current is produced by the photomultiplier. When using the scintillation counter we are actually counting single individual photons and we cannot use the radioactive standard. Luminol chemiluminescence is used by some workers but in bioluminescence assays we are working at a much lower light level.

Dr. Birks Way back in 1952 we had to make an absolute measurement of the scintillation efficiency of anthracene crystals. At that time we were unable to measure single photons and we used an attenuation procedure in which a lamp of known intensity, from the National Standards Laboratory, was reduced in intensity by using the inverse square law, utilizing apertures and grey filters until we got down to comparable levels. Perhaps you need not go that far now since Hastings and Weber (J. Opt. Soc. Amer. 53: 1410, 1963) have now established a scintillation standard light source.

Dr. Kobayashi I have made up a ^{14}C-standard in a butyl-PBD fluor for a colleague to use as a standard for the firefly assay. Is that a bad standard to calibrate an instrument?

Professor Schram This works well with analogue or current measuring devices. The source gives a known number of scintillations per unit time but the quantum counter cannot resolve the individual photons. What is needed is a source emitting single photons that can be resolved in time by the counter.

Dr. Ross I would like to second Dr. Birks's suggestion about the technique of using a very low voltage on a tungsten bulb. This procedure has been applied but I cannot recall the references. The beauty is that from the emissivity of tungsten and the electrical parameters one can quite accurately calculate the rate of photon emission.

Mr. Laney In the laboratory we use a green light emitting diode (LED) with a light pulser and an attenuating filter. The pulses are generated *via* an oscillator. But the source has to be calibrated.

Dr. Lee We have proposed the luminol chemiluminescence reaction as an absolute light standard (J. Lee *et al.*, Bioluminescence in Progress, Princeton University Press, 1966, p. 35). Its calibration is traceable to the National Bureau of Standards standard lamp and is also consistent with the fluorescence yield of diphenylanthracene and the photochemical quantum yield of the ferrioxalate actinometer (J. Lee and H.H. Seliger, Photochem. Photobiol. 15: 227, 1972). It is also in agreement with the gas phase NO-O chemiluminescence standard (A. Fontijn and J. Lee, J. Opt. Soc. Amer. 62: 1095, 1972). The luminol reaction can be used down to 10^{-11} M concentration, that is about 10^8 total photons, which allows it to be applied to the bioluminescence assay calibration. Reynolds and Hastings (Bioluminescence in Progress, p. 45) have found that the scintillation standard of Hastings and Weber differs from the luminol chemiluminescence standard by about three times. The scintillation standard deteriorates with time and this makes it less useful as a standard.

SUBJECT: *SEALED GLASS AMPOULES*

Mr. Downes I refer to Mr. Sharry's paper. Most people think sealing glass ampoules is a dangerous procedure and indeed we used to cool these ampoules in dry ice and ethanol before sealing. Then Drs. Everett and Gibbs at Packard said they considered this unnecessary, so since 1970 we have been sealing ampoules at room temperature and by now have dealt with in excess of 50,000 with no real troubles. I have discussed with manufacturers of ampoule fillers and sealers the problem of applying the technique with toluene and other flammable materials. Evidently this has been done in industry with no real problems. The machines cost about a thousand pounds sterling in England and they can seal over 10,000 ampoules per day. The ampoules can be readily flushed with inert gas so that oxygen quenching is minimized and we could get higher counting efficiencies. The procedure is easy to learn and a thousand ampoules can be sealed in a morning by hand. I cannot really see how any glass vial with an imperfect cap can compete with the sealed glass ampoule and we expect to save several thousand dollars in the coming year, which is more than sufficient to purchase the automatic filler and sealer.

Dr. Painter I agree very much with your comments. We prepare thousands of liquid scintillation standards each year in 20 ml vials and never have a problem since the concentration of oxygen and toluene has to be just right to get an explosion. Occasionally one starts to burn but it's not a real problem.

SUBJECT: *LOW LEVEL COUNTING AND SCINTILLATION VIALS*

Dr. Calf We are mainly interested in dating of water - in other words, determining how long the water has been removed from the hydrological cycle or how long it has been underground. The maximum activity we have to deal with is a fraction of a picocurie and so our counting sys-

tem must be one with a maximum E^2/B or figure of merit. You can improve this value by either increasing E (efficiency) by better light collection or photomultipliers, or by reducing B (background). As Professor Noakes has shown us, the background can be reduced by electronic means and massive shielding or by changing the material from which the vial is constructed. Teflon vials have a much lower background than glass vials since they contain no potassium-40 and they also bring about a reduction in cross-talk between the two photomultipliers.

Dr. Underwood In our radioimmunoassay laboratories at the Peter Bent Brigham Hospital we dispense 2-3000 vials a week. Because of economy we are using plastic vials. I would like to ask Dr. Painter about a problem we have noticed with the background count rate of polyethylene vials. If we add PCS to polyethylene vials and store them in the counter at 7°C, then the background count rate can increase up to 7-fold over a 2-3 month period. In contrast, the background count in glass vials remains constant. Although this is a rather academic point because background vials are dispensed at the same time as those for samples, I wonder whether Dr. Painter has noticed this in his experiments with polyethylene vials.

Dr. Painter I alluded to this problem in my paper the other day. Dave Hansen of Nuclear Chicago in some unpublished studies showed that if you take polyethylene vials, some with no activity and some with activity, you get permeation of substances like ^{3}H- and ^{14}C-toluene and ^{3}H- and ^{14}C-hexadecane into the atmosphere within the counter. Diffusion then occurs into the other polyethylene vials which are being used as backgrounds and their count rate goes up dramatically in quite a short time. This would be aggravated in an ambient temperature counter. This is one possibility; I don't know whether it is the answer to your problem.

Mr. Polach I would like to make some comments on background and the causes of background in liquid scintillation counters. To me, three things stand out. Firstly, there is incident radiation as a single cause of background. Commercial liquid scintillation units are all top loading and thus susceptible to cosmic radiation and neutron 'skyshine'. Bottom-loading equipment with complete shielding on the top will lower background due to 'zeni-

thal' distribution of cosmic ray mesons. The second factor is cross-talk between the photomultiplier tubes. At the Liquid Scintillation Symposium at North-West University held in 1967, Professor Aron Nir of Israel established from phototube masking experiments that some of the cross-talk was due to events in one tube being seen by the other tube. I think Dr. Calf's remarks on the advantages of teflon vials are interesting because here we have an opaque substance, non-light-transmitting teflon, acting as an obstacle between the phototubes which will lower background. In Dr. Sharry's example he has taken 1, 5 and 20 ml vials and placed them in lucite or perspex containers. This is one way of maintaining the background at a high level and this is what he found; the background was not related to the volume of the vials. I could suggest ways of lowering the background of the 1 and 5 ml vials. This brings us back again to the vial design. It is obvious that vials and vial materials will influence background. I think that we should endeavour to design a vial with thin walls to reduce the Cerenkov effect. Thin-walled plastic or thin film vials held in an aluminium container which exposes only the volume of the vial to the phototubes arranged in an aluminized reflector where the phototubes cannot see each other might achieve the lowest possible background.

Dr. Coghlan Commented that the Nuclear Enterprises machine loads from the bottom.

Mr. Hartley With teflon vials the actual light cross-talk reduction can be very high. In the counter I described earlier I calculated that only 4% of the light emitted from one photomultiplier reaches the other one. This is due to a good fit of the vial into the light guide. Even a small space round the vial will lose a lot of the advantage.

Mr. Laney In 1971 (Organic Scintillators and Liquid Scintillation Counting, ed. D.L. Horrocks and C.-T. Peng, Academic Press, New York, 1971, p. 991) I presented a paper on electronic rejection of optical cross-talk which showed that three-quarters of the cross-talk between photomultiplier tubes can be eliminated. This is done by recognizing that there is a dysymmetry between pulses that occur in one phototube that are seen by the other.

Dr. Stanley One point that needs to be raised is that of total internal reflection. You do lose quite a lot of light from the vial because of the big changes in refractive index going from toluene scintillant (RI 1.50) and glass (RI 1.50) into air, which is essentially RI 1.0. As you go from glass directly through to a perspex or lucite light pipe, the RI is a great deal higher. This may cut down the light lost by total internal reflection.

Dr. Birks If you have an air gap between the perspex and the glass and if they are not welded together, you will still get the same reflection losses.

Dr. Scoggins I would like to make a few comments of a practical nature on Australian-made counting vials. Because of the large number of different batches of AGM (Australian Glass Manufacture Co.) vials that we use, we carefully check their background and counting efficiencies. Recently we had a batch which had a 10% reduction in ^{14}C efficiency when compared with other vials from the same company. This shows that the glass quality used in vial manufacture in this country may vary and emphasizes that each different batch of vials should be checked prior to use.

SUBJECT: *MEASUREMENT OF* ^{125}I

Professor Bransome When we measure a gamma-emitter in a liquid scintillation counter, what contributes to the observed spectrum? What is the contribution of the Auger electrons and what are the relative quantum yields of the various contributions and do they vary with energy?

Dr. Kobayashi When ^{125}I decays, it does so by electron capture. Here the nucleus captures an electron and its position is filled by an electron from the outer orbit and during this process gives off an X-ray. This X-ray is similar to a gamma ray. Dr. Horrocks recently pointed out that the efficiency of counting this X-ray is very poor but the efficiency of measuring ^{125}I is around 60-70% and he considers this high efficiency to be the result of the ejection of Auger electrons which act just like β-particles.

Professor Bransome We can measure technetium-99m, which is a weak gamma-emitter, although at lower efficiency. The K - X-ray of ^{125}I is supposed to be able to interact with good efficiency with scintillators. What about other gamma-emitters?

Dr. Painter In the case of the pure weak gamma-emitter, we are not counting Auger electrons but conversion electrons which are very similar. We are counting electrons and not photons. If you work out the probability of a photon-toluene interaction, it is very small, and if we had a pure gamma-emitter which did not produce conversion electrons or a pure X-ray emitter which did not result in Auger electrons, we would be able to see this very nicely. In the case of ^{131}I, the decay scheme is complex and there are several betas and you can calibrate the pulse height spectrum. Horrocks has taken an alpha-emitter which is almost mono-energetic in the liquid scintillation spectrometer (as was shown in Dr. Fardy's paper on plutonium) and actually calibrated this spectrum to show that the peak we set is not equal or close to the energy of the gamma. Unless we have loaded the scintillator with lead or tin, we do not have a good probability for gamma interaction.

Dr. Birks Chapter 1 of my book gives all the information necessary for this discussion. There are three ways in which gamma rays interact, and in the case of a hydrocarbon in the liquid scintillator the one operative at low energy is the photoelectric effect and it cuts out around 20 Kev. There is also Compton interaction and this falls off to about half when you reach 500 Kev and keeps going down. The Compton cross-section merely depends on the number of electrons per unit volume and since there are electrons in hydrocarbons you get such interaction; otherwise you could not use external standards unless you are one of these people who believe the secondary electrons all come from the walls of the vial, and they don't! Above 1 Mev you have the possibility of pair production and at higher energies it becomes the major process. In the photoelectric process almost all the gamma energy goes into the photoelectron but with Compton's only a portion of the energy is given up - the rest is released as a lower energy gamma. This low energy gamma has a better possibility of being absorbed and this will depend on its pathlength through the scintillator. The two may add up to give you an energy which looks like the total energy.

Professor Bransome What about the yield of the K - X-ray in electron capture?

Dr. Birks It depends on the nuclei.

Professor Bransome Let's say ^{125}I.

Dr. Birks I'm sorry, I can't quote the value.

Dr. Kobayashi Are you saying that for ^{125}I what is being counted is the X-ray?

Dr. Birks No, I'm not taking a view on that. I was merely answering the question on the interaction of gamma rays with matter.

Dr. Lowenthal ^{125}I decays so that the gamma is 93% converted internally and that means you get conversion electrons and it's these which are being counted.

Professor Bransome I'm still not happy about the situation. Where do those two photopeaks come from?

Professor Noujaim Speaking as a non-physicist, I understand that there are two gammas emitted for ^{125}I. One is at 27 Kev and the other at 30 Kev. You get more than one gamma per disintegration, so the efficiency should be higher than 100% if you are counting the two individual events. ^{125}I is a special isotope since the half-value layer of the 30 Kev gamma is about one inch, which is the approximate size of the scintillation vial. For higher energy gammas you need much more than an inch. Most gamma-emitters also emit beta particles with them, either as such or as conversion electrons together with the Compton electrons produced by gamma interactions. I cannot see how you can quench correct in this case.

Professor Bransome The gamma you refer to I believe is the 35 Kev X-ray, which I cautiously assumed to be the second peak in my paper yesterday. It is pulse height shifted much more rapidly than the first so I think it's important to get the theory straight.

Dr. Kobayashi I believe Dr. Horrocks has very carefully analysed the peaks for ^{125}I and although they are close to where the gamma peaks might lie, they are not in the right place and turn out to be due to the conversion electrons.

(The editors considered that this matter had not been sufficiently clear cut in its outcome and decided to ask Dr. D.L. Horrocks, who did not attend the conference, to add an addendum to this section on what he considers to be the situation in counting ^{125}I by liquid scintillation. The following are his comments.)

Dr. Horrocks (Beckman Instruments Inc., Fullerton, California, U.S.A.) The counting efficiency of ^{125}I dissolved in a liquid scintillator solution (without a metal such as Pb or Sn to act as an electron increaser) is about 76% when there is no quenching. This has been determined with a ^{125}I standardized source purchased from Amersham/Searle and has been cross checked by a source standardized by the X-ray-X-ray coincidence methods (Eldridge and Crother, Nucleonics 22(6): 57, 1964). There is no way that the efficiency can be greater than 100% because when an X-ray is stopped in the solution it will be coincident with any other event which excited the solution, such as Auger electrons or conversion electrons. The probability of stoppage of a 27.5 keV X-ray in a 15-ml volume of toluene is only about 8%. Thus the high counting efficiency has to be the result of excitations by processes other than the stoppage of the Te K-X-rays.

The primary source of electrons to excite the LSS occurs by the internal conversion of the 35.5 keV transition and the Auger process. The process of gamma emission competes with the process of internal conversion. The internal conversion occurs in 93% of the ^{125}I decays and gamma emission occurs in only 7% of the decays. The internal conversion process will produce a conversion electron with energy equal to the energy of the gamma transition, E_{γ}, less the binding energy, E_{BE}, of the electron. Thus the conversion electrons will have energies dependent upon the shell of the conversion:

K shell 35.5 - 31.8 = 3.7 keV
L shell 35.5 - 4.8 = 30.7 keV

The other source of electrons is the release of energy as a result of filling the electron vacancies produced by the electron capture and internal conversion processes. When the vacancy is in the K shell, 80% of the time an X-ray will be produced and 20% of the time the energy will go into the production of one or more Auger electrons.

The total energy deposited in the LSS for any single disintegration will determine the response measured (i.e. the pulse height output of the system). The total energy is made up of any X-rays which are stopped, the conversion electrons and the Auger electrons, but since all of these are coincident only one pulse (count) will be produced. Thus again it can be stated that the counting efficiency cannot be greater than 100%. Another way to consider the excitation of the LSS is to say that the energy for excitation from the decay is equal to the total energy released minus that which escapes from the solution. The total energy will be a function of which electron shells are involved in the electron capture and the internal conversion. But the energy which escapes from the solution will be either 27.5 keV (Te K-X-rays), 35 keV (the unconverted gamma ray) or 55 keV when two coincident X-rays escape. Consider the case of K capture followed by internal conversion in the K shell. The total energy released would be:

31.8	K edge
35.5	excited state
67.3 keV	Total energy

If all energies stopped in the LSS, the response would be equivalent to excitations by 67.3 electrons. The probability of this occurs in about 5% of the decays. When one of the two X-rays escapes (carries away 27.5 keV), the response would be equivalent to excitation by 39.8 keV electrons. The probability of this occurs in about 29% of the decays. If both X-rays escape (carries off 55 keV), the response would be equivalent to excitation by 12.3 keV electrons. The probability of this occurs in about 41% of the decays. One has to go through all the combinations (i.e. EC(L) and L conversion and combinations with EC(K) and K conversion) to obtain the energy released and the probabilities of the various pathways. Having done this one arrives at the following summarized table.

Energy Deposited	Probability
4.3 - 4.8	.054
12.3 - 12.8	.537
31.8 - 40.3	.360
67.3	.049
	1.000

Figure 8 shows a plot shows a plot of the pulse height distribution of ^{125}I compared to the ^{3}H spectrum and Figure 9 shows ^{125}I compared to the ^{109}Cd-^{109m}Ag spectrum. All samples are prepared in an emulsion containing LSS (Ready-Solv VI). These figures show that the response produced in the LSS is not energetic enough to be due to stoppage of X-rays and γ-rays.

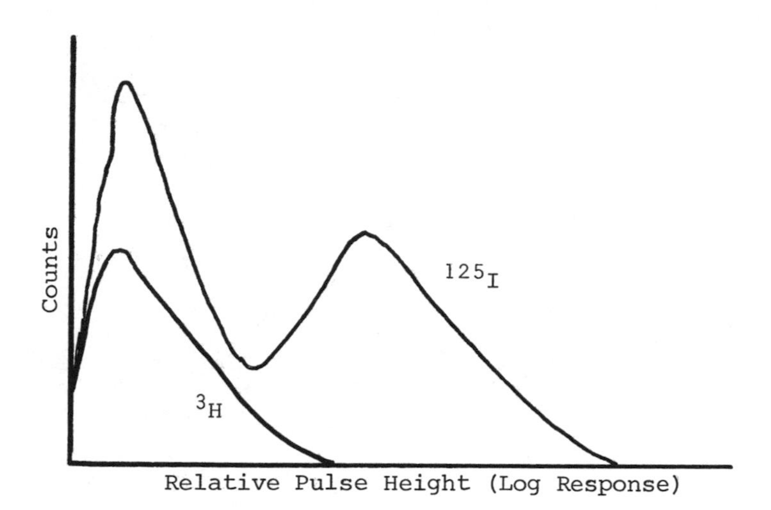

Figure 8

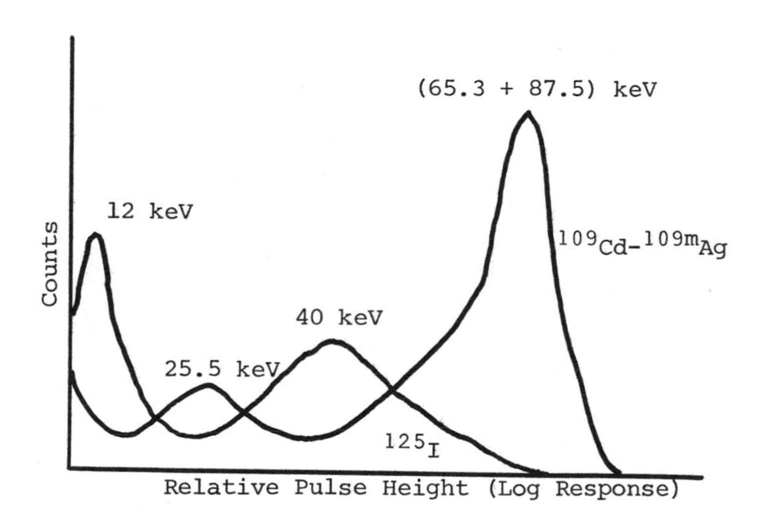

Figure 9

CLOSING REMARKS

J. B. BIRKS

THE SCHUSTER LABORATORY

UNIVERSITY OF MANCHESTER, MANCHESTER, U.K.

Over the years I have attended fifteen conferences or symposia devoted to various aspects of scintillation counting. On several occasions I have been invited to present the opening paper. This is the first time that I have been asked to make the concluding remarks. I find myself in the same position as the rich heiress' seventh husband on their wedding night. I know what is required, but I'm not sure that I can make it sufficiently interesting.

Sometimes matters are clarified by adopting an alternative standpoint. During the first few hours of this Symposium, it became clear to me that we were not just attending a scientific meeting, we were also participating in a religious convention. The plenary lecturers were the visiting bishops expressing their theological beliefs. The other contributors and delegates were other members of the clergy preaching or believing similar different doctrines, sometimes heretical, but a common religious purpose linked us all together. We were all striving to see the light, feeble though it might be.

Practically all the assembly were Unitarians, devoted to searching for the truth in one sample at a time. After the Bishop of Manchester's opening address, with its dire warning that some are more susceptible to impurity than others, Dr. Laney attempted to convert us to the doctrine of the Trinity, the three-in-one. Despite electronics of devilish ingenuity, he had found that the three talked to each other. Older theologians were heard to mutter, "One God indivisible".

The assembly then moved to the age-old doctrine, "rid yourself of sin". Sin comes in two forms, the external world of mammon, known technically as "background", and the internal blemish of impurity. Dr. Noakes had a simple recipe for preserving his virgin specimen from the external world: a thick sodium iodide crystal chastity belt and a few tons of shielding on top. Dr. Hartley recommended a less rigorous regimen: Teflon vials and a reduced photomultiplier voltage to minimise noise. Dry ice/acetone mixtures simply applied to the photomultiplier represent an alternative method of "cooling it". This is a modern version of the English public school doctrine of cold showers as an antidote to temptation or sin.

Dr. Polach stressed the necessity of extreme purity for anyone who was involved in dating. Dr. Painter entertained us with lurid examples of some extremely impure commercial specimens that he had encountered, and of the Salvationist efforts he had made to rid them of their contamination. At this point of the meeting, a rationalist viewpoint was presented, when mathematician Phil Malcolm described how he had taught a computer to perform the sacred rite of liquid scintillation counting. This was the most original and significant contibution, and it was certainly the one involving the greatest number of equivalent man-years.

By now the assembly was ready for a new religion, and Dr. Noujaim preached "lesser pulse height analysis", one which had worked, at least in Alberta. New hope was offered for those whose ardours were being quenched, though there was an element of faith, as well as reason, in the new doctrine. A local preacher, Dr. Downes, then gave a sermon on the text "The good shepherd takes care of his sheep, and the good Australian takes care of his wool". The Monday session closed with a lamentation from Dr. Gresham. Paraphrasing Irving Berlin, his message was "My computer's too small, my users are too many, and I don't know where I am".

After the opening address by Dr. Kobayashi, Bishop of Massachusetts (the only State that didn't vote for Tricky Dick), the Tuesday session turned from the pure to the vial. Dr. Painter reviewed the whole range of vial things that may tempt us. For the rich man Dr. Calf advocated the Teflon model, for the middle-income group

Dr. Sharry praised glass ampoules, and for the poor man our resident Guru, Dr. Gupta, offered plastic minibags in a variety of sizes, hardly conforming to the requirements of the Population Council.

Dr. Bransome told us that surfactants behave as scintillators, thus providing the first indication that some of the cocktail components are beginning to rise above their station. Our hearts went out to Dr. Oades when he displayed vividly coloured pictures of his specimens. They were all soiled, and some of them so deeply stained, that the inner light could hardly get out. Dr. Lowenthal, the man responsible for Commonwealth radioactivity standards, proved an agnostic, who was sceptical of the use of scintillation counting for standardization.

Dr. Ross, Bishop of Tennessee (of which State I am proud to be an honorary citizen) addressed us on the parallel religion of Cerenkovism. Dr. Apelgot, described in today's issue of the newspaper "The Australian" as a modern Madame Curie who disapproves of the French nuclear tests, charmed us into believing that she had found the eternal truth that we lesser mortals were still searching for. During the following three sermons, I was absent discussing Phil Malcolm's next computer programme. Later comments suggest that Dr. Bransome had some difficulty with self-conversion in Iodine-125. Self-conversion is always a difficult exercise for the religious.

Wednesday morning opened with a session from the hot gospellers, the preachers of hell fire and combustion. Dr. Gupta provided us with a stirring text. "We should be thankful to God that everything that is simple is true, and that everything that is not simple is not true". As candidates for the test of truthful simplicity, may I offer quantum mechanics, DNA, myoglobin, man and liquid scintillation counting. He showed us pictures of combustion coils in a variety of sizes (apparently related to other activities of the Population Council) and he advocated combustion as the solution of the world's problems. Stop your children becoming hippies, was the message, buy them a combustion kit. Dr. Noakes, then described how mice could be eliminated by continuous spark ignition. Mr. Kreveld told us of the Downers Grove

(Illinois) solution to the combustion problem, a handsome instrument which would do credit to any crematorium.

The Church Assembly, or Workshop session, under Dr. Coghlan recorded some surprising areas of ignorance among the congregation. Short sermons from the panel of bishops and from other clergy went some way to dispel this ignorance, but it was clear that many had been guilty of inadequate study of the "Bible", describing the theory and practice of the subject.

In the afternoon, Dr. Schram, Bishop of Brussels, told us of a further branch of the religion, devoted to bioluminescence. Dr. Lee and Dr. Stanley, as two disciples of this particular schism, applied it to bacteria and enzymes, respectively. Finally Dr. Underwood and Dr. Coghlan drew our attention to an important application of the LSC religion, namely radioimmunoassay. The results may be critical to the life of a guinea pig, or even a human being; however, it is clear that new instrumental techniques, incorporating conveyor belt methods and Dr. Gupta's miniaturization ideas, are required for saturation analysis.

Throughout the meeting, the visiting bishops and some others preached the "old-time religion". According to John's gospel, "the more you know about how the scintillator and the counting instrument works, the more likely you are to generate accurate numbers". To view any of the elements in an experiment as mysterious "black boxes" is unscientific. If the numbers generated relate to human beings and their welfare, it is downright sinful.

The committee who organised this scientific-cum-religious symposium did a grand job, and I wish to express the thanks of the whole congregation. Thanks to Philip Stanley for his fine chairmanship, to Bruce Scoggins for his excellent programme secretaryship, to Peter Annand for his generous financial advice, and to Convention Organisers Pty. Ltd., for their official execution.

All that remains, my brethren, is to send you back to your individual altars to worship, with minds cleansed of unclean thoughts, resolved to distinguish the true from the false. Our Guru has told us that the truth lies in meditation. To my old friends and to my many new friends, go in peace, continue to strive to see the light, and here's to our next meeting. Let us join in the LSC hymn, "Lead kindly light amid the encircling doom".

SUBJECT INDEX

A 4
B 5
C 6
D 7
E 8
F 9
G 0
H 1
I 2
J 3